Blockchain Technologies

Series Editors

This book series aims to provide details of blockchain implementation in technology and interdisciplinary fields such as Medical Science, Applied Mathematics, Environmental Science, Business Management, and Computer Science. It covers an in-depth knowledge of blockchain technology for advance and emerging future technologies. It focuses on the Magnitude: scope, scale & frequency, Risk: security, reliability trust, and accuracy, Time: latency & timelines, utilization and implementation details of blockchain technologies. While Bitcoin and cryptocurrency might have been the first widely known uses of blockchain technology, but today, it has far many applications. In fact, blockchain is revolutionizing almost every industry. Blockchain has emerged as a disruptive technology, which has not only laid the foundation for all crypto-currencies, but also provides beneficial solutions in other fields of technologies. The features of blockchain technology include decentralized and distributed secure ledgers, recording transactions across a peer-to-peer network, creating the potential to remove unintended errors by providing transparency as well as accountability. This could affect not only the finance technology (crypto-currencies) sector, but also other fields such as:

Crypto-economics Blockchain
Enterprise Blockchain
Blockchain Travel Industry
Embedded Privacy Blockchain
Blockchain Industry 4.0
Blockchain Smart Cities
Blockchain Future technologies
Blockchain Fake news Detection
Blockchain Technology and It's Future Applications
Implications of Blockchain technology
Blockchain Privacy
Blockchain Mining and Use cases
Blockchain Network Applications
Blockchain Smart Contract
Blockchain Architecture
Blockchain Business Models
Blockchain Consensus
Bitcoin and Crypto currencies, and related fields

The initiatives in which the technology is used to distribute and trace the communication start point, provide and manage privacy, and create trustworthy environment, are just a few examples of the utility of blockchain technology, which also highlight the risks, such as privacy protection. Opinion on the utility of blockchain technology has a mixed conception. Some are enthusiastic; others believe that it is merely hyped. Blockchain has also entered the sphere of humanitarian and development aids e.g. supply chain management, digital identity, smart contracts and many more. This book series provides clear concepts and applications of Blockchain technology and invites experts from research centers, academia, industry and government to contribute to it.

If you are interested in contributing to this series, please contact msingh@endicott.ac.kr OR loyola.dsilva@springer.com

Gunjan Chhabra · Keshav Kaushik
Editors

Understanding the Metaverse

Applications, Challenges, and the Future

Springer

Editors
Gunjan Chhabra
Department of Computer Science
and Engineering
Graphic Era Hill University
Dehradun, Uttarakhand, India

Keshav Kaushik
Amity School of Engineering
and Technology
Amity University Mohali
Punjab, India

ISSN 2661-8338 ISSN 2661-8346 (electronic)
Blockchain Technologies
ISBN 978-981-97-2280-8 ISBN 978-981-97-2278-5 (eBook)
https://doi.org/10.1007/978-981-97-2278-5

This Springer imprint is published by the registered company Springer Nature Singapore Pte Ltd.
The registered company address is: 152 Beach Road, #21-01/04 Gateway East, Singapore 189721, Singapore

To my family, thank you for encouraging me in all my pursuits and inspiring me during this journey. I am especially grateful to my parents and wife, who supported me emotionally. I always knew that you believed in me and wanted the best for me.

—Dr. Gunjan Chhabra

This book is dedicated to my beloved Parents-Sh. Vijay Kaushik, Smt. Saroj Kaushik,
Wife-Priyanka, and daughter Kashvi. May god always bless us, Har Har Mahadev!!

—Keshav Kaushik

Foreword

Welcome to the cutting edge of the digital era, where virtual and real reality collide and infinite possibilities beckon. Throughout the book, readers are going to take a voyage into the metaverse on the pages that follow; this is a place that exists outside the boundaries of our physical reality and has the power to drastically alter it.

The concept of the metaverse has captured the imagination of visionaries and technologists alike, offering a glimpse into a future where digital environments seamlessly blend with our everyday lives. This book serves as a guide through this immersive landscape, offering insights into its origins, evolution, and transformative potential.

As we delve into these pages, readers will learn about the complex network of technologies that support the metaverse and various implementations, from blockchain to AI to augmented and virtual reality. Readers will investigate the ramifications of this digital frontier in a variety of fields, including business, entertainment, healthcare, and education.

But this journey is not merely one of exploration; it is a call to action. As we confront the challenges of security, privacy, and identity within the metaverse, we must also envision its vast potential for positive change. From revolutionizing industries to fostering global collaboration, the metaverse holds the key to unlocking a future limited only by our imagination.

So, Dear Readers, I invite you to join this intellectual odyssey as we unravel the complexities, envision the possibilities, and navigate the uncharted territories of the metaverse.

Aryan Chaudhary
Chief Scientific Advisor
Bio Tech Sphere Research, India
Chair, Meerut ACM Chapter
FIOASD, SE-(T&F), USTPC, MIACC
IEEE CTSoc-CSH TC
Kolkata, India

Preface

This book serves as a guide through the immersive landscape of the metaverse, a paradigm-shifting concept that is reshaping the way we perceive and engage with the digital world. Within these pages, we embark on a journey that unravels the origins and evolution of the metaverse. From its conceptual foundation to the intricate web of integrated technologies driving its development, we explore the transformative forces of augmented reality (AR), virtual reality (VR), mixed reality, and the convergence of technologies that define the metaverse.

The narrative unfolds systematically, shedding light on the interconnectedness of the metaverse, non-fungible tokens (NFTs), and the evolution toward Web 3.0. Chapters delve into the depths of digital twins, metaverse services, and the far-reaching implications for education, healthcare, business, and entertainment. As we navigate this digital landscape, we confront pressing issues of security, privacy, and identity. From the intricacies of social interactions to the challenges of combating cyber frauds and safeguarding data, this book provides a nuanced understanding of the complexities inherent in the metaverse.

The journey extends into the future, exploring the metaverse's potential applications in travel, tourism, smart cities, and military defense. We examine the role of blockchain in shaping the metaverse's futuristic landscape, along with the legal and jurisdictional considerations that accompany this emerging digital frontier. This book is more than a mere exploration; it is a roadmap for those eager to comprehend the intricacies of this transformative digital universe. The narrative looks beyond the present, addressing the challenges and future prospects that await us as we embrace the metaverse. Embark on this intellectual journey with us, as we unravel the complexities, envision the possibilities, and navigate the uncharted territories of the metaverse. Your understanding of this digital frontier commences here.

Dehradun, India — Gunjan Chhabra
Punjab, India — Keshav Kaushik

Contents

About the Editors

Dr. Gunjan Chhabra is a Ph.D. in computer science and engineering, working as Associate professor in the Department of Computer Science and Engineering, Graphic Era Hill University, Dehradun. He has a teaching experience of 10 years, and an area of his expertise is algorithms, image processing, and machine learning. He has published ten patents, with four granted patents from IPR. Also, he has published several research papers in various renowned journals. Additionally, he has authored three textbooks on the domain Internet of Things, smart technologies, and machine learning. He has also mentored many projects in different domains of computer science to solve real-world problems. Under his guidance, his students have successfully incubated their ideas.

Keshav Kaushik is an accomplished academician, cybersecurity, and digital forensics expert currently serving as an Assistant Professor at the Amity School of Engineering and Technology, Amity University Mohali, Punjab, India. As a key member of the Cybersecurity Centre of Excellence, he has been instrumental in advancing the field of cybersecurity through his dedicated teaching and innovative research. In addition to his academic role, he holds the prestigious position of Vice-Chairperson for the Meerut ACM Professional Chapter, highlighting his leadership and commitment to the professional community. His academic journey includes a notable stint as a Faculty Intern during the Summer Faculty Research Fellow Programme 2016 at the Indian Institute of Technology (IIT) Ropar, reflecting his continuous pursuit of knowledge and professional development. His scholarly contributions are extensive and impactful, with over 135 publications to his credit. This includes 25 peer-reviewed articles in SCI/SCIE/Scopus-indexed journals and 50+ publications in Scopus-indexed conferences. He is also an inventor, holding one granted patent and six published patents, alongside five granted copyrights. His editorial expertise is showcased by publishing 30 books and 25 book chapters, further cementing his reputation as a thought leader in the field. His professional certifications are a testament to his expertise and commitment to excellence. He is a Certified Ethical Hacker (CEH v11) by EC-Council, a CQI and IRCA Certified

ISO/IEC 27001:2013 Lead Auditor, a Quick Heal Academy Certified Cyber Security Professional (QCSP), and an IBM Cybersecurity Analyst. His recognition as a Bentham Ambassador by Bentham Science Publishers and his role as a Guest Editor for the IEEE Journal of Biomedical and Health Informatics underscore his influence and authority in cybersecurity. He is a dynamic speaker, having delivered over 50 national and international talks on cybersecurity and digital forensics topics. His mentorship was acknowledged during the Smart India Hackathon 2017, under the aegis of the Indian Space Research Organization (ISRO), with a certificate of appreciation from AICTE, MHRD, and i4c. A two-time GATE qualifier with an impressive 96.07 percentile (2012 and 2016), he has also received accolades from the Uttarakhand Police for his significant contributions to cybercrime investigation training. With a career marked by significant achievements and a profound impact on cybersecurity and digital forensics, he continues to inspire and lead in both academic and professional circles.

Introduction to Metaverse

Umesh Gupta and Abhinendra Singh

Abstract Sharing information has been one of the main goals for ages. The need for more efficient and interactive sharing methods increased as time passed. This gave birth to the "World Wide Web", which enabled searching for information amongst different documents. Initially, everything was static, meaning users could only read the information from the web. With the growth of social media, a new Web model emerged: a two-way flow of information. Now, users not only read information from the web but can also share it. The productivity level can be increased tremendously by integrating various technologies like Artificial Intelligence, Virtual Reality, Augmented Reality, Blockchain and Internet of Things (IoT). With these new technologies, a new era of the web is emerging in which, instead of traditional 2D websites, a 3D virtual space will be created where users can meet their friends face-to-face via virtual avatars, create, buy and sell digital artefacts or tokens (NFTs). This chapter aims to cover these new emerging technologies and what difference they can make in lives and the whole working structure of the Internet.

Overview

It all started on October 29, 2021, when Facebook changed its logo and name to "Meta" boosting. Some governments even started taking various initiatives to accelerate metaverse adoption. Metaverse sounds like science fiction in which there is a virtual world, and everything is like a game, but there is an increasing trend of its application in commerce. But till now, metaverse is more of a concept, so this chapter will take the reader on a journey to all the aspects of metaverse, what benefits it can provide, and will there be any future consequences? With this imagined virtual space, there will be an increase in the use of digital currencies like Non-Fungible Tokens (NFTs), creating opportunities for various artists to commercialize their products like art pieces, music, memes,

U. Gupta (✉) · A. Singh
SCSET, Bennett University, Times of India Group, Gr. Noida, Uttar Pradesh 201310, India
e-mail: er.umeshgupta@gmail.com

G. Chhabra and K. Kaushik (eds.), *Understanding the Metaverse*, Blockchain Technologies, https://doi.org/10.1007/978-981-97-2278-5_1

etc. This space also comes with lots of responsibilities and requirements, first and foremost is safety and another quite important need is the end of centralization i.e., decentralized surfing on the web in which flow of information is not restricted and controlled by only one party. Users can roam around space and buy land and properties as NFTs. This version is called "Web 3.0", in which the user will own the internet and can regulate it on their information record. This Web version will be centred on blockchain technologies, thus allowing the decentralized ownership of information, services, or platforms. Metaverse and Web 3.0 will be omnipresent in the future; we do not know many things about these new buzzwords since they are in the very early stages. There might be many changes in what we are expecting from them. There can be various problems faced while building them. But we can only say that these new technologies can drastically change how we surf the internet and buy something. It is also possible that the schools will be shifted to this virtual space, one can travel around the world and many more things. This chapter will try to answer all the questions in the readers' minds regarding these new techs.

> the Web as I envisaged it, we have not seen it yet. The future is still much bigger than the past.—Tim-Berners Lee.

The web has come a long way; it was in 1989 when Tim Berners-Lee, a computer scientist working at CERN, made an automated information-sharing system that enabled a connection between various universities and institutes, thus helping scientists to share information. It was marked as the beginning of the World Wide Web or "Internet". From this point, there is no looking back; with many new technologies like Artificial intelligence, Blockchain, Internet of Things (IoT) and Augmented Reality, a new web is emerging with a new level of user interaction and security. From traditional 2D websites, we are moving towards a virtual 3D space where one can interact with friends. Let's dive into the detail of these new emerging technologies.

1 Web 3.0

1.1 Brief History

The Web can be defined as sharing information, documents and resources between users through an international network (Internet). It can also be considered an ample storage of information stored on the servers and accessed by the clients through the internet. Since its development, the web has evolved a lot. The earlier web was only the collection of some text designed and formatted on Hyper-Text Markup Language

(HTML) [1]. Firstly, it was only intended for sharing information and was in read-only mode, meaning it was not interactive. It was March 1989 when Berners-Lee discussed a system called Mesh. The main idea of Berners-Lee behind the web was a common information space in which one can communicate with each other by sharing information [2]. Till now, we have two eras of the web, which are as follows:

- **Web 1.0**: It was the first unidirectional web model, i.e., the user can only read the information and not upload anything. It used HTML, HTTP and URLs. Some other protocols like XML, XHTML and CSS were also used—some merged technologies between the server and client, like ASP, PHP, JSP and CGI. JavaScript and VBScript were used on the server, and Flash on the client side. This web was very slow; it was not responsive. Users must refresh the page unnecessarily to see any new information uploaded.
- **Web 2.0:** With the growth of social media in the 2000s, a new model of the web emerged, which was Web 2.0, also called Participative and Social web; this allowed the user to read as well as upload information on the web. Web 2.0 uses AJAX technologies like JavaScript, XML, DOM, REST and CSS. This web concentrates more on user interaction, User Generated Content (UGC) and connectivity with other systems. Hacking is one of the key and main issues of this web as this web has more interaction and less control.

1.2 What is Web 3.0?

Until now, the web was just an information-sharing platform where users read and upload their information. But now there was a need for an intelligent web to understand the user's words and give them more efficient and accurate results. Another issue with the past web version was hacking, where an unauthorized third party can manipulate and steal any user's data. Thus, Web 3.0, a decentralized, open to everyone, built on the top of blockchain technologies, is emerging as the new generation of the web. It aims to end centralized companies, where users control their data and transactions.

1.3 Web 3.0 Technologies

This web is an intelligent and innovative web that can interpret and process information like humans. Thus, many technologies are being used in developing this web, such as human-like interpretation ability, which will be achieved by Natural Language Processing, which aims to process and handle text data and help to find meaning.

i. Blockchain technology

It is one of the recent technologies which aims to provide a trusted data-sharing and record-management system. It is a colossal transaction ledger, replicated across multiple users in which no third-party company has any intervention. Thus, it is a decentralized transaction and data management technology without third-party control for data and user trust [3]. The system is made such that every transaction is given a time stamp; thus, without approval, the data cannot be altered or modified. This will ensure the credibility of the information when performing any transaction over a network.

Let us try to understand with an example; let us assume that a student graduated, and the recruiter requests an official document or proof of graduation. This transcript can be directly collected from the university, where the university acts as a trusted intermediary between the student and the employer, ensuring the information is authentic and trustworthy. There can be one question: why can't the employer ask the student to submit a copy of the transcript? The answer is "Trust", as the student can change the content for their benefit. So, briefly, the true service provided by a third party is trust, which is precisely the Blockchain proposition.

Definition Blockchain is a transactional database technology that is a decentralized way to validate and allow reliable and consistent transactions across the participants, also known as nodes [4].

Before moving forward, there are a few things one should know. As discussed previously, blockchain is a distributed technology. Each peer maintains a copy of the ledger to prevent a single-point failure. Now, if a new thing comes to the blockchain network, each node must agree or disagree on whether it is valid or invalid. This process is used to get agreement on its validity among a distributed system. This is called the *Consensus algorithm.* One can think of it as a group of friends trying to decide which place they can go for their holidays.

There are major Blockchain Key elements, which are as follows:

a. *Distributed ledger:* A shared database in the decentral network stores all the transaction information. It is a replicated, shared and synchronized spread across many sites, countries, or institutions. On the other hand, unlike a centralized database, a distributed ledger has no controlling authority or administrator since it is replicated across a distributed computer node, so it requires a peer-to-peer (P2P) computer network and Consensus algorithms [5].

The distributed ledger based on the P2P network which is spread on multiple nodes. Each node duplicates and stores a similar copy of the ledger and independently updates itself, thus ensuring no central authority. Let us see if it works; a ledger has been updated (Fig. 1).

Now, it will be broadcasted to the P2P network in which each node will process a new update transaction independently, and then, using the consensus algorithm, a correct copy of the updated ledger is determined.

- *Smart contracts*: These are executable codes that run on the decentralized network Blockchain that enforce an agreement between the fraudulent parties; this is done without any central authority [6]. It replaces paper contracts with digital contracts and gives network automation. These contracts can guarantee the reduction of

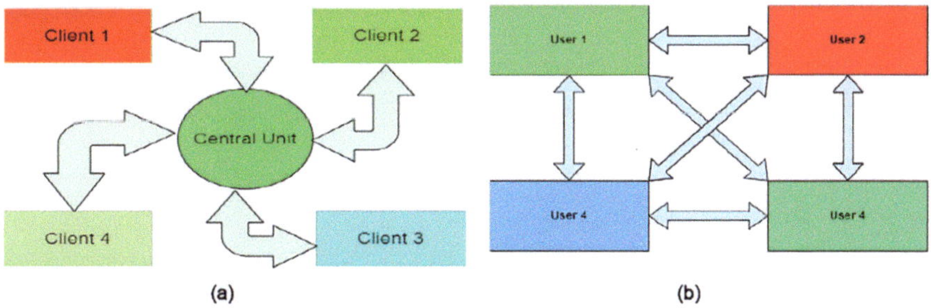

Fig. 1 **a** Centralized system, **b** Decentralized system

human errors to avoid disputes in these contracts. To prevent alteration by any party, these contracts are copied to each node of the network. Programmatically, these contracts store information, process inputs and write outputs using some pre-defined functions. Some examples of pre-defined functions:

- Constructor function: This enables intelligent contract creation. So, to host a new contract, this function is invoked through a transaction, and the transaction's sender becomes the contract's owner.
- Self-Destructor function: The owner can only invoke this function to destroy the contract.

These contracts contain state variables, functions, function modifiers, events and structures that control and execute various actions. One contract can also be called by another contract.

- *Public key cryptography*: This security feature gives the Blockchain network members a key that gives them unique identification. There are two types of keys: the public key common to everyone on the network and the private key unique to the members. These keys are used to access any data in the ledger.

Blockchain integration in Web 3.0 will enable secure transactions since data will be managed and validated in this decentralized peer-to-peer network. Various layers of blockchain are shown in Fig. 2.

ii. Artificial intelligence/Machine learning

Artificial intelligence is a machine mimicking how humans think, understand and implement things. It is one of the definitions that has surfaced worldwide in the past decades. Simply artificial intelligence (AI) combines computer science and user data, thus making the computer machines learn and solve problems. It has sub-fields, which are Machine learning (ML) and Deep learning (DL):

- *Machine Learning (ML)*: This algorithm teaches the machine to handle the data more efficiently. Since finding patterns in the data and extracting relevant data is often very complex, we use Machine learning algorithms that understand the complex data easily in such scenarios. With abundant available data, Machine

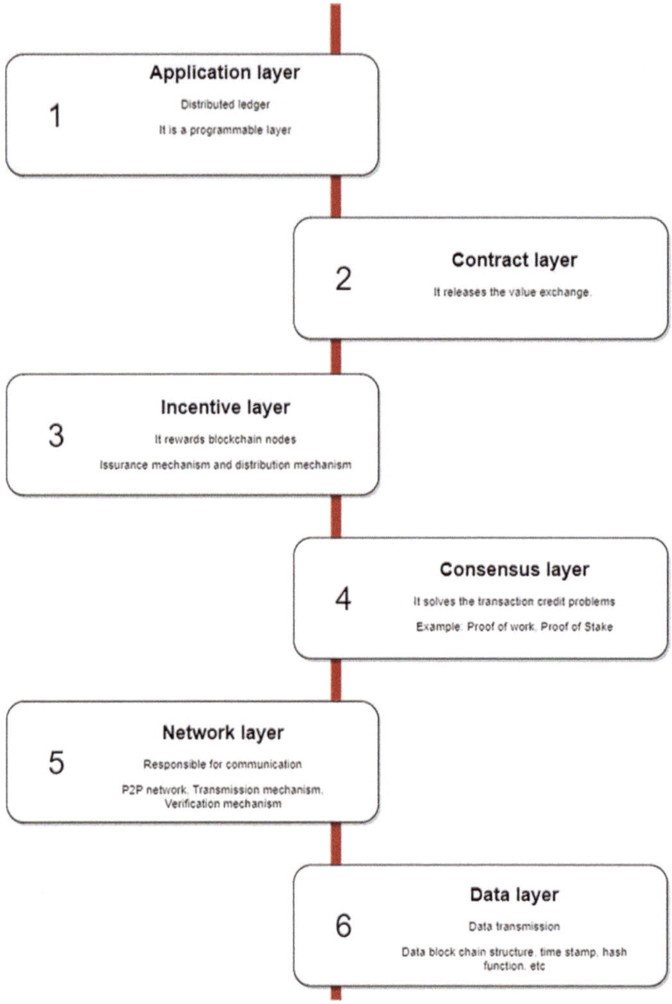

Fig. 2 Layers of blockchain

learning algorithms are now applied in various fields. Machine learning models solve multiple problems, such as classifying the data or classes. There are many types of Machine learning algorithms, as shown in Fig. 3.

- *Deep learning:* This method trains computers to process data like a human brain. This model can recognize complex patterns in images, text, sounds and other forms of data. This model contains multiple processing layers, enabling the machine to learn from the data with different levels of abstraction. Human neural networks inspire deep learning algorithms.

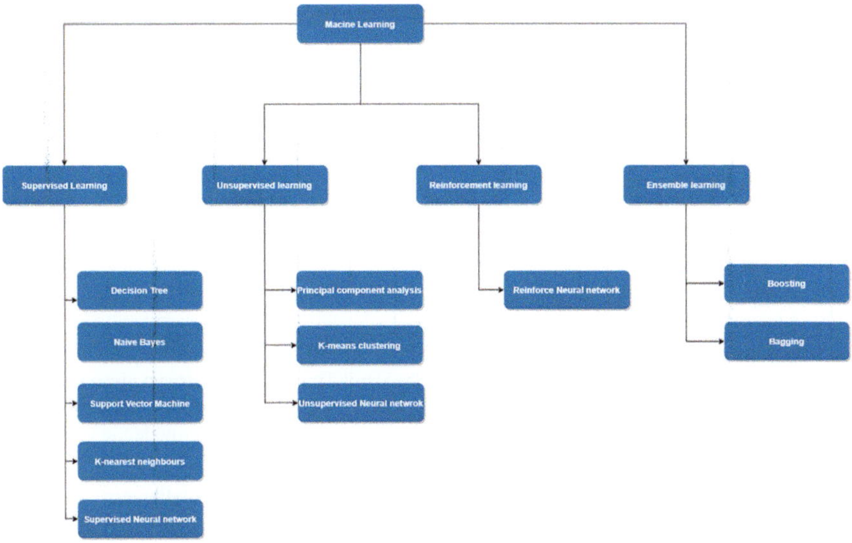

Fig. 3 Various machine learning algorithms

Web 3.0 aims to understand the "meaning" and "emotions" of the data. Thus, using an AI algorithm will enable us to create a web not concerned with data structure but with "*Meaning*".

1.4 How Will Web 3.0 Work?

The conventional layouts and delivery of the webpages defined using HTML will be same in Web 3.0. Still, changes can be expected in how they connect with the data sources and where they reside. This web with different data connection techniques also aims towards a Semantic web, which means it will study the relationship between the words. Let us see an example,

I Love Decentralization.
I < 3 Decentralization.

The above two sentences seem different, but their semantics are the same, or both are saying the same thing. Since Web 3.0 will be the future of the web, it will be better to understand how Web 3.0 works by comparing it with the existing web form, web 2.0, refer Table 1.

Table 1 Difference between Web 2.0 and Web 3.0

S. no.	Web 2.0	Web 3.0
1	**Centralized** All the services and platforms are governed by a central authority that always acts as an intermediary between the users	**Decentralized** A peer-to-peer network with a distributed system will be used where users control their data and transactions
2	**Flat currency** The currency that the government issues is used, like the Indian Rupee	**Cryptocurrency** Using blockchain technology (as mentioned in Sect. 3.1), digital currencies such as Bitcoin are used
3	**Cookies** This is used to see and track the users who visit any site, to understand the pattern in which the user is surfing, and to provide personalization accordingly	**NFTs** This is a unique token given to a user; these tokens are assigned some values that can provide the users with some perks
4	**CSS and Ajax** This web is defined using this layout that sees which form data is used and not the meaning	**AI (Artificial Intelligence)** This smart web can understand the meaning and emotion of the data and not only the structure
5	**Relational database** These central databases store the data in one or more tables of columns and rows. The relationship is logically established between different tables	**Blockchain** A distributed ledger will be used to store the data (Sect. 3.1)
6	**Social Network** This web created a new space where each user relates to other users and can share their information	**Metaverse world** This emerging world is a virtual space that is comprised of Virtual Reality and Augmented Reality
7	**IPv4 addressing space** Web 2.0 uses a 32-bit address space, which provides 2^32 unique addressing space	**IPv6 addressing space** In Web 3.0, the address space is increased to 2^128

2 Non-fungible Tokens

Earlier, when blockchain technologies did not exist, a system for verifying and securing digital assets was not very safe, and there was always a chance of tampering by unauthorized parties. Owing to this problem, NFTs were introduced, tokens that represent digital assets and are made above blockchain technology, giving them a unique ID and a decentralized ledger thus preventing tampering by any third party.

Definition These are assets that are based on the blockchain. Each token has been given a unique identification code and metadata, distinguishing every token from one other.

Let us understand NFT, word by word:

- Non-fungible: In terms of economics, the item is unique and distinguishable; any other item cannot replace it. For example, the Rupee bill is fungible as other rupee bills can interchange it.
- Tokens: According to the Oxford Dictionary, it is a piece of an item of any form that can be used to buy something of a specific value.

2.1 *Brief History*

In the world of digital currencies, in May 2014, the first NFT was developed by Kevin McCoy and Anil Dash; it was then registered on the Namecoin blockchain and sold for $4. At that time, it was not called NFTs, but they called it "monetized graphics".

In 2015, the first NFT project was developed and demonstrated as DEVCON 1 in London. This project's name was Etheria, 457 purchasable and tradable hexagonal tiles. In the first four years of its launch, Etheria remained unsold. After all this year, it was March 13, 2021, since by this time, interest in digital currencies and NFTs increased, so within 24 h, all the tiles were sold for US$ 4 million. Thus marked the beginning of the NFTs.

In 2017, CryptoKitties, an online game, was profitable because of tradable Cats. It was in 2021 that a JPEG digital picture was sold for $69 million, even at the time of the pandemic. This boosted the interest of the whole world in this technology and thus started a new form of currency (one can say), NFT. It is a type of currency derived from the smart contract of Ethereum.

2.2 *Ethereum*

NFT is based on blockchain technologies, and NFTs are a new emerging digital asset that people can sell and buy, so there is a need for a blockchain above which all the stated things can be done securely. Thus, **Ethereum,** a community-run technology software that provides a platform for developing and deploying Dapps (Decentralized Applications) [7, 8], is designed to enable NFT transactions.

It is a blockchain network with Turing-complete programming and various abstract layers to allow everyone to own ownership, transaction formats and state transition methods. It is a platform where anyone can build and deploy a decentralized application. This is the only blockchain with a programming language that enables users to write smart contracts and DAPPs to create rules for ownership and transactions.

2.2.1 Ethereum Features

a. Ether

Like the real-life scenario, one needs payment to do a task. Ether is Ethereum's currency, used as payment for resources and fees for performing any transaction on the network. It is used to pay for transactions and buy GAs, which is a payment for the computation of any transaction; we will explore "Gas" in the later part of this chapter. The value of Ether works similarly to a stock market, but the only difference is that the prices are more liquid, which means prices change very frequently. Overall, we can say that it is a fuel of the Ethereum network.

b. Smart contracts

As mentioned in Sect. 3.1, these are executable codes that facilitate the exchange of any info or assets (in the case of NFTs) between the users. It consists of various terms and conditions that should be agreed upon by the parties in the network. It is far more secure than traditional contracts, as they can't be altered once they are executed, and payment done on the top of a contract is permanently registered, so if in future there is any change in the contract, transactions related to the original contract will not be altered. One more thing that makes it more beneficial is that its verification process is carried out by anonymous parties, making it decentralized.

c. Ethereum Virtual Machine (EVM)

As mentioned in the previous sections, smart contracts are programs written in a programming language (Solidity in Ethereum) executed to enable and register it to the network. So, for any programming language, there needs to be an engine that understands the language of smart contracts. Thus, Ethereum Virtual Machine (EVM) is a runtime environment that provides a compiler for Ethereum-based smart contracts and provides deployment features.

The language used to develop the smart contracts is converted to bytecode, which is understood and executed by the EVM. Another thing about the EVM is that it is a sandbox environment, which means it is a stand-alone environment for testing and development. Before deploying the smart contract in the leading Ethereum network, one can verify and test the contract several times.

Let Us Now Understand the Working of the EVMs:

Let us consider a scenario where Ansh wants to pay 20 Ethers to Rahul. This transaction will be executed using a fund transfer smart contract between Ansh and Rahul; this contract will be sent to the EVM for validation. Validation is done by the Proof-of-work consensus algorithm (Sect. 3.1). Miner nodes in the Ethereum network check and validate the transaction. All the nodes on the network will execute these contracts using their respective contracts.

- **Proof-of-work**: It is a technique to verify any new transaction added to the blockchain network. Since blockchain has no centralized authority, this technique ensures the integrity of the new data in the network.

The miners in the network are the participants who are allowed to check and verify the transactions. Their goal is to validate the blocks. They use various hashing algorithms like the Ethash algorithm, which returns an appropriate hash value. A hash value is considered suitable when it is less than the pre-defined target per the proof-of-work consensus. Miners in the network compete to get a block verified, and once a miner succeeds in proving a block, they are rewarded.

- **GAs**: It is the fuel of the Ethereum network; users need to pay "GAs" to get any transaction done. The transaction fees are calculated using Eq. 1.

$$\text{Ether} = T * \text{fees} = \text{Gas limit} * \text{Gas price} \tag{1}$$

where, (i) **Gas limit** = Amount of gas fuel used for the computation.
(ii) **Gas price** = Amount of Ether a user is required to pay.

d. Decentralized applications (DApps):

It is a decentralized application that is developed on blockchain. Let us understand by comparing it with the traditional applications currently used. When we log in to a conventional application, the web app is rendered using HTML, and it will call the APIs, which will access the centrally hosted data.

In the case of the DAPPs, the APIs are replaced with smart contract-based APIs that fetch data from the blockchain. These are more secure applications as it is not controlled by a single supreme central authority but rather by a decentralized and more transparent system.

e. Decentralized Autonomous Organizations (DAOs)

It is a system based on the blockchain that gives people the facility to coordinate and self-govern themselves using self-executing rules (Smart Contracts) [9, 10].

- **D—Decentralized**: It is called decentralized because it works over a blockchain, which is a decentral server-less infrastructure. The decision-making depends not on any authority but on various collective agreements of the multiple members. This considerable dependence is achieved by a voting mechanism in which network members can participate. Let us see how this voting system works.

People add funds to the DAO as it requires payments to execute and make decisions. Now, each member is given tokens that talk about that member's share in the DAO. Now, as the traditional voting systems work, the maximum votes decide the fortune of the decision. Thus, if anyone tries to tamper with the information in any network node, it will not affect the information. Once the information is deployed to the network, it is copied to every node, and each node validates it.

- **A—Autonomous**: It is called so because it is not dependent on its creator unless explicitly declared in the code. One more thing about autonomous is that since it is built on a blockchain, no authority can turn OFF the DAO. If its members are active, DAO will operate all the facilities.

f. A project on Ethereum

Now, we will make a simple, smart contract using a language called Solidity. Solidity is a statistically typed curly braces programming language used to develop smart contracts and runs on the EVM (Sect. 2.2.1).

This project is a primary **counter-smart contract**, built on solidity language and remix IDE and is deployed on Remix VM (Merge). The basic structure of the solidity program is as follows.

- The first line of the program is always the declaration of the version of the solidity used; in this version, 0.5.1 version is used.
- The main smart contract code is enclosed in a contract counter function. We will start declaring all the state variables and functions inside this.
- We have first declared a count variable with a uint data type (unsigned integer) with an initial value of 0.
- Now in Solidity, any new function is declared using the function keyword, so an Increment function is displayed in which the counter is increased. One can see that there is one more thing in the function ***emit Increment (count)***; this line calls the event named Increment, events in the solidity and Ethereum allows anyone in the blockchain to subscribe to the smart contract and can see the value of the count.
- In a similar way, the decrement function is declared.
- A getCount function is declared, which will display the value of the count.
- Lastly, project is first compiled to see whether there are any errors; once compiled, it is deployed in the blockchain, which comes with the remix IDE, i.e., Ethereum.

Here is the program:

```
//declaring the version of the solidity
pragma solidity 0.5.1;
contract Counter
    event Increment(uint value);
    event Decrement(uint value);
  uint count = 0;
    function increment() public{
        count = count + 1;
        emit Increment(count);
    }
    function getCount() view public returns(uint){
        return count;
    }
    function decrement() public{
        count = count - 1;
        emit Decrement(count);
    }
}
```

2.3 NFT Marketplace

As discussed earlier, NFTs are digital assets and can be anything.mp3 file, digital art, etc. But just creating these will not make it an asset; one must sell it somewhere from where others can buy it. Thus, the NFT marketplace comes into the picture; it is like the traditional shops where NFTs are stored and can be sold and brought.

Once NFT is created, it is "Minted" in the network. *Minting is a process by which digital art is uploaded on the Ethereum blockchain*. Briefly, the NFT marketplace is an NFT exchange platform. Like a normal marketplace, there are some prerequisites for the NFT marketplace.

- **A crypto wallet**: It is a storage of all the private keys (passwords that give access to the cryptocurrencies the user has) that allows safe access to the crypto. One can send, receive and spend crypto. It is like a bank account in the physical world. Various crypto wallets are available; one must choose the wallet according to its compatibility with different blockchain networks. Earlier crypto wallets were compatible with only one cryptocurrency. Now, we have various wallets compatible with more than one cryptocurrency.

  There are types of crypto wallets, let's compare these.

- **Some amount in the wallet**: The wallet should be pre-funded before you create, mint or buy any NFT from the marketplace.
- **A user account**: One must create an account on the marketplace; this account will handle the details about NFTs the user has brought, minted, etc.

Thanks to the increased interest in NFTs shown by users worldwide. Owing to such huge interests and users buying and selling it, numcrous NFT art marketplace platforms exist where one can create, sell and buy NFT. Some of the famous marketplaces are being compared in Table 2.

Table 2 Hot wallet and cold wallet differences

	Hot wallet	Cold wallet
Price	It is a free wallet that can have some pay interest on stored crypto	These are not free
Use	These are more useful for trading as they are convenient to access	They are more suited for long-term storage
Cybersecurity	This wallet is connected to the internet, so they are vulnerable to hacking	It is the most secure type of wallet as it's not connected to the internet and has various layers of security
Loss protection	It has a backup and recovery system, though it could be better	The backup and recovery options are for lost passwords but not for lost devices
Efficiency	Since they are already connected to the internet, they are easily accessible. Thus, the exchange is easy	Connecting them to the internet requires extra work like USB, Wi-Fi, etc.

2.4 NFT Applications

A. Digital Art

These are any creative content uploaded on any virtual or digital medium; it can be anything: music, films, paintings, images and many more. Just like physical arts, anyone can sell and buy these. If people can buy and sell physical and digital art, why are these decentralized tokens and NFTs needed? The answer is that in normal art, be it physical or digital, the original art can be stolen or copied and sold. In NFT, each art is given a unique signature, which allows it to be differentiated from any other art. Thus, NFTs enable the authenticity of the produced item, which is also very helpful for the artists, not only securing the art from theft but giving them a massive profit as it has a concept of royalty; in this, the artist receives an amount whenever his art is transferred to a new owner. The most notable sale of digital art in history till now is of art made by Mike Winkelman (also known as Beeple), which caused a sale worth $69 million (Table 3).

B. Licenses and Certification

This is a very time-saving application of NFTs, as it helps minimize the time and effort of the companies who spend much time verifying critical documents. It can also support various institutions in tracking the records of all the certificates issued since each certificate is assigned a unique signature, which can be used to check its authenticity.

C. Virtual World

A new emerging world in the internet will be a virtual place where people can buy and sell assets in the physical world. NFTs in this world will help Maintain and track

Table 3 NFTs marketplaces

S. no.	NFT marketplace	Blockchain	Crypto	Launch	Accepted type
1	OpenSea	Ethereum	$ETH $WETH $DAI	2018	Image, audio, video
2	Variable	Ethereum	$ETH $WETH $DAI	2020	Image, audio, video
3	SuperRare.Co	Ethereum	$ETH	2018	Image
4	Foundation	Ethereum	$ETH	2020	Image
5	Marketplace	Ethereum	$ETH $DEFI	2018	Image
6	Hic et Nunc	Tezos	$XTZ	2021	Image, audio; video
7	KnownOrigin	Ethereum	$ETH	2020	Image
8	CryptoPunks	Ethereum	$ETH	2017	Image

financial transactions and also authenticate them. Since NFT is developed on the blockchain, hiring an external auditor to evaluate the assets is unnecessary.

Until now, we have learnt about many new technologies like Web 3.0, a decentralized and intelligent web that will remove every third party's intervention in any transaction between the two users. It will be handled by the self-made rules of the users in the form of a smart contract. Then we explored the NFTs, which are non-fungible tokens that can be of any form of digital assets with a unique signature that anyone can sell and bring. Together, these technologies lead to another big emerging technology, or a world just like the physical world, but fully virtual. Let us explore this new virtual world, also known as Metaverse.

3 Metaverse

In this world of social networks, people share their activities or things happening with everyone on a platform from which anyone can see and react. Now let us imagine a space where people can interact with each other by becoming virtual people, also known as Avatars; this imagination is now becoming a reality and thus opening a new area for research and development, which is famously known as **Metaverse**.

3.1 What is Metaverse?

Metaverse is a world that is present virtually where billions of people can do daily life chores and interact with each other while sitting in one place in the physical world. In this world, the people will be in their virtual form, also called "*Avatars*". The term "Metaverse" originated from the science fiction novel *Snow Crash* [11] by Neal Stephenson. Metaverse is comprised of two words: "*Meta*", which means beyond, and "*verse*" is taken from the word "*Universe*".

Definition It is a single, shared, immersive, persistent, 3D virtual space where humans can perform all the real-world things, but without actually moving.

This definition was just one way of describing Metaverse; here are some more:

- Metaverse is a mixture of virtual worlds and augmented real worlds.
- Metaverse is a persistent platform, which means it is available whenever someone wants to visit it.
- Metaverse is an integrated system that uses X-tended Reality (XR) and other technologies.
- Metaverse is tightly related to the Reality, actions and interactions that can be exchanged with the real world. Everyone will have a digital twin, which will allow more possible communication.

The metaverse is nascent, or you may say it does not exist. But with the development of various technologies like Blockchain, Artificial intelligence, Augmented Reality and Virtual Reality, it is believed that the metaverse will happen one day. It will be a space that will change how people live, shop, learn, etc.

3.2 Why Metaverse?

As mentioned, this term was first found in a science fiction novel; at that time, no one believed it would be proclaimed the future of the internet. It was October 2021 when Facebook, the social giant, changed its name to "*Meta*" and announced plans to invest around $10 billion in this technology or a concept. This grabbed the attention of everyone towards this emerging, fascinating world.

Just before the Facebook move in 2020, since the world was hit with COVID and everyone was locked in their rooms, no one was allowed to go out, this was the time when people realized that there is a need for a virtual world in which they can roam around freely, learn, work and earn just by sitting at a single place in their house.

This concept has been known for many years, but till now, it was just an idea. The development of various technologies like Virtual Reality, Augmented Reality, Artificial intelligence, etc., gave it some life, though still there is no Metaverse. But computer scientists are predicting or claiming it to be the future of the internet.

3.3 Metaverse Technologies

The development of any technology is dependent on various other technologies that are used to operate it. For example, high-rise buildings can only be built with the invention of lift technology. Similarly, Metaverse is also dependent on various technologies with which it will be fully functional and can provide an immersive experience to the user.

- **Hardware**: Metaverse is a virtual space that requires some intermediate device that will connect the user in the physical world with the virtual world. These are various types of VR/AR/MR headsets; mobile phones also need to be developed so that they can navigate the Metaverse. Different enterprise hardware is also required to help the industry use AR-based environments. Some hardware like Hand-Based input devices, Motion Input devices and non-hand-based input devices.
- **Immersive technologies**: Extended Reality is a comprehensive technology under which there are a series of immersive technologies, electronic and an environment where the data is projected. It also includes AR, VR and MR.
  - **Augmented Reality (AR)**: This is a mixture of physical and virtual worlds in which the devices spatially project the devices' digital artefacts. These devices can be mobile phones, tablets, or glasses. It is a technology that merges the

real and virtual worlds. An example of augmented Reality is the new feature in Google, which shows various digital artefacts of animals on the physical world surface (see Fig. 4b).

- **Virtual Reality (VR)**: It is a separate digitally created space in which the users are immersed and are in a different world located somewhere else. Users can operate things similarly to the physical world using multisensory equipment such as VR headsets and immersion helmets. These can be anything from interacting with a virtual object. There are many developments in this field; now, mobile phones have a 360° gyroscope sensor, which helps see around the virtual world just by moving it in different directions.
- **Mixed Reality (MR)**: This technology is very complex, and it's still not clear how it will shape or what it is exactly. This technology has many definitions and keeps changing with the continuous advancement of various other technologies. It is an advanced AR iteration in which the digital artefact can be projected on the real-world surface, examine the surroundings and interact with them, like hiding behind an object, etc. [12].
- **Computing power**: As mentioned earlier, Metaverse will be a persistent world, which means it can be accessed or viewed at any time; so, to accomplish this, the computing power should be very high to ensure the smooth functioning of all the functions from complex mathematical calculations, data synchronization to Artificial intelligence. The requirement of advanced computation is essential for the seamless functioning of the Metaverse.

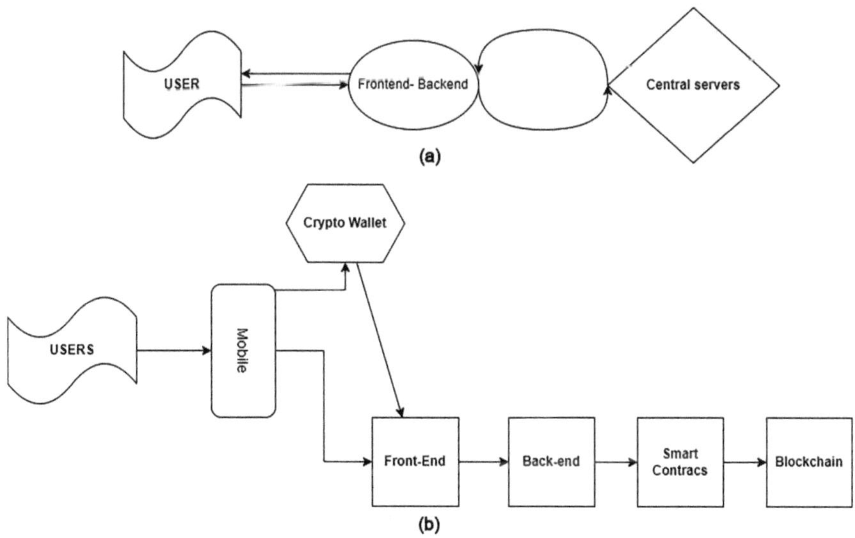

Fig. 4 **a** Centralized app flow; **b** decentralized apps (Dapps)

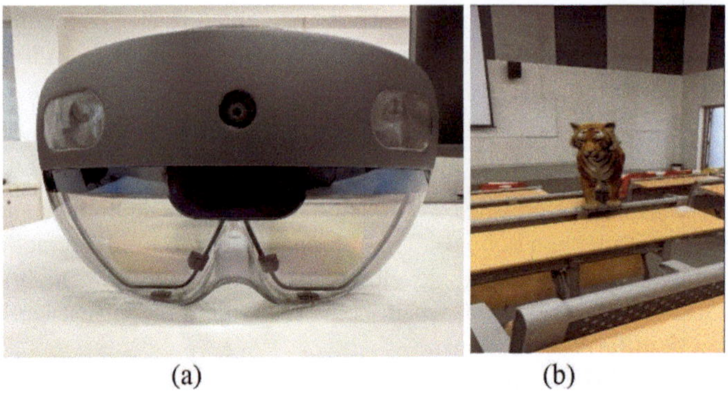

Fig. 5 **a** AR glasses (Bennett University Lab); **b** Google AR search feature (Bennett univ. hall with Google AR feature)

Fig. 6 Oculus Quest 2 VR headset (Bennett University Lab)

There are some more technologies with which Metaverse can become a highly efficient, flexible and secure world (Figs. 5 and 6).

3.4 The Ecosystem of the Metaverse

The economy is one of the backbones of a real world society, people manufacture products, sell them to various distributors who take them forward and sell them in the different markets, where everyone can buy them. In the same way, the virtual world (Metaverse) component is its economic system, which will make the Metaverse just like the real world. This ecosystem consists of four parts: Digital creation, digital asset, digital market and digital currency; this will form the basis of this ecosystem (Table 4).

Table 4 Difference between the conventional economy and Metaverse economy system

S. no.	Conventional	Metaverse
1	The value of anything is always defined based on the time or effort put into the work	It will be a more equitable economy in which the value will be decided based on unique skills and contributions brought by the individual
2	This economy is more bounded by various things like limited cost and availability of resources, etc.	It is a more accessible economy since it is virtual; there is no limitation of resources, availability of space, etc.
3	The cost of the products is high as they depend on various factors like resource costs, shipping and transporting costs and many more	Because of digital distribution, the products will be distributed digitally, and there is no need for shipping; the cost of products will decrease
4	The products could be more scalable as there are many things in between, like shipping, reachable, etc.	Once a product is digitally created, it can be distributed to an unlimited number of users without any extra cost of production

- **Digital Creation**: It is one of the most basic and one of the foundations of the Metaverse. It is a virtual space, means a space of various Graphics, 3d avatar, etc. There is a need for creators and graphic designers who can create them. A tool kit for creating these digital artworks can be anything ranging from immersive arts with immersive technologies (refer to the previous section) to other robotic crafts.
- **Digital Assets**: These are the digital elements that users can buy and sell. These assets can be anything in the Metaverse, from a digital land to a small digital artwork, each with a unique signature. Blockchain technology must help the system verify and protect the assets from theft, piracy, etc.
- **Digital Market**: As discussed in the earlier sections, this place will be available for selling all digital assets. To make this market work seamlessly and control the trading, use of various automation and AI technologies that will study the market trends and will take well analysed decisions to grow the financial market.
- **Digital currency**: There are products to sell and buy, but it can't be done without a safe, acceptable and well-distributed currency. Digital currency will form the base of Metaverse technology.

3.5 Artificial Intelligence in Metaverse

Artificial intelligence is a technology that enables the computer to learn, think and work like humans or even beyond their ability. This technology consists of three sub-technologies or algorithms: Machine learning, Deep learning and Reinforcement learning.

AI in Metaverse can be imagined as a central controlling unit or a user support system. It will be responsible for controlling and guiding and will help the virtual world interact with the physical world seamlessly. It will also help in deciding the financial and market trades by studying the market trends. It can also be used as a technical support system because of its high learning ability.

Let us learn about the three main algorithms of AI that can be used in the Metaverse.

- **Machine Learning in Metaverse**: This is a collection of various algorithms enabling the computer to learn from experience and data humanly. It is subdivided into two segments, i.e., supervised and unsupervised. In supervised learning, the computer is given some labelled data to train itself and deliver results accordingly. For example, if we are making a system for detecting fraud messages, then to prepare the machine, we have to give some examples of fraud messages and some examples of non-fraud messages, which the machine will study and find patterns of what is expected in fraud messages and non-fraud messages and will give the result accordingly.

Support Vector Machine (SVM) is one of the leading candidates that can be used for problems of pattern recognition, regression, etc. This method finds a hyperplane that divides the entire dataset into different segments so that the machine can classify all these segments. The SVM classifier optimizes the Eq. (2) object function.

$$\text{Superscript } L(w) = \sum \max\left(0, 1 - yi\left[w^T x_i + b\right] + \lambda \|w\|^2\right) \tag{2}$$

Here w represents the weight vector, b denotes the threshold and λ denotes a Lagrangian factor; it determines the trade-off between margin maximization and regularization.

However, Machine learning algorithms are not much encouraged as they require selecting the feature manually, and in the case of Metaverse, the feature space is vast, and thus, it will not be efficient to use (Table 5).

Table 5 Application of AI in Metaverse

Use case	Types	ML approaches	Years
Play game tracking	AI objects	RL-based Bayesian network graph [13]	2005
Animate and control avatars	AI objects	State action RL [14]	2006
Face recognition	AI objects	Markov random field [15]	2012
Smart city services	Virtual worlds	Semi-supervised RL [16]	2018
Multi-people tracking	AI objects	CNN [17]	2018
Collaborative learning	Virtual worlds	Mobile edge computing [18]	2022
Action detection	Virtual worlds	[19–21]	2019–2020, 2022
Augmented reality	Virtual worlds	[22–24]	2017, 2020
Avatar moving	AI Objects	RL [25]	2012
Digital twin	VR-interaction	[26]	2022

- **Deep learning in Metaverse**: As discussed earlier, this algorithm works like a human brain. Convolutional Neural Network (CNN) is an algorithm mostly encouraged in Metaverse, as it can solve various computer vision problems like facial recognition, speech recognition, AR technologies, etc. This algorithm uses a weight-sharing method between the neurons, which helps decide which features are essential for the results. In CNN, C stands for Convolution, which consists of various convolutional layers or functions that help extract the elements from the input. These are pooling and padding. This layer is also known as filters or kernels, which float around the input image. In Metaverse, it can help in the immersive technologies, as in MR, the virtual object can interact with the surroundings. So, it can achieve this using object detection and recognition techniques, which are developed using Deep learning.
- **Reinforcement learning (RL)**: It is an algorithm that is based on self-learning and self-punishing method, which means that using this algorithm, the machine first learns from the data, then produces results, and accordingly rewards itself if the results have fewer errors or false alarms and punishes itself if the errors are high. So, it is just like trial and error, as we humans do. Deep learning and RL are mixed, which can bring a massive change in the field of AI.

3.6 Blockchain in Metaverse

As discussed earlier in this chapter, this is a transactional database management technology that uses a decentralized transaction ledger in which every data is replicated and sent to all the nodes of the network or chain, thus removing the dependence on anyone to control and verify it [27–31].

In the Metaverse, this technology can be beneficial in maintaining the economic system, thus forming the base of the Metaverse. One of the most popular applications of the blockchain is cryptocurrency, and since the cryptocurrency is digital and very trustable, this can be used in Metaverse. One can clearly see in Fig. 7 that many cryptocurrencies have been created till now and its growing continuously.

Thus, due to its continuous development and increasing popularity, cryptocurrency is encouraged to be used in the metaverse.

Blockchain technologies are used in cryptocurrency, as discussed in the previous sections of this chapter. It provides a decentralized space where the cryptocurrencies are stored, maintained and monitored on how many are sold, the cost, etc. (Table 6).

4 Conclusion

This chapter takes the reader on a small tour of Web 3.0, NFTs and Metaverse. Also discusses how all the technologies are brought together to create a world that was once imagined and read in fictional stories. This chapter started with a discussion of

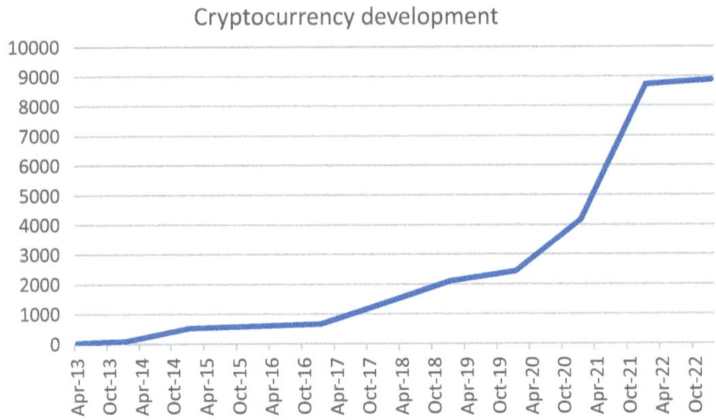

Fig. 7 Development of cryptocurrency

Table 6 Application of blockchain technology in the metaverse

Use	Types	Approaches	Years
Transaction	Blockchain architecture	[32]	2020
PoW, PoS	Consensus mechanism	[33]	2018
Bitcoin, ethereum	Cryptocurrency	[34, 35]	2021
Supply chain	Chain network	[36]	2019
IoT	Blockchain in IoT	[37]	2018

the various ages of the World Wide Web; from Web 1.0 to Web 3.0, we saw various technologies shift and concepts that changed how people think of the Web. Next, this it discusses about digital currency called NFT, which aims to provide safe, smooth and tamper-less transactions in the virtual worlds or web. Lastly, the chapter discusses Metaverse by listing some key features of this new uncertain world. There are many theories, assumptions and uncertainty regarding these topics. Any new technology or concept can replace or change the whole structure, working and execution of the Metaverse. But one thing is sure, "*this will be the future of Internet*".

References

1. Hung A (2023) Assessing the potential involutionary effects of new copyright laws: a techno-legal analysis based on the impact of Web 3.0 on copyright protection. Seattle J Technol Environ Innov Law 14(1):3
2. Berners L. https://www.w3.org/People/Berners-Lee/ShortHistory.html. Accessed 26 Jun 2023
3. Ali O, Jaradat A, Kulakli A, Abuhalimeh A (2021) A comparative study: blockchain technology utilization benefits, challenges and functionalities. IEEE Access 9:12730–12749. https://doi.org/10.1109/ACCESS.2021.3050241

4. Abou Jaoude J, Saade RG (2019) Blockchain applications: usage in different domains. IEEE Access 7:45360–45381
5. Sadeghi M, Mahmoudi A, Deng X (2022) Adopting distributed ledger technology for the sustainable construction industry: evaluating the barriers using ordinal priority approach. Environ Sci Pollut Res 29(7):10495–10520
6. Khan SN, Loukil F, Ghedira-Guegan C, Benkhelifa E, Bani-Hani A (2021) Blockchain smart contracts: applications, challenges, and future trends. Peer-to-Peer Netw Appl 14:2901–2925
7. Mahesh B (2020) Machine learning algorithms: a review. Int J Sci Res 9(1):381–386
8. Rehman W, Zainab H, Imran J, Bawany NZ (2021) NFTs: applications and challenges. In: Proceedings of the 2021 22nd international Arab conference on information technology (ACIT). IEEE, pp 1–7
9. Hildenbrandt E, Saxena M, Rodrigues N, Zhu X, Daian P, Guth D, Moore B et al (2018) Kevm: a complete formal semantics of the ethereum virtual machine. In: Proceedings of the 2018 IEEE 31st computer security foundations symposium (CSF). IEEE, pp 204–217
10. Faqir-Rhazoui Y, Arroyo J, Hassan S (2021) A comparative analysis of the platforms for decentralized autonomous organizations in the Ethereum blockchain. J Internet Serv Appl 12(1):1–20
11. Duan H, Li J, Fan S, Lin Z, Wu X, Cai W (2021) Metaverse for social good: a university campus prototype. In: Proceedings of the 29th ACM international conference on multimedia, pp 153–161
12. Ritterbusch GD, Teichmann MR (2023) Defining the metaverse: a systematic literature review. IEEE Access
13. Chen J-F, Lin W-C, Bai H-S, Yang C-C, Chao H-C (2005) Constructing an intelligent behavior avatar in a virtual world: a self-learning model based on reinforcement. In: IRI-2005 IEEE international conference on information reuse and integration, conference, 2005. IEEE, pp 421–426
14. Lee J, Lee KH (2004) Precomputing avatar behavior from human motion data. In: Proceedings of the 2004 ACM SIGGRAPH/eurographics symposium on computer animation, pp 79–87
15. Yampolskiy RV, Klare B, Jain AK (2012) Face recognition in the virtual world: recognizing avatar faces. In: Proceedings of the 2012 11th international conference on machine learning and applications, vol 1. IEEE, pp 40–45
16. Mohammadi M, Al-Fuqaha A (2018) Enabling cognitive smart cities using big data and machine learning: approaches and challenges. IEEE Commun Mag 56(2):94–101
17. Fabbri M, Lanzi F, Calderara S, Palazzi A, Vezzani R, Cucchiara R (2018) Learning to detect and track visible and occluded body joints in a virtual world. In: Proceedings of the European conference on computer vision (ECCV), pp 430–446
18. Jovanović A, Milosavljević A (2022) VoRtex Metaverse platform for gamified collaborative learning. Electronics 11(3):317
19. Wu K, Wang H, Esfahani MA, Yuan S (2019) BND*-DDQN: learn to steer autonomously through deep reinforcement learning. IEEE Trans Cognit Develop Syst 13(2):249–261
20. Huang J, Li N, Zhang T, Li G, Huang T, Gao W (2018) Sap: self-adaptive proposal model for temporal action detection based on reinforcement learning. In: Proceedings of the AAAI conference on artificial intelligence, vol 32
21. Zhang J, Zi L, Hou Y, Wang M, Jiang W, Deng D (2020) A deep learning-based approach to enable action recognition for construction equipment. Adv Civil Eng 2020:1–14
22. Park K-B, Kim M, Choi SH, Lee JY (2020) Deep learning-based smart task assistance in wearable augmented reality. Robot Comput Integ Manuf 63:101887
23. Alhaija HA, Mustikovela SK, Mescheder L, Geiger A, Rother C (2017) Augmented reality meets deep learning for car instance segmentation in urban scenes. In: British machine vision conference, vol 1
24. Lampropoulos G, Keramopoulos E, Diamantaras K (2020) Enhancing the functionality of augmented reality using deep learning, semantic web and knowledge graphs: a review. Vis Inform 4(1):32–42

25. Kastanis I, Slater M (2012) Reinforcement learning utilizes proxemics: an avatar learns to manipulate the position of people in immersive virtual reality. ACM Trans Appl Percept 9(1):1–15
26. Zhu H (2022) Metaaid: a flexible framework for developing metaverse applications via AI technology and human editing. arXiv preprint arXiv:2204.01614
27. Bhardwaj A, Krishna CR (2018) Performance evaluation of bandwidth for virtual machine migration in cloud computing. Int J Knowl Eng Data Min 5(3):139–152
28. Bhardwaj A, Krishna CR (2019) Improving the performance of pre-copy virtual machine migration technique. In: Proceedings of 2nd international conference on communication, computing and networking: ICCCN 2018, NITTTR Chandigarh, India. Springer, Singapore, pp 1021–1032
29. Bhardwaj A, Gupta U, Budhiraja I, Chaudhary R (2023) Container-based migration technique for fog computing architecture. In: Proceedings of the 2023 international conference for advancement in technology (ICONAT). IEEE, pp 1–6
30. Bhardwaj A, Singh AP, Sharma P, Abid K, Gupta U (2023) Performance evaluation of virtual machine and container-based migration technique. International conference on data analytics and management. Springer, Singapore, pp 551–558
31. Kaur G, Shrivastava R, Gupta U (2023) Blockchain integration with internet of things (IoT)-based systems for data security: a review. International conference on data analytics and management. Springer, Singapore, pp 617–625
32. Daian P, Goldfeder S, Kell T, Li Y, Zhao X, Bentov I, Breidenbach L, Juels A (2020) Flash boys 2.0: frontrunning in decentralized exchanges, miner extractable value, and consensus instability. In: Proceedings of the 2020 IEEE symposium on security and privacy (SP). IEEE, pp 910–927
33. Duong T, Chepurnoy A, Fan L, Zhou HS (2018) Twinscoin: a cryptocurrency via proof-of-work and proof-of-stake. In: Proceedings of the 2nd ACM workshop on blockchains, cryptocurrencies, and contracts, pp 1–13
34. Ethereum (2023) Welcome to Ethereum. https://bitcoin.org/en/bitcoin-for-businesses. Accessed 16 Jun 2023
35. Bitcoin (2023) Bitcoin for businesses. https://bitcoin.org/en/bitcoin-for-businesses. Accessed 16 Jun 2023
36. Malik S, Dedeoglu V, Kanhere SS, Jurdak R (2019) Trustchain: trust management in blockchain and IoT supported supply chains. In: Proceedings of the 2019 IEEE international conference on blockchain (Blockchain). IEEE, pp 184–193
37. Novo O (2018) Blockchain meets IoT: an architecture for scalable access management in IoT. IEEE Internet Things J 5(2):1184–1195

The Metaverse: Disclosing a Computerized Wilderness of Conceivable Outcomes

Mukesh Choubisa, Nachiket Patel, and Dhruv Patel

Abstract Through the use of VR, AR, and other immersive technologies, users can interact with computer-generated surroundings in real time in the metaverse, a virtual world made possible by the fusion of physical and virtual reality. The concept has gained significant attention in recent years, with many companies and organizations working towards creating their own versions of the metaverse. The metaverse offers a unique, immersive, and potentially transformative online experience that allows people to connect, collaborate, and create in ways that were previously impossible in the physical world. As the metaverse continues to evolve, it is likely to play an increasingly important role in our lives, both online and offline. Examples of current metaverse platforms include Second Life, Minecraft, and Fortnite. As technology continues to develop, the metaverse promises to become more immersive, interactive, and ubiquitous. It has the potential to transform industries such as entertainment, education, and business, providing new opportunities for social interaction, learning, and business transactions while also addressing challenges related to privacy, security, and ethics.

Keywords Virtual reality · Metaverse · VR · Environments · Metaverse space · Cold storage · AR · Extended · Mixed reality · Replicate · Gaming · Education · Entertainment

1 Introduction

In the era of Artificial Intelligence, metaverse is a promising concept that refers the virtual world. The separation of physical and non-physical (virtual) reality creates a virtual world. In the metaverse world, users/client can interact with a system generated environment in real-time with the help of VR, AR and other technology. VR stand for virtual reality and AR stand for augmented reality. It is essentially a virtual

M. Choubisa (✉) · N. Patel · D. Patel
Department of Computer Science and Engineering, SoE, Indrashil University-Mehesana, Rajpur, Gujarat, India
e-mail: mukesh24sa@gmail.com

G. Chhabra and K. Kaushik (eds.), *Understanding the Metaverse*, Blockchain Technologies, https://doi.org/10.1007/978-981-97-2278-5_2

world where client/users can directly interact with each other and communicate with digital objects [1] as if they were in a real-world environment.

The new interesting concept of the metaverse has more focused in recent years by technocrat, Scientist and Researchers, particularly as advancements in technology have made it possible to create more immersive and engaging virtual experiences. While there is no single metaverse yet, many companies and organizations are working towards creating their own versions of the metaverse, and it is seen as a potential future direction for the internet as a whole.

The metaverse gives a unique, immersive, and potentially transformative online experience that allows different-2 people to connect at one place, collaborate for multidisciplinary work, and create different ways that were not possible in offline or physical mode. Metaverse has the potential to create a path we connect with each other with technology, communication, creativity and collaboration for research, education, etc. As the metaverse continues to progress, it is to be expected play an important role in our offline and online lives.

1.1 What is Metaverse?

A virtual environment or cosmos that is shared in real time by millions of people is referred to as the "metaverse." It is a three-dimensional environment where users may interact with digital items and environments while also communicating with one another online. Though the idea of a metaverse has its roots in science fiction, it has grown in significance with the development of virtual and augmented reality. Users can design their own avatars, explore virtual worlds, and take part in a variety of activities in the metaverse, including trading, socializing, and gaming.

Some supporters of the Metaverse envision it as a brand-new type of internet where users may easily and completely immerse themselves in a variety of services and experiences. Some see it as a potential hub for creativity and innovation, where fresh genres of art, entertainment, and social interaction may flourish. Although the Metaverse is still mostly a vision for the future, there are already a lot of virtual worlds and platforms, such as Second Life, Minecraft, and Fortnite that provide certain aspects of the Metaverse experience. The metaverse is projected to become a more significant and influential part of our digital life as augmented reality and virtual reality technologies advance.

In Metaverse, users can interact in real time, explore virtual environments, participate in virtual events, and perform activities that simulate real experiences. Some examples of current Metaverse platforms include virtual games, online communities (Minecraft), and various social-media platforms. However, as technology continues to develop, the Metaverse promises to become more immersive, interactive, and ubiquitous. The metaverse has the would-be to renovate many industries, including entertainment, education, and business [2]. This can provide new opportunities for social interaction, learning, and business transactions, as well as new challenges related to privacy, security, and ethics. As such, it is a rapidly developing area of interest for entrepreneurs, technologists, and policy makers.

2 Existing Modern Prototypes of Metaverse Applications

Several modern prototypes of Metaverse applications exist, including:

- Second Life: A virtual environment called Second Life was introduced in 2003. It enables users to make avatars, connect with one another, and carry out tasks like purchasing attend performances and online classes.
- Minecraft: Popular sandbox video game Minecraft allows players to build and explore virtual worlds. Due to the fact that it enables students to take part in group learning exercises online, it is frequently employed in education.
- VRChat: Users of the social virtual reality network VRChat can make and personalize avatars, communicate with one another virtually, and attend events like concerts and comedic performances.
- Decentraland: A virtual environment called Decentraland was created using the Ethereum network. Users can build and personalize avatars, exchange virtual goods and services, and interact with one another in virtual environments.
- Somnium Space: Users of the virtual reality platform Somnium Space can explore and engage in virtual environments. Users can also construct and sell virtual property, go to events, and participate in activities like playing games and seeing art exhibits.

These prototypes show some of the potential uses and functions of the metaverse, including social, educational, entertaining, and commercial ones. They are still in the early stages of development, though, so there is still a ton of space for creativity and development [3].

3 Enabling Technologies of Metaverse

In order to build a flawless virtual world, Metaverse depends on a number of supporting technologies.

The following are some of the key technologies that make up the Metaverse:

- Virtual reality (VR) and augmented reality (AR) technologies: VR and AR enable users to interact with virtual items and other users while being immersed in virtual surroundings [1]. In contrast to augmented reality, which uses cameras and screens to superimpose digital data on the real environment, virtual reality primarily uses a headset and hand controls.
- 3D Modelling and Animation: Through the use of 3D modeling and animation, virtual characters, settings, and objects can be created. With the aid of these

technologies, users can design and develop their own avatars, virtual homes, and other things they can use in the metaverse.

- Blockchain and cryptography: In the metaverse, a decentralized and safe virtual economy is being built using blockchain technology. It enables users to purchase, sell, and exchange virtual goods and money while maintaining the security and transparency of all transactions [3].
- Artificial Intelligence (AI) and Machine Learning (ML): AI and ML technology provide sophisticated virtual identities and individualized user experiences in the metaverse. They enable virtual characters to interact with users in a more genuine and human manner and to modify their behaviour to match user preferences.
- Cloud Computing: To store and manage the massive amounts of data and processing power required to enable Metaverse, cloud computing is used. It enables users to connect to the metaverse from any location and engage in real-time interactions with other users and virtual items [3].

These enabling technologies are essential to the development and operation of Metaverse, and they will continue to evolve and improve as Metaverse develops and matures.

3.1 Threats and Countermeasures to Authentication and Access Control in Metaverse

Access control, authentication and authorization are keys to ensuring user security and privacy within Metaverse platform. However, there are also threats to these systems that must be addressed by appropriate countermeasures. Some common threats and countermeasures for authentication and access control in metaverse [4] include:

- Spoofing: Hackers can steal user credentials, such as usernames user and passwords, to gain unauthorized access to user accounts. Social Engineering: Hackers can use social engineering tactics to trick users into revealing their credentials or other sensitive information.
- Malware: It can be used to steal user credentials or other sensitive information, such as credit card numbers.
- Insider Threats: Insider threats, such as employees or contractors with authorized access, can abuse their privileges to gain unauthorized access to user accounts or sensitive data.
- Countermeasures: Strong authentication: Strong authentication measures, such as multi-factor authentication (MFA) and biometric authentication, can help prevent identity theft by requiring proof of identity extra in addition to passwords.
- User Education: Educating users about the dangers of social engineering tactics and how to avoid them can help prevent users from falling victim to these attacks. Virus and malware protection: Installing virus and malware protection

software can help prevent malware from stealing user credentials or other sensitive information.
- Role-Based Access Control: RBAC can help prevent insider threats by limiting access to sensitive data and systems based on role and responsibilities. Overall, it is important to implement a layered approach to security and to regularly review and update authentication and access control measures to address emerging threats in Metaverse.

3.2 Economy-Related Threats and Countermeasures in Metaverse

As the Metaverse grows in popularity, it will likely become a target for various economic-related threats. Here are some common threats and countermeasures related to the economy in the Metaverse:

- **Fraud**: Fraudulent activities, such as scams and Ponzi schemes, can be a significant threat to the Metaverse economy. Fraud can take many forms, including fake ICOs, fraudulent virtual assets, and phishing attacks.
- **Money Laundering**: The anonymity provided by virtual currencies can be used for money laundering purposes. Money laundering can involve buying virtual currencies with illicit funds and then converting them back into fiat currency.
- **Hacking**: Hacking attacks can target virtual currencies and digital wallets, which can result in the loss of funds and damage to the Metaverse economy.
- **Price Manipulation**: Price manipulation can occur when a small group of individuals controls a significant portion of a virtual asset. These individuals can artificially inflate or deflate the price of the asset to their advantage.

Countermeasures:

- **Regulatory Frameworks**: Regulatory frameworks can help to prevent fraudulent activities and money laundering in the Metaverse. Regulations can provide oversight and ensure that operators and users of the Metaverse comply with ethical and legal practices.
- **Smart Contracts**: Smart contracts can be used to automate transactions, reduce the need for intermediaries, and ensure transparency [3]. Smart contracts can help to reduce the risk of fraudulent activities by enforcing pre-defined rules and conditions.
- **Anti-Money Laundering (AML) and Know Your Customer (KYC) Regulations**: AML and KYC regulations can be implemented to prevent money laundering and fraudulent activities. AML regulations require virtual currency exchanges to verify the identities of their customers, while KYC regulations require exchanges to verify the source of the funds used to purchase virtual currencies.

- **Cold Storage**: Cold storage can be used to store virtual currencies offline, reducing the risk of hacking attacks.
- **Decentralization**: Decentralization can help to reduce the risk of price manipulation by preventing a small group of individuals from controlling a significant portion of a virtual asset. Decentralized networks can ensure that virtual assets are distributed more evenly across the network.
- By implementing these countermeasures, Metaverse operators can help to ensure the security and stability of the Metaverse economy, and protect users from fraudulent activities, money laundering, and other economic-related threats [1].

4 Extended Reality, Mixed Reality and Virtual Reality

1. **Extended Reality (XR)**: VR, AR and MR are included in the emerging technology is known as XR. With the use of these technologies, consumers can be transported to brand-new virtual worlds or given digital experiences that can be superimposed on the actual world.

VR = Virtual Reality, AR = Augmented Reality, MR = Mixed Reality, XR = Extended Reality.

2. **Virtual Reality (VR)**: VR is a used to express an entirely immersive digital experience that immerses people in an entirely virtual setting. When using VR, a motion controller and headset are often used to interact with the virtual environment.
3. **Augmented Reality (AR)**: The term "augmented reality" (AR) refers to a technology that uses the camera of a smart-phone or tablet to superimpose digital content over the physical world. AR can be utilized for a multitude of things, such adding details about a real-world object or surroundings or improving the gaming experience.
4. **Mixed Reality (MR)**: In mixed reality (MR), which combines VR and AR, digital content is superimposed on the physical environment and fixed to particular places or items. With MR technology, users often wear a headset or special glasses that let them view both the real and virtual worlds simultaneously.

All three of these immersive technologies have significant potential for an extensive range of applications, including gaming, education, training, and even healthcare. As these technologies continue to evolve and improve, we can look forward to see even more novel and exciting use cases emerge in the future (Fig. 1).

Now let's see each of the technology in detail.

Virtual Reality (VR) can be categorized into several types based on the level of immersion and the hardware and software used. Here are the main types of VR in detail.

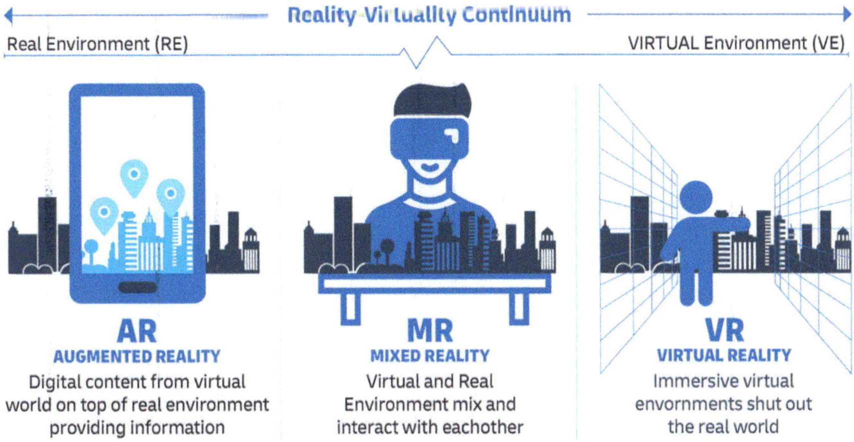

Fig. 1 Extended reality, mixed reality and virtual reality

Completely Immersive VR:

Fully immersive virtual reality (VR) is defined by the usage of head-mounted displays (HMDs), which completely enclose the user's field of vision and frequently come with coordinating headphones for sound immersion. Spatial Following: High-quality sensors and following frameworks are utilized to distinguish the user's head developments and alter the virtual environment accordingly.

Motion Controllers: In expansion to HMDs, movement controllers may be utilized to track the user's hand developments and signals, permitting for interaction with virtual objects.

Importance: Completely immersive VR gives the foremost practical and immersive encounter, making it appropriate for applications like gaming, reenactments, and training.

Partially Immersive VR:

Screen-Based VR: This sort of VR does not completely encase the user but employments screens to form a virtual environment. It can be experienced through gadgets like PCs, smart- phones, or projectors.

Augmented Reality (AR): AR overlays computerized data or objects onto the genuine world. Whereas it doesn't completely inundate clients in a virtual environment, it improves the real-world encounter with advanced elements.

Importance: In part immersive VR is more available and can be utilized for a wide run of applications, from portable gaming and virtual visits to information visualization and training.

360° VR:

360° Recordings: These recordings are shot with specialized cameras that capture all encompassing to see the environment. Clients can see these recordings utilizing VR headsets to see around and investigate the captured environment.

Limited Interaction: 360° VR regularly offers constrained interaction since it's basically utilized for inactive encounters like observing documentaries or encountering virtual tourism.

Importance: 360° VR gives a sense of nearness and submersion for detached substance utilization, making it perfect for travel encounters, documentaries, and virtual tours.

Web-Based VR:

WebVR/WebXR: These advances permit clients to get to VR encounters straightforwardly through web browsers, disposing the required for devoted VR apps or program installations.

Accessibility: Web-based VR makes VR encounters more available to a broader group of onlookers by rearranging the get to process.

Importance: Web-based VR brings down boundaries to section for VR substance utilization and can be utilized for promoting, instructive substance, and intuitively web encounters.

Significance of VR:

Immersive Learning: VR is utilized in instruction and preparing to make reasonable recreations for aptitude advancement. Restorative understudies can hone surgeries, and pilots can prepare in realistic flight simulations.

Entertainment: VR gaming could be a critical industry, advertising immersive and intelligently gaming encounters. VR moreover empowers virtual concerts, motion picture theatres, and other excitement experiences.

Healthcare: VR is utilized for torment administration, presentation treatment, and physical restoration. It can give diversions for patients amid restorative procedures.

Architecture and Plan: Planners and architects utilize VR for virtual walkthroughs of buildings and models. It helps in plan visualization and client presentations.

Virtual Tourism: Clients can investigate removed areas for all intents and purposes, giving a travel-like involvement from the consolation of their homes.

Drawbacks of VR:

Cost: High-quality VR equipment and computer program can be costly, constraining openness to a few users.

Motion Affliction: A few clients may encounter movement affliction or distress when utilizing VR, particularly amid fast developments in virtual environments.

Isolation: Completely immersive VR can be confining as clients are cut off from the genuine world whereas utilizing VR headsets.

Technical Challenges: VR frameworks require effective equipment, and making practical virtual situations can be in fact demanding.

Future Angles and Scope of VR:

Education and Training: VR's potential for immersive learning encounters will proceed to develop. It'll be utilized in schools, colleges, and corporate preparing programs.

Healthcare: VR will be progressively coordinates into treatment and restoration programs. It may moreover be utilized for restorative meetings and inaccessible surgeries.

Entertainment: VR gaming and amusement will advance with more reasonable design and intuitively encounters. Virtual reality topic parks and arcades will likely gotten to be more common.

Social Interaction: Social VR stages will permit individuals to associated in virtual spaces, making long-distance communication more engaging.

Work: VR will play a bigger part in farther work and collaboration, empowering virtual gatherings, conferences, and office spaces.

Accessibility: As innovation progresses and gets to be more affordable, VR will ended up more available to a broader extend of users.

In conclusion, Virtual Reality could be a transformative innovation with a wide extend of applications and a promising future. As equipment gets to be more reasonable and program more advanced, VR's significance and scope will proceed to grow, affecting different businesses and changing the way we learn, work, and engage ourselves.

4.1 Economy-Related Threats and Countermeasures in Metaverse

1. **Marker-Based AR:**
    - How it Works: Marker-based AR depends on physical markers or pictures that are recognized by AR computer program. These markers serve as reference focuses for the framework to get it where to overlay advanced content.
    - Illustrations: Common cases incorporate utilizing QR codes, printed pictures, or particular designs as markers. When a user's device's camera recognizes the marker, it triggers the show of advanced substance, such as 3D models, activities, or data, superimposed onto the marker in real time.
    - Applications: Marker-based AR is frequently utilized in instructive apps, intuitively exhibition hall shows, publicizing campaigns, and limited time materials. It gives an intuitively and locks in way to communicate data or engage users.
2. **Markerless AR (Location-Based or Positional AR):**

- How it Works: Markerless AR depends on a device's sensors, counting GPS, accelerometers, and spinners, to decide the user's area and introduction. This data is utilized to put advanced substance relative to the real-world environment.
- Cases: Location-based AR apps, like Pokémon GO, utilize GPS to position virtual animals within the genuine world. Other applications can utilize sensors to put virtual objects on surfaces or inside particular physical spaces.
- Applications: Markerless AR is utilized broadly in location-based gaming, route apps (e.g., Google Maps AR), and tourism guides. It empowers users to investigate and connected with computerized substance that's consistently coordinates with their surroundings.

3. **Projection-Based AR:**
    - How it Works: Projection-based AR includes the utilization of projectors to show computerized substance onto physical objects or surfaces within the genuine world. These projectors can be mounted on gadgets or stationary fixtures.
    - Cases: Envision anticipating advanced craftsmanship onto the dividers of a room or showing intelligently controls on a tabletop. Projectors can moreover be utilized for virtual shopping shows or intuitively gallery exhibits.
    - Applications: Projection-based AR is commonly utilized in craftsmanship establishments, publicizing (e.g., intelligently storefronts), and intelligently shows in galleries and galleries.
4. **Superimposition-Based AR:**
    - How it Works: Superimposition-based AR replaces or increases the user's real-world see with computer-generated substance. It can include overlaying objects, characters, or data on the genuine world, making it show up as on the off chance that they coexist.
    - Illustrations: In mechanical applications, superimposition-based AR can overlay repair informational onto a bit of apparatus. In customer applications, it might include attempting on virtual clothing or setting virtual furniture in a real room.
    - Applications: Superimposition-based AR is broadly utilized in mechanical settings for errands like upkeep, gathering, and preparing. Within the buyer space, it's utilized for virtual try-ons, insides plan visualization, and intelligently advertising.

5 Significance of AR

Enhanced Client Encounters: AR improves the genuine world with advanced data and intelligently components, giving clients with locks in and instructive experiences.

Education and Preparing: AR is utilized in instruction for intuitively learning encounters and in preparing for reenactments and aptitude development.

Navigation and Wayfinding: AR-based route apps improve users' capacity to discover headings and find focuses of intrigued within the genuine world.

Entertainment: AR gaming and excitement apps offer intuitively and immersive encounters, obscuring the lines between the computerized and physical worlds.

Retail and Promoting: AR is utilized for virtual try-ons, item visualizations, and intuitively showcasing campaigns to lock in consumers.

Drawbacks of AR:

Technical Confinements: AR depends on sensors, cameras, and handling control, which may change over gadgets and influence the quality of the AR experience.

Privacy Concerns: AR can raise security issues, particularly when capturing and utilizing information almost the user's surroundings.

Dependency on Portable Gadgets: AR encounters are overwhelmingly gotten to through smartphones and tablets, constraining their scope in certain applications.

Future Viewpoints and Scope of AR:

Wearable AR Gadgets: The improvement of AR glasses and headsets (e.g., Microsoft HoloLens, Google Glass) will increment the predominance and comfort of AR.

Enterprise and Mechanical Utilize: AR is anticipated to play a developing part in areas like fabricating, healthcare, and upkeep, where it can move forward productivity and decrease errors.

5G and Cloud Integration: Quicker web speeds and cloud computing will upgrade AR capabilities by empowering real-time information preparing and collaboration.

Education and Inaccessible Collaboration: AR will gotten to be more unmistakable in instruction and inaccessible work, offering imaginative ways to memorize and collaborate.

Consumer Selection: AR applications for shopping, gaming, and social interaction will proceed to pick up notoriety, extending its part in day by day life.

Augmented Reality in Healthcare: AR will be utilized for restorative preparing, telemedicine, and making strides quiet results through visualization and guidance.

AR in Design and Plan: Planners and originators will utilize AR for 3D modeling, visualization, and client presentations.

Augmented reality's potential is endless, and its development is driven by innovative headways and a wide run of applications. As AR gadgets ended up more open and program more advanced, its significance and scope will proceed to extend, bridging the crevice between the physical and computerized universes.

5.1 Let's Dig into Blended or Mixed Reality (MR) in Detail, Covering Its Sorts, Significance, Downsides, Future Perspectives, and Scope

Mixed Reality (MR):

1. **Expanded Reality (AR) + Virtual Reality (VR) Spectrum:**
   - MR exists on a range between AR and VR. It consistently mixes the genuine world with virtual components, permitting clients associated with both simultaneously.
   - Depending on the adjust of real-world and virtual substance, MR encounters can incline more toward AR (more real-world substance) or VR (more virtual content).

Importance of MR:

1. **Improved Client Encounters**: MR gives more immersive and intuitively encounters than AR alone by consolidating the computerized and physical universes seamlessly.
2. **Mechanical and Proficient Utilize**: MR is profitable for errands such as inaccessible help, item plan, upkeep, and preparing in businesses like fabricating, healthcare, and aviation.
3. **Instruction and Preparing**: MR offers practical reenactments and intelligently learning encounters, upgrading instruction and proficient training.
4. **Design and Plan**: Designers and architects utilize MR for 3D modeling, visualization, and client introductions, permitting clients to see virtual representations of their ventures in real- world environments.
5. **Inaccessible Collaboration**: MR encourages collaboration by permitting clients to connected with computerized substance and each other inside shared virtual spaces, in any case of physical location.

Drawbacks of MR:

1. **Complexity and Fetched**: Creating MR applications can be complex and expensive due to the require for specialized equipment and software.
2. **Specialized Challenges**: Accomplishing consistent integration of virtual and real-world components can be challenging and may require exact following and mapping technologies.
3. **Client Adjustment**: Clients may require time to adjust to MR encounters, especially in case they are modern to the innovation or on the off chance that it includes wearing headsets.

Future Angles and Scope of MR:

1. **Progressions in Equipment**: The improvement of lightweight, comfortable, and more reasonable MR headsets will drive selection and extend MR's reach.

2. **5G and Cloud Integration**: Quicker and more solid organize network will empower real-time information handling, improving MR encounters and inaccessible collaboration.
3. **Mechanical Applications**: MR will proceed to be received in businesses such as fabricating, healthcare, and upkeep for assignments that require real-time data overlay.
4. **Instruction and Preparing**: MR will play an expanding part in instruction and proficient preparing, advertising immersive and intuitively learning experiences.
5. **Inaccessible Work and Collaboration**: MR will encourage inaccessible work by giving virtual office situations and intuitively assembly spaces.
6. **Customer Amusement**: MR gaming and amusement applications will advance to offer more intuitively and immersive experiences.
7. **Healthcare**: MR will be utilized for restorative preparing, telemedicine, and progressing understanding results through visualization and guidance.
8. **Retail and Commerce**: MR will be utilized for virtual try-ons, item visualizations, and intuitively shopping experiences.
9. **Structural Visualization**: MR will gotten to be fundamentally in design and plan for making practical 3D visualizations and walkthroughs.

As innovation progresses and MR gets to be more open, it'll proceed to convert businesses and improve lifestyle by seamlessly blending the advanced and physical universes. MR's potential for advancement and its capacity to form locks in and enlightening encounters make it an energizing field with considerable future prospects.

6 Limitations of 2D Learning Environment

Two-dimensional (2D) learning environments have several limitations that may hinder the effectiveness of the learning process. Some of these limitations include:

Lack of realism: 2D learning environments are limited in their ability to replicate real-life situations, which may reduce the degree to which learners can transfer what they have learned to practical, real-world settings.

Limited interactivity: 2D environments generally offer limited interactivity compared to three-dimensional (3D) environments, which can hinder the learners' engagement and immersion in the learning experience.

Limited spatial cues: 2D environments lack the spatial cues provided by 3D environments. This can make it difficult for learners to orient themselves in the environment and understand the relationships between different objects and elements.

Limited sensory cues: 2D learning environments do not provide the same sensory cues as real-life or 3D environments. This can reduce the effectiveness of learning related to skills that require sensory feedback, such as physical coordination and balance.

Limited ability to simulate complex systems: 2D learning environments may not be suitable for simulating complex systems, such as ecological systems, due to their limited ability to represent the relationships and interactions between different components of the system.

Overall, while 2D learning environments can be effective in certain contexts, they are generally less effective than 3D learning environments in providing an immersive and realistic learning experience.

6.1 *Brief History of Virtual Media and XR Technologies*

The history of virtual media and extended reality (XR) technologies can be traced back to the mid-twentieth century. Here are some significant milestones (Fig. 2):

1950—The first attempts at creating virtual reality (VR) technology began with the Sensorama, an arcade-style machine that provided multisensory experiences such as smell and vibration.

1960—Ivan Sutherland invented the first head-mounted display (HMD) and created the first VR system called the Sword of Damocles.

1970—The term "virtual reality" was given by Jaron Lanier, who later founded VPL Research, which was the first company to produce VR software and hardware.

1980—The first commercially available VR system, the VPL Data Glove, was released, allowing users to manipulate virtual objects with their hands.

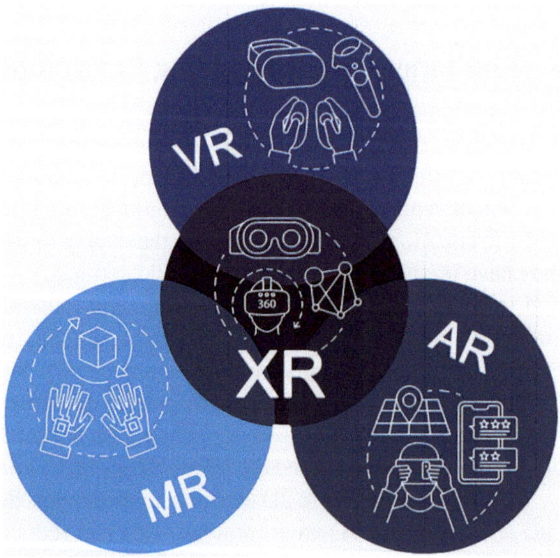

Fig. 2 VR, XR, AR, MR

1990—The introduction of the World Wide Web led to the development of new forms of virtual media, massively multiplayer online role-playing games (MMORPGs), such as World of War-craft, and virtual worlds like Second Life.

2000s—Mobile technology and advancements in graphics processing allowed for the development of augmented reality (AR) and mixed reality (MR) technologies, which integrate digital elements into the real world.

2010s—The launch of the Oculus Rift in 2012 and subsequent release of other consumer VR headsets marked a new era of accessible and immersive VR experiences. The rise of smartphones also led to the proliferation of AR technology through mobile apps such as Pokémon Go.

2020s—The development of XR technologies continues to accelerate, with applications ranging from entertainment and gaming to education, healthcare, and industrial training.

Overall, the development of technology and software has led to increasingly immersive and lifelike virtual experiences throughout the history of virtual media and XR technologies.

7 Virtual Worlds (VW) and Virtual Reality (VR) in Education

VW and VR offer significant potential for education, providing immersive and interactive learning environments that can supplement or replace traditional classroom instruction. Here are some ways in which virtual worlds and VR are being used in education:

Simulations: Virtual worlds and VR can be used to create simulations of real-world environments or scenarios that are too expensive, dangerous, or impractical to replicate in real life. Examples include flight simulators for aviation training, medical simulations for healthcare professionals, and engineering simulations for training and design purposes.

Experiential learning: Virtual worlds and VR can provide experiential learning opportunities that allow students to explore and interact with content in new ways. For example, students can explore ancient ruins in a virtual world, dissect a frog in a VR simulation, or experience the effects of climate change on an ecosystem.

Remote learning: Virtual worlds and VR can facilitate remote learning, allowing students to participate in classes and collaborate with peers and teachers from anywhere in the world. This can be especially useful for students who are unable to attend traditional classes due to geographical or other constraints.

Enhanced engagement: Virtual worlds and VR can increase student engagement by providing a more interactive and engaging learning experience. This can be especially beneficial for students who may struggle with traditional classroom instruction or have different learning styles.

Accessibility: Virtual worlds and VR can provide accessible learning opportunities for students with disabilities, allowing them to participate in educational activities that may be challenging or impossible in real life.

Overall, virtual worlds and VR offer a wide range of possibilities for enhancing and transforming education, providing new and innovative ways to engage students and improve learning outcomes.

VR Affordances

Virtual Reality (VR) provides several affordances that make it a unique and powerful tool for a wide range of applications, including education, training, entertainment, and more. Here are some key affordances of VR:

Immersion: VR provides a highly immersive experience that can fully engage users' senses and transport them to another world. This can be especially useful for creating simulations or training scenarios that closely mimic real-world experiences.

Interactivity: VR allows for highly interactive experiences, enabling users to manipulate objects and interact with the environment in ways that are not possible with other media. This can be especially useful for creating experiential learning opportunities that enable learners to explore and experiment with content in new ways.

Presence: VR can create a strong sense of presence, making users feel as if they are truly present in the virtual environment. This can be especially useful for creating social experiences or training scenarios that require a high degree of realism and immersion.

Flexibility: VR provides a flexible platform for a wide range of applications, from gaming and entertainment to education and training [3]. This can be especially useful for creating personalized learning experiences that cater to the needs and preferences of individual learners [1].

Data collection: VR provides a rich source of data that can be used to track user behaviour and gather insights into learning and performance. This can be especially useful for creating personalized learning experiences that adapt to individual users' needs and preferences.

Overall, VR provides a range of unique affordances that make it a powerful tool for a wide range of applications, providing new and innovative ways to engage users and enhance learning, training, and other experiences.

7.1 Metaverse Contemporary Development

The perception of the metaverse has been around for decades, but it is currently experiencing a resurgence in interest and development [5]. Metaverse is basically a virtual world where person/users can communicate with each other and with digital content in a flawed and immersive way [3]. Here are some of the contemporary developments in the metaverse space:

Virtual worlds: Virtual worlds such as 2nd life and VR Chat have been around for over a decade, providing users with immersive social and gaming experiences. These virtual worlds are now being augmented with new technologies such as blockchain, which can provide a decentralized and secure infrastructure for virtual economies and transactions.

Social media and gaming: Social media platforms such as FB and Snapchat are also investing heavily in the metaverse, with plans to construct virtual EVS, in this platform users can communicate with each other and with digital content in new and innovative ways. Gaming companies such as Roblox and Fortnite are also exploring the metaverse space, providing users with immersive gaming experiences that incorporate social elements.

Blockchain and cryptocurrencies: Blockchain and cryptocurrencies are being used to power virtual economies in the metaverse, permitting client/users to buy and sell virtual goods and services using digital currencies such as Bit-coin and Ethereum. These virtual economies are becoming increasingly sophisticated, with some users making significant amounts of money by buying and selling virtual real estate, art, and other assets.

Virtual concerts and events: The pandemic (COVID-19) has accelerate the development of virtual concerts and events, providing users with immersive experiences that replicate the feeling of attending a live event. These virtual events are becoming increasingly sophisticated, with some artists and performers using motion-capture technology to create realistic avatars and interactions with the audience.

Overall, the metaverse is an electrifying and speedily growing space that is likely to have a considerable impact on the way we cooperate with digital content and with each other in the coming years. As the technology continues to progress and grown up, researcher can anticipate seeing even more innovative and exciting developments in the metaverse space.

8 Future Research Directions of Metaverse

The Metaverse is an evolving concept, and there are still many research directions that can be explored. Here are some potential future research directions for the Metaverse:

Interoperability: The Metaverse consists of many different virtual environments and platforms, each with its own standards and protocols. Research can explore ways to improve interoperability between these platforms, enabling users to move seamlessly between them.

Governance: The Metaverse is a decentralized network, and research can explore ways to govern this network effectively. Governance structures can help to ensure that the Metaverse operates fairly and transparently, and that users' rights are protected.

Privacy: As the Metaverse becomes more popular, the privacy of users is likely to become a significant concern. Research can explore ways to ensure that user data is protected and that users can control how their data is used.

Accessibility: Metaverse has the latent to provide new opportunities for individuals with disabilities, but research can explore ways to make the Metaverse's more reachable and inclusive for all users.

Ethics: The Metaverse raises many ethical questions, such as the impact of virtual experiences on the physical world and the ethics of creating virtual environments that simulate real-world experiences [1]. Research can explore these questions and provide guidelines for ethical practices in the Metaverse.

Sustainability: Metaverse has the impending to consume significant amounts of energy, and research can explore ways to make the Metaverse more sustainable and reduce its environmental impact.

Security: As the Metaverse becomes more valuable, it will likely become a target for hackers and other malicious actors. Research can explore ways to recover the security of the Metaverse and protect user data and virtual assets [4].

Overall, the Metaverse presents many exciting opportunity and challenges for researchers in various fields [2], together with computer science, economics, sociology, and philosophy. Future research can help to shape the development of the Metaverse and ensure that it operates in a way that is fair, beneficial, and transparent for all users [5].

9 Conclusion

The metaverse, a concept once limited to the domains of science fiction, is presently quickly advancing into a substantial and transformative advanced reality. It speaks to a joining of immersive innovations such as VR, increased AR, and blended reality MR, making interconnected virtual universes where clients can live, work, play, and socialize. As we peer into the metaverse's skyline, a few key takeaways emerge.

First and first, the metaverse hold the latent to reform how we interface with one another and lock in with computerized situations. Social intelligent inside the metaverse rise above geological boundaries, advertising modern roads for worldwide collaboration, communication, and shared encounters. Whether it's going to virtual concerts, collaborating on a extend, or basically hanging out with companions in a computerized space, the metaverse offers uncommon openings for association and community.

Moreover, the metaverse has the control to disturb conventional businesses and rethink how we conduct commerce, learn, and engage ourselves. In instruction, immersive learning encounters can transport understudies to chronicled occasions or complex logical marvels, improving comprehension and maintenance. Within the work environment, virtual workplaces and collaborative stages empower farther groups to collaborate consistently, introducing in a modern period of adaptable work arrangements.

However, the metaverse isn't without its challenges. Security and security concerns linger huge, requiring vigorous shields to ensure users' information and characters. Issues of openness and inclusivity must moreover be addressed to guarantee that the metaverse may be a welcoming space for all, notwithstanding of physical capacities or financial circumstances.

As we look toward long haul, the metaverse's scope shows up boundless. Innovative headways will proceed to refine the client involvement, making it more immersive and natural. The advancement of interoperable measures and conventions will cultivate a more open and interconnected metaverse, empowering consistent moves between distinctive virtual universes and experiences.

In conclusion, the metaverse is balanced to gotten to be an indispensably portion of our computerized lives, reshaping how we associated, learn, work, and engage ourselves. Its potential for positive change is tremendous, but it too carries noteworthy duties. As we explore this courageous modern computerized wilderness, it is officeholder upon us to guarantee that the metaverse is built on standards of inclusivity, protection, and moral stewardship, enabling people to investigate and shape their virtual predeterminations whereas protecting the values that characterize our shared humankind.

References

1. Zhang Q (2023) Secure preschool education using ML and metaverse technologies. Appl AI
2. Srivastava AK, Tyagi S, Sharma D, Dubey A (2023) Metaverse is the future of influencer marketing. IGI Global
3. Ray PP (2023) Web3: a review on background, applications, technologies, architectures, zero-trust, challenges and future directions. IoT Cyber Phys Syst 3:213–248
4. Su Z, Xing R, Zhang N, Wang Y, Liu D, Shen X, Luan TH (2022) A survey on metaverse: fundamentals, security, and privacy. IEEE Commun Surv Tutor
5. Kouchih A, Lyoussi D (2022) Metaverse and financial inclusion opportunities and risks for the banking ecosystem. IGI Global

Metaverse, NFTs, and Web 3.0: Applications, Challenges, and Future Opportunities

A. Mansurali, K. Nikitha Reddy, Tania Chakraborty, and K. Jyothish

Abstract The emergence of Metaverse, NFTs (non-fungible tokens), and Web 3.0 technologies has created a new ecosystem for digital content ownership and online interaction. This book chapter aims to explore the applications of this new paradigm by analyzing the technical, economic, social, privacy, and security aspects of Metaverse, Web 3.0, and NFTs. The book chapter first presents a comprehensive overview of Metaverse, including its applications and challenges. It then discusses the role of NFTs in Metaverse and how they can facilitate ownership and monetization of digital assets. It then discusses that Metaverse and NFTs have the potential to disrupt existing business models and create new opportunities for creators and consumers alike. This book chapter explores the combination of Metaverse, NFTs, and Web 3.0 and presents an architecture that integrates all three. Furthermore, the paper discusses social implications of Metaverse and NFTs, including issues of identity, privacy, security, and governance. It further discusses that the decentralized nature of Web 3.0 technologies presents both opportunities and challenges for creating a fair and equitable Metaverse ecosystem. Finally, book chapter discusses the current trends and future of Metaverse, NFTs, and Web 3.0 which has potential in shaping the future of online interaction and content ownership.

Keywords Metaverse · NFTs · Web 3.0 · Privacy · Security · Virtual world · Applications

---

A. Mansurali
Central University of Tamil Nadu, Thiruvarur, Tamil Nadu 610005, India

K. N. Reddy · T. Chakraborty · K. Jyothish (✉)
Christ University, Bengaluru, Karnataka 560029, India
e-mail: k.jyothish@mba.christuniversity.in

G. Chhabra and K. Kaushik (eds.), *Understanding the Metaverse*, Blockchain Technologies, https://doi.org/10.1007/978-981-97-2278-5_3

1 Introduction

The Metaverse, Web 3.0, and NFTs are all interrelated and are expected to have a significant impact on the future of the internet. The Metaverse is a virtual world that aims to offer a fully immersive experience for its users. Web 3.0 is the internet's next evolution, and it is projected to be more decentralized, secure, and transparent. NFTs, on the other hand, are one-of-a-kind digital assets recorded on a blockchain that provide proof of ownership and authenticity.

In the context of the Metaverse, NFTs are expected to play a crucial role in creating a virtual economy. Users will be able to purchase, sell, and exchange virtual items including apparel, homes, and even virtual land. NFTs will secure ownership of these assets, ensuring that they are unique, valuable, and cannot be reproduced. This virtual economy is expected to be constructed on a decentralized blockchain, giving individuals ultimate control over their assets and allowing peer-to-peer transactions without the necessity of intermediaries.

Web 3.0 is expected to be the backbone of the Metaverse, providing the infrastructure necessary for creating a fully immersive and decentralized virtual world. It will enable users to interact with each other in real time, create their own content, and have complete control over their data and assets. Web 3.0 will also enable the integration of various technologies such as AI, IoT, and blockchain, which will enhance the user experience and enable new use cases.

Overall, the combination of the Metaverse, Web 3.0, and NFTs is projected to alter the way people interact with the internet and open up new avenues for creativity, innovation, and economic growth.

1.1 Metaverse

The word Metaverse is a combination of the prefix "meta" (meaning transcendence) and the suffix "verse" (word universe), which is a shared virtual world linked to the physical world. The term was first coined by science fiction author Neal Stephenson in a novel Snow Crash. The author described the Metaverse as a massive interconnected network of virtual environments in which users interact with each other. The major technology drivers of Metaverse are virtual reality (VR) and augmented reality (AR), 3D modeling and animation, game engines, blockchain, artificial intelligence (AI), cloud computing, and Internet of Things (IoT). Individual users in the Metaverse build digital versions of themselves called avatars, which are used to explore and experience the Metaverse [1].

1.2 Non-fungible Tokens

Non-fungible tokens (NFTs) are one of the types of crypto-assets, and it is a building block of Metaverse economy. It is a "cryptographic asset on a blockchain containing unique identifying information and codes that separate them from one other", as defined by Peres et al. Since NFTs are stored on a blockchain, they cannot be duplicated or destroyed, and their ownership can be easily verified. This makes them an ideal way to manage ownership of virtual assets in the Metaverse, where the value of these assets can be significant [2, 3].

One of the examples of usage of NFTs in Metaverse is through digitally collectible real estate property ownership. Users can purchase land on Metaverse using any of the following platform's native currency such as MANA, SAND, Axie Infinity Shards (AXS), WAX (WAXP), and ENJ, and once the users owns their virtual property, it can be utilized for many purposes such as building virtual homes, businesses, and any other structures. Here, NFTs represent the ownership of the bought land on the Metaverse platform by creating a unique LAND token for the land which is bought [2, 3]. Users can easily establish the existence and ownership of land or other digital assets using this method.

NFTs have garnered significant attention in recent years especially in art and entertainment industries. As per the report, the 24-h trading volume on average of the NFT market is 35,534,962,685 USD and the 24-h trading volume of the entire cryptocurrency market is 1,074,288,222,081 USD. The total value of NFT sales as of March 2023 is led by Ethereum with 37,386,616,034 USD with the total volume of NFT sales 49,663,893,752.65 USD. The total NFT transactions all time are 154,015,693 with total NFT buyers reaching to 9,976,443. Such rapid growth and the humongous amount of NFT sales occurring now are an indication of NFTs likely to continue to play an important role in the digital economy.

1.3 Web 3.0

Web 1.0 refers to the earliest version of the World Wide Web, which was developed in the late 1980s and early 1990s. Tim Berners-Lee described Web 1.0 as "read-only web". In Web 1.0, the web pages were static and were not interactive. The web pages developed in Web 1.0 generation were static and were not changing frequently. Producers and service providers started publishing of online catalogues for the advertisement of their products or services. The main goal of the websites was to publish the information for anyone at any time and establish an online presence.

Web 2.0 is the second generation of the internet. During a conference in 2004, Dale Dougherty, a web pioneer and O'Reilly VP, coined the phrase. It changed the way individuals on the internet connect with one another and exchange content. Web 2.0 has become a platform for users to create and share their content; notable social media platforms such as Twitter, Facebook, and Instagram were emerged; and highly responsive, interactive, and dynamic web applications were born.

Web 3.0 is the third iteration to the internet which is a Decentralized Web running on top of blockchain technologies. In Web3.0, data is stored in decentralized and distributed storage structure, which means that there is no centralized server nodes. In a blockchain network, data is stored across a distributed network of nodes, each of which has a copy of the data. This means that there is no single point of failure, and the network can continue to function even if some nodes go offline.

Users have control over their data and identities on Web 3.0. Each user will manage their own data using blockchain tokens in which organizations will have no say, such as the usage of non-fungible tokens (NFTs). Tim Berners-Lee quoted Web 3.0 as the Semantic Web (Semantic Web Technologies enable people to create data stores on the web, build vocabularies, and write rules for handling data) [4].

2 Metaverse, Web 3.0, and NFTs: Privacy, Governance, and Opportunities

The Metaverse uses data collected from the real world to provide immersive experiences, and that sensors attached to users, such as gyroscopes, can be used to control their avatar movements in the virtual world [5]. However, it is vital to emphasize that the usage of sensors and data collection in the Metaverse can generate privacy and security problems. Users, for example, may not wish to submit some sorts of personal data, such as biometric information or location data, which could be used to monitor or identify them. In addition, the Metaverse's vast virtual worlds and communities can also create new challenges in terms of privacy and security. For example, users may be subject to eavesdropping or surveillance by other users or platform operators, which could compromise their privacy and personal information. To address these challenges, Metaverse developers and operators will need to prioritize privacy and security protections, such as strong encryption, user consent mechanisms, and secure data storage and management [6]. This will be essential to building trust and ensuring that the Metaverse is a safe and secure environment for all users.

The Metaverse has the ability to transform our civilization by opening up new avenues for expression and engagement. Ethical design, on the other hand, must be a primary priority. When developing the Metaverse, it is vital to address the creation process, privacy and security, inclusivity and diversity, and governance.

2.1 Privacy and Security in the Metaverse

Privacy and security are crucial aspects to consider when designing the Metaverse. With so much personal information shared and saved in the Metaverse, it is critical to protect users' data. The adoption of blockchain technology can provide a secure and transparent method of storing and managing user data, ensuring that it cannot be tampered with or stolen.

However, blockchain technology alone is not enough to ensure privacy and security in the Metaverse. Additional measures such as encryption, multi-factor authentication, and biometric identification can also be used to protect users' data and prevent unauthorized access. Furthermore, the implementation of privacy policies and transparency in data handling can also enhance user trust in the platform [7].

2.2 *Inclusivity and Diversity in the Metaverse*

The Metaverse has the potential to break down barriers and promote inclusivity and diversity. It provides a platform where individuals can interact and express themselves without limitations imposed by their physical appearance, location, or social status. However, this potential can only be realized if the Metaverse is designed with inclusivity and diversity in mind [8].

To achieve this, designers should ensure that the platform is accessible to individuals with disabilities and those from different socioeconomic backgrounds. The Metaverse should also be designed to promote gender and racial diversity and avoid perpetuating existing biases and stereotypes.

2.3 *Governance in the Metaverse*

The Metaverse presents unique challenges for governance due to its decentralized and open nature. To ensure the ethical use and development of the Metaverse, a governance framework must be established [9]. This framework should involve all stakeholders, including users, creators, and platform owners, and be based on principles of transparency, accountability, and inclusivity.

One possible approach to governance is the use of decentralized autonomous organizations (DAOs), which are self-governing entities that operate based on smart contracts. DAOs can provide a transparent and democratic way for stakeholders to make decisions and manage resources in the Metaverse.

Non-fungible tokens (NFTs) are used to confirm the ownership and existence of digital assets, including videos, photographs, works of art, event tickets, and more. NFTs are created using smart contracts on the Ethereum blockchain. NFTs provide a promising solution for intellectual property (IP) protection, as their full-history tradability, deep liquidity, and convenient interoperability allow creators to earn royalties each time of a successful trade on any NFT market or by peer-to-peer exchanging. Although NFTs are essentially just code, they have ascribed value due to their comparative scarcity as a digital object [2]. NFTs have gained popularity in the art world as they allow for the ownership and transfer of digital artwork in a unique and secure way. NFTs provide artists with the ability to monetize their digital creations, as each NFT represents a unique piece of art that can be bought and sold in a marketplace. The ownership of an NFT proves the authenticity and ownership of the digital artwork, providing a way for artists to protect their work from piracy and theft. The market for NFT art has grown rapidly, with some pieces selling for millions of dollars at auctions [2, 10]. NFTs have also been used in the music industry, with musicians selling NFTs representing unique versions of their songs or albums. The use cases for NFTs continue to expand, and it is expected that they will play an increasingly important role in the future of intellectual property and digital asset ownership.

Many recent research work discussed various opportunities presented by NFTs in different industries. One such domain is gaming, where NFTs have the potential to change the gaming experience by giving ownership records for things in games and establishing an economic marketplace in the ecosystem that benefits both creators and users. NFTs also offer great potential in virtual events, such as ticketing, where blockchain-based smart contracts can provide a transparent ticket trading platform for stakeholders such as event organizers and customers, eliminating the need for third-party intermediaries. In addition, NFTs can protect digital collectibles such as artworks by transforming them into digital formats with integrated identities, thereby providing an efficient way to manage and protect digital masterpieces while enabling artists to receive predetermined royalty fees each time their artwork exchanges in the markets. Finally, by offering a perfect decentralized environment for the virtual online world, where users can enjoy games, display their own works of art, trade assets and virtual properties, and even make money from the virtual economy, NFTs can inspire the Metaverse, a collective virtual shared space that permits all types of digital activities.

Web 3.0 is considered as the next phase of the internet, operating on top of blockchain-related technologies, which offers a decentralized approach to data storage and management, significantly reducing service costs. Web 1.0 and Web 2.0 have played their respective roles in shaping the internet, with Web 1.0 providing a single-directional approach to accessing information, while Web 2.0 enabling user-created content and personalized recommendations [11]. Web 3.0, marked by decentralization and intelligence, operates on blockchain and artificial intelligence technologies, allowing websites to learn autonomously, employ DAO for digital identity and asset management, and utilize extended reality and blockchain distributed storage technologies to form a decentralized autonomous network [4]. Web 3.0 can provide unified consensus data valorization services and decentralized services, laying a solid platform for the growth ecology of the digital economy.

The security and network structure are crucial components of Web 3.0. To enhance the security of the network, various efforts have been made. The original centralized network protocols are to be replaced with the fully decentralized communications layer called Mainframe. The foundation of Web 3.0 is thought to be blockchain, and to maintain privacy, data should be stored in a distributed fashion with many copies. For reducing the consumption of resources, Kan et al. proposed a new PRE scheme. In terms of smart contracts, researchers have proposed various methods to characterize and detect gas-inefficient patterns and cross-contract vulnerabilities. To make blockchain technology healthier, researchers employed parallel-fork symbolic execution to expedite smart contract vulnerability detection and proposed proper permission constraints to prevent Ponzi schemes on Ethereum.

The emergence of Web 3.0 has led to the development of new network architectures that impact the internet and information technology industries. The adoption of blockchain technology has enabled the creation of digital currencies such as Bitcoin, which has transformed the digital economic system. The decentralization, privacy, and security preserving features of blockchain have also made it a solution for various applications. Non-fungible tokens (NFTs) have enabled unique digital assets to be sold on the market via blockchain technology. Web 3.0 technology has also been investigated for its possible use in the Metaverse and decentralized storage systems.

In the context of industry digitization, the industrial internet has become a central infrastructure. The industrial internet intends to connect machines to networks, collect production data, and enable remote control. The abstraction of software has enabled the optimization of entire factories that are larger than a single machine. The rise of the Internet of Things (IoT) has created a solid academic foundation for the interaction of physical items with networks. Blockchain technology has provided a good way to organize an Industrial Internet of Things (IIoT) that addresses challenges related to interoperability, heterogeneity, confidentiality, and safety vulnerabilities. Blockchain techniques have been used to provide solutions with enhanced safety and dependability. Implementing blockchain-based IIoT has already been utilized in numerous industries, such as the food supply chain, as an effective way of monitoring food quality and safety by offering a more traceable and transparent chain of custody. Specific models, such as the M2M communication model and the UAV-assisted M2M system, have been employed to improve the IIoT's stability and to enable data computation and decision-making.

The Metaverse relies on Web 3.0 technologies such as blockchain to create a decentralized and secure environment where users can interact and transact with each other. NFTs are also powered by blockchain technology, and their unique properties make them a natural fit for use in the Metaverse, where users can own and trade virtual assets. Additionally, the Metaverse can serve as a platform for NFT-based experiences, such as virtual art galleries or concerts.

3 Understanding Metaverse

A Metaverse is a virtual world that is totally immersive and interactive, allowing users to communicate with one another in real time. It is the concept of a collective virtual shared environment that is not restricted by physical reality and is produced and altered by its users' collective participation. The term Metaverse was coined by science fiction author Neal Stephenson in his 1992 novel Snow Crash, which referred to a virtual reality space where people could interact [12].

The Metaverse is a vast, ever-evolving digital universe. It is a place where people can explore and create their own realities. It is a place of infinite possibilities and endless exploration. People can create virtual worlds, build businesses, and engage in social activities. The Metaverse is constantly changing and evolving. It is a living, breathing digital ecosystem constantly being updated and improved. It is a place where people can communicate and exchange ideas, experiences, and stories with one another. It is a place where people can learn, create, and grow. A Metaverse is a place of endless possibilities. It is a place where people can express themselves and explore their creativity [12]. It is a place where people can come together and collaborate on projects and ideas. It is a place where people can connect and build relationships.

The Metaverse is a vast and intricate digital world that offers endless possibilities for exploration, creation, and connection. However, navigating and understanding the Metaverse can be a challenging task, as it encompasses various virtual worlds, economies, currencies, goods, laws, governments, communities, environments, tools, networks, identities, opportunities, investments, markets, services, technologies, regulations, analytics, data, systems, lifestyles, cultures, events, activities, relationships, and experiences. To explore and create within the Metaverse safely, it is crucial to comprehend the rules and regulations that govern it [1]. It is necessary to understand the different types of virtual worlds and how they are structured, ranging from massive multiplayer games to social media platforms. Similarly, grasping the diverse virtual economies and how they work, including barter, auction, subscription, and freemium models, is crucial to take full advantage of the Metaverse's potential. Furthermore, comprehending the various virtual currencies and how they are used, from Bitcoin to Ethereum, is essential to make informed financial decisions. Similarly, it is necessary to participate in the Metaverse's vibrant marketplace to understand the different types of virtual goods and how they are traded, such as virtual real estate, fashion items, and digital art.

Moreover, understanding the different types of virtual laws and how they are enforced, ranging from terms of service to copyright laws, is vital to avoid legal issues. Additionally, comprehending the different types of virtual governments and how they are structured, from decentralized autonomous organizations (DAOs) to virtual nations, is essential to understanding how power is distributed in the Metaverse. Furthermore, understanding the different types of virtual communities and how they interact, from fan clubs to online forums, is crucial to building meaningful relationships in the Metaverse [1]. It is also essential to understand the different types of virtual environments and how to navigate them, from virtual reality headsets to web browsers, to move seamlessly between different parts of the Metaverse.

To thrive in the Metaverse, it is also necessary to understand the different types of virtual tools and how to use them, such as content creation software, digital wallets, and communication platforms. Similarly, understanding the different types of virtual networks and how to interact with them, such as blockchain and peer-to-peer networks, is crucial to leverage the Metaverse's potential. Moreover, protecting virtual identities and understanding the different virtual security and privacy types are crucial to avoid fraud and data breaches [1]. Also, managing virtual currencies and understanding the different types of virtual communication and how to use them is essential to communicate with other users and participating in the Metaverse's economy.

3.1 Metaverse Architecture and Supporting Concepts

See Fig 1.

The concept of the Metaverse is still evolving, and there is ongoing debate about what exactly it encompasses. However, some common themes have emerged, and there are generally two key aspects that are seen as critical to its development: technology and ecosystem. Under the technology aspect, the fourteen focused areas for the Metaverse can be broadly categorized as follows:

- **Infrastructure**: The infrastructure required to support the Metaverse is significant, as it requires a massive amount of computational power and bandwidth to create and operate virtual worlds on a large scale. This includes both the hardware and software components, such as servers, data centers, cloud computing, and networking protocols. The development of more advanced hardware, such as GPUs and high-speed processors, will also be crucial for the Metaverse's growth.
- **Interoperability**: It refers to the ability of many platforms, apps, and devices to coexist in the Metaverse, independent of their origin or ownership. This is crucial for the Metaverse's success since it will let users to freely roam between virtual worlds and experiences, as well as enabling developers to design programs that can interact with numerous platforms.
- **Identity**: Identity is essential for users to establish a persistent, secure, and unique identity within the Metaverse. This includes creating and managing virtual identities, linking them to real-world identities, and maintaining control over personal data and information. Using blockchain technology, digital signatures, and biometrics can help establish secure and reliable identity systems within the Metaverse.

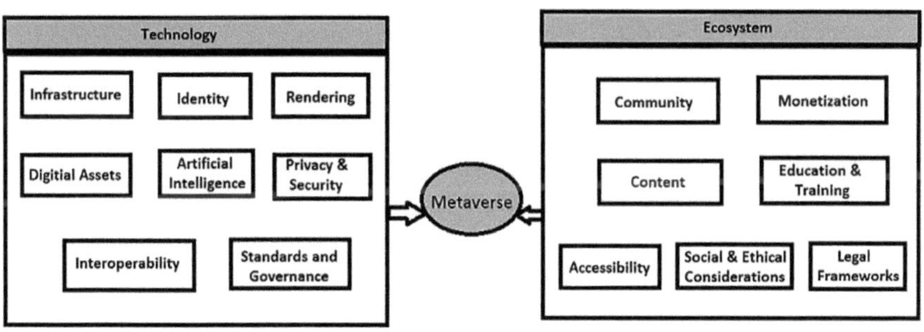

Fig. 1 Metaverse architecture

- **Digital Assets**: These assets are the lifeblood of the Metaverse, and they include virtual currency, virtual goods, and virtual real estate. The ability to create, own, trade, and use digital assets within the Metaverse is critical for its economic and social development. This includes developing standards and protocols for asset creation, exchange, and management, as well as ensuring their security and authenticity.
- **Rendering**: It is the process of creating realistic and immersive virtual environments, and it requires advanced rendering techniques, such as real-time ray tracing, physics simulation, and AI-driven content creation. This is essential for creating engaging and believable virtual worlds that can attract and retain users.
- **Artificial Intelligence**: These technologies can enhance the functionality and user experience of the Metaverse in various ways. This includes natural language processing, computer vision, and machine learning, which can help to create more interactive and intelligent virtual environments, as well as improve user interfaces and user engagement.
- **Privacy and Security**: Privacy and security are critical aspects of the Metaverse, as they involve the protection of user data and digital assets within virtual environments. This includes encryption, authentication, and fraud prevention, as well as the development of privacy policies and user consent mechanisms.
- **Standards and Governance**: The development of industry standards and best practices for the Metaverse is crucial for its equitable and sustainable growth. This includes the establishment of governance models that can ensure fairness and transparency in decision-making, as well as the development of ethical and legal frameworks that can address social and ethical issues related to the Metaverse's development.

Under the ecosystem aspect, the fourteen focused areas for the Metaverse can be broadly categorized as follows:

- **Community**: Building a vibrant and engaged community is essential for the Metaverse's success, as it will help to drive growth and innovation. This includes creating social networks, forums, and other channels for users, creators, developers, and investors to connect and collaborate.
- **Content**: Creating and curating compelling content is crucial for attracting and retaining users within the Metaverse. This includes developing a diverse range of experiences, such as games, social networks, education, and entertainment, as well as providing tools and resources for creators to develop their own content.
- **Monetization**: Developing sustainable and scalable business models is critical for the Metaverse's long-term success. This includes advertising, subscriptions, microtransactions, and virtual real estate, as well as exploring new revenue streams, such as NFTs and blockchain-based transactions.
- **Education and Training**: Providing resources and training for users, creators, and developers is essential for supporting the Metaverse's growth and development. This includes providing access to tutorials, documentation, and other educational resources that can help users to build their skills and knowledge.

- **Accessibility**: Ensuring that the Metaverse is accessible and inclusive for all users is crucial for promoting diversity and equity. This includes addressing issues related to physical accessibility, such as hardware requirements and interfaces, as well as cognitive and socioeconomic accessibility, such as language and cultural barriers.
- **Social and Ethical Considerations**: Addressing the social and ethical implications of the Metaverse is essential for ensuring its responsible and sustainable development. This includes issues related to privacy, identity, behavior, and governance, as well as exploring the potential social and cultural impacts of the Metaverse.
- **Regulation and Legal Frameworks**: Developing regulatory and legal frameworks is critical for ensuring the safety, security, and fairness of the Metaverse, while also fostering innovation and growth. This includes establishing standards and guidelines for user behavior, content moderation, data privacy, and intellectual property rights.

3.2 Leveraging the Potential of Metaverse

For those looking to capitalize on the Metaverse's potential, it is crucial to understand the different types of virtual opportunities and how to pursue them, such as starting a virtual business or investing in digital assets. Understanding the different types of virtual investments and how to maximize them, such as NFTs and virtual real estate, is also crucial to take advantage of the Metaverse's potential returns. Furthermore, it is essential to understand the different types of virtual markets and how to manipulate them, from decentralized exchanges to trading platforms, to create and capture value in the Metaverse's economy [1]. Similarly, leveraging the different types of virtual goods and services and how to monetize them, such as selling virtual merchandise or offering virtual experiences, is necessary to create sustainable revenue streams. To utilize the Metaverse's potential, it is crucial to understand the different types of virtual platforms and how to use them, such as social media and gaming platforms. Also, developing virtual applications and understanding the different types of virtual analytics and how to interpret them, such as user behavior and market trends, is essential to create innovative solutions in the Metaverse. Additionally, it is vital to integrate virtual services and understand the different types of virtual experiences.

Metaverse has several characteristics and features that make it unique from other digital platforms. Some of the key characteristics and features are:

- **Immersive and Interactive**: Metaverse is a fully immersive and dynamic digital reality where users can connect with one another in real time. It lets users to engage with the environment and with one another, generating a sense of presence in the virtual world.
- **Shared Space**: Metaverse is a collaborative virtual shared place generated and altered by its users' collective activity. It is not limited by physical reality and can be accessed from anywhere in the world.

- **Persistent**: Metaverse is a persistent digital universe, meaning that the environment and objects within it exist and persist even when users are not present.
- **Scalable**: Metaverse is a scalable platform that can support many users simultaneously. Millions of users can access it worldwide, and the environment can be expanded and scaled up as needed.

Metaverse has the potential to impact several areas of our lives, including entertainment, socialization, education, and commerce. Some of the potential applications and impacts are:

- **Entertainment**: Metaverse has the potential to revolutionize the entertainment industry by creating immersive and interactive experiences for users. It can be used to create virtual concerts, gaming, and other forms of entertainment.
- **Socialization**: Metaverse can be used as a platform for socialization, allowing users to interact with each other in real time in a virtual space. It can create virtual communities, online events, and social gatherings.
- **Education**: The Metaverse has the ability to alter the education industry by providing students with immersive and interactive learning experiences. It has the ability to construct virtual classrooms, simulations, and educational games.
- **Commerce**: Metaverse can be used as a platform for e-commerce, allowing users to shop and make purchases in a virtual environment. It can create virtual storefronts, digital marketplaces, and virtual advertising.

Overall, the Metaverse has the potential to alter the way we connect with one another and with the digital world, opening us previously unimaginable opportunities and experiences. The Metaverse can extend the physical environment by utilizing extended, augmented, and virtual reality technology. The Metaverse's ancestors include interactive virtual environments and immersive gaming. The transformative impact of the Metaverse on society, particularly in terms of its effects on social interactions as it becomes more widely adopted. Sectors potentially impacted by the Metaverse include marketing, education, tourism, and healthcare.

4 Understanding NFTs

NFTs, or non-fungible tokens, are digital assets that represent ownership of a unique piece of content, such as art, music, videos, or any other digital file. Unlike traditional cryptocurrencies, which are interchangeable with one another and have the same value, NFTs are one-of-a-kind and cannot be replicated or exchanged for another asset [3, 13]. NFTs are built on blockchain technology, allowing their authenticity and ownership to be verified, tracked, decentralized, and transparent. This means that anyone on the blockchain network can easily verify the ownership and history of an NFT [14]. The working of NFTs is based on blockchain technology, a decentralized and transparent ledger that allows for verifying and tracking ownership of digital assets. Here is a simple working example of how an NFT works:

- A creator uploads a digital asset, such as a piece of artwork, onto a marketplace platform that supports the creation and sale of NFTs.
- The marketplace platform creates a unique smart contract representing digital asset ownership. This smart contract contains a set of rules that define the ownership and transfer of the NFT.
- The creator lists the NFT for sale on the marketplace platform, setting a price and specifying any other conditions for the sale.
- A buyer purchases the NFT directly from the marketplace platform. The purchase price is typically paid in cryptocurrency, such as Ethereum.
- The smart contract representing the NFT is updated to reflect the transfer of ownership to the buyer. The blockchain records the ownership history of the NFT, which anyone on the network can easily verify [14].
- The buyer can then hold onto the NFT as a collectible or sell it on the marketplace platform to another buyer.

4.1 Properties of NFTs

- **Verifiability**: Verifiability is a crucial property of NFTs, as it allows for the public confirmation of ownership and metadata associated with the token. This is made possible through the use of blockchain technology, which provides an immutable and transparent record of all transactions on the network. By making ownership and metadata publicly verifiable, NFTs enable greater trust and transparency in digital transactions.
- **Interoperability**: Interoperability refers to the ability of NFTs to be designed to be interoperable with other blockchain networks, allowing for greater flexibility and use cases. Interoperability enables NFTs to be transferred seamlessly between different networks, providing users with more options and enabling them to access new markets.
- **Transparent Execution**: The transparency of NFT execution means that all activities related to NFTs are publicly accessible. This includes the creation or "minting" of the NFT, as well as all subsequent transactions, such as buying and selling. This level of transparency allows for greater accountability and enables users to easily track the history and ownership of a particular NFT.
- **Tradability**: The tradability of NFTs means that they can be traded and exchanged freely, allowing for maximum flexibility and value in the digital marketplace. NFTs are designed to be fully transferable, enabling users to easily buy, sell, or exchange them with other parties. This property of NFTs has enabled the creation of new digital marketplaces and the emergence of a vibrant and dynamic digital economy.
- **Availability**: The availability of NFT systems refers to the fact that they never go down, and all tokens and issued NFTs are always available for selling and buying.

This is made possible through the use of decentralized blockchain networks, which are designed to be resilient and available 24/7. This high level of availability makes NFTs an attractive option for those seeking to engage in digital commerce and create unique digital assets.

- **Usability**: Usability is an important property of NFTs, as it ensures that ownership information is up-to-date, user-friendly, and easy to understand. This is critical for facilitating the smooth and efficient transfer of NFTs between different parties. NFTs must be designed with usability in mind, ensuring that they are accessible to users of all technical backgrounds.
- **Tamper-resistance**: The tamper-resistance of NFT metadata and trading records means that once a transaction is confirmed, it cannot be manipulated or altered in any way. This ensures the authenticity and integrity of the NFT and provides a high level of trust and confidence in the digital marketplace. Tamper-resistance is a critical property of NFTs, as it helps to prevent fraud and ensure that all transactions are conducted in a fair and transparent manner.
- **Atomicity**: The atomicity of NFT transactions means that they can be completed in a single ACID transaction, ensuring that they are consistent, isolated, and durable. This is important for ensuring that all transactions are completed successfully and that there are no inconsistencies or errors in the process. Atomicity also ensures that all parties involved in the transaction are able to achieve their desired outcome, making NFTs a reliable and trusted digital asset.
- **Immutability**: The immutability of NFTs means that they cannot be modified, altered, or deleted once they are created. This ensures that the authenticity and ownership of the NFT remains intact over time, providing users with a high level of confidence in their digital assets. Immutability is a critical property of NFTs, as it helps to prevent fraud and ensure that all transactions are conducted fairly and transparently.

4.2 Properties of NFTs

Some of the use cases of NFTs are:

- **Gaming**: In gaming, NFTs can be used to represent unique in-game things like weapons, skins, or characters. This allows gamers to own and sell their digital goods, making the gaming experience more immersive and engaging.
- **Music**: NFTs can represent ownership of unique and limited-edition music releases, concert tickets, and other exclusive merchandise. This enables artists to monetize their work directly and bypass traditional intermediaries like record labels or ticketing platforms.
- **Virtual Real Estate**: NFTs can represent ownership of virtual real estate in online worlds, such as Decentraland or Somnium Space. This allows investors to purchase and own digital land and developers to create and sell virtual buildings and other structures.

- **Collectibles**: Digital valuables like sports trading cards, priceless artwork, and other unique goods can be made with NFTs. As a result, collectors can possess and exchange digital assets in the same way they do real treasures.
- **Identity and Reputation**: NFTs can represent identity and reputation in digital communities, such as social media platforms or online forums. This allows users to own and control their online identities and to be recognized and rewarded for their contributions to the community.
- **Royalties and Revenue Sharing**: NFTs can create revenue-sharing agreements between creators and investors. This allows creators to share in the profits generated from the sale of their digital assets, providing a new way to monetize their work.

4.3 *Potential Trends that Shape the Future of NFTs*

The future of NFTs is a topic of much discussion and speculation, as the technology is still relatively new and rapidly evolving [10, 15]. However, here are a few potential trends that may shape the future of NFTs:

- **Expansion into New Industries**: While NFTs have already seen major acceptance in the art, gaming, and music industries, there is still room for NFTs to be embraced in a variety of other industries, including real estate, fashion, sports, and others. As technology advances, new applications for NFTs may develop.
- **Increased Interoperability**: Currently, most NFTs are created on the Ethereum blockchain, which limits their interoperability with other blockchain networks. However, efforts are underway to create cross-chain NFTs that can be used across multiple blockchain networks, potentially increasing their utility and value.
- **Evolution of Marketplace Platforms**: Currently, most NFT marketplaces are relatively new and decentralized, with a limited set of features and services. In the future, we may see the emergence of more sophisticated and user-friendly NFT marketplaces that provide a broader range of services, such as curation, authentication, and resale support.
- **Integration with Other Technologies**: NFTs may be integrated with other emerging technologies, such as virtual and augmented reality, to create even more immersive and engaging experiences for collectors and users.
- **Sustainability and Environmental Concerns**: The energy consumption required for NFT transactions on the blockchain has come under scrutiny, and there is increasing focus on finding more sustainable and environmentally friendly solutions for NFT creation and transactions.

5 Understanding Web 3.0

Web 3.0, also known as the "Semantic Web" or the "Decentralized Web", is the next generation of the internet that is being developed to overcome the limitations of the current web, which is often referred to as Web 2.0.

Web 2.0 is characterized by the dominance of large, centralized platforms, such as Google, Facebook, and Amazon, that collect and control vast amounts of user data. While Web 2.0 has enabled many powerful applications, it has also raised concerns about data privacy, censorship, and the power of centralized platforms [16, 17]. Web 3.0 aims to address these concerns by creating a more decentralized and user-centric web. Some of the key features of Web 3.0 include:

- **Decentralization**: By enabling peer-to-peer interactions without the use of centralized intermediaries, Web 3.0 seeks to build a more decentralized internet. It does this by utilizing distributed systems such as blockchain technology. As a result, centralized platforms would have less influence and would be more autonomous, transparent, and resilient.
- **Interoperability**: Web 3.0 enables data and applications to be transferred across many platforms and systems in order to increase internet interoperability. By lowering entry barriers, this would encourage innovation by making it simpler for users to move their data and apps between different providers.
- **Privacy**: Web 3.0 aims to create a more private internet by using technologies such as encryption and zero-knowledge proofs to protect user data and enable anonymous interactions. This would give users greater control over their data and reduce the risk of surveillance and data breaches.
- **Personalization**: Web 3.0 aims to create a more personalized internet by using artificial intelligence and machine learning algorithms to understand user preferences and provide personalized experiences. This would enable more targeted and relevant content and services, as well as empowering users to create their own custom experiences.
- **Openness**: Web 3.0 aims to create a more open internet by using open source software and open standards to enable anyone to build and contribute to the web. This would promote collaboration and innovation, as well as ensuring that the web remains accessible and inclusive for everyone.

5.1 Characteristics of Web 3.0

- **Virtualization**: The first characteristic of Web 3.0 is virtualization, which means that it will have high-speed internet bandwidths and high-end 3D graphics that can be used for virtualization. This will allow for the creation of virtual 3D environments, which can be used in various applications. For example, Second Life is one of the most popular 3D web applications of Web 3.0. With virtualization, users can create and experience immersive digital environments that are closer to real life.

- **Decentralization**: Web 3.0 is expected to be more decentralized than the current Web 2.0, which is dominated by a few large corporations. With Web 3.0, users will have more control over their data and will be able to participate in decentralized applications and networks, such as blockchain-based platforms.
- **Trust**: Trust is a critical aspect of Web 3.0, as users will need to trust the decentralized networks and applications they use. Technologies such as blockchain, smart contracts, and decentralized identity management systems will help to establish trust in Web 3.0.
- **Interoperability**: The second characteristic of Web 3.0 is interoperability, which refers to the ability of different systems, devices, and applications to communicate and exchange data with each other seamlessly. In Web 3.0, interoperability will be achieved through the use of open standards and protocols, which will allow for collaboration and reuse of data and information. Web 3.0 applications will be easy to modify and will be able to run on different types of devices, including hand-held devices, TVs, and automobiles, among others.
- **Privacy**: With the increasing amount of personal data being generated and shared online, privacy is becoming a significant concern. Web 3.0 is expected to prioritize privacy by enabling users to control their data and decide who has access to it. Decentralized identity management systems and encryption technologies will play a crucial role in ensuring privacy in Web 3.0.
- **Intelligence**: Web 3.0 will also be characterized by intelligence, which means that applications will work intelligently with the help of human–computer collaboration. Various artificial intelligence tools and techniques, such as neural networks, rough sets, fuzzy sets, and machine learning, will be integrated into applications to work intelligently. This means that applications will be able to analyze data intelligently and provide optimum output without much user interference. Web 3.0 will also allow for intelligent conversion of documents in other languages into any language, allowing users to communicate with others in their native language.
- **Semantic Web**: Web 3.0 is expected to have a more Semantic Web, which will enable machines to understand the content and meaning of information on the web. This will facilitate more intelligent search and discovery of information, as well as the development of intelligent applications and services.
- **Personalization**: The fourth characteristic of Web 3.0 is personalization, which means that applications will be tailored to individual preferences. In Web 3.0, personalization will be achieved through the use of Semantic Web Technologies. This means that applications will be able to understand and interpret the meaning of data and information, allowing for better personalization. For example, search engines will be able to provide more relevant search results based on individual preferences, and personalized portals can be created on the web.
- **Scalability**: As the web grows and more people and devices come online, scalability becomes increasingly important. Web 3.0 is expected to address scalability challenges through the use of decentralized networks and peer-to-peer protocols,

which can handle large volumes of data and transactions in a more efficient and scalable manner.

5.2 Implications of Web 3.0

The potential implications of Web 3.0 are far-reaching and could fundamentally reshape the internet and our society [18–20]. Some potential implications include:

- **Greater user control**: Web 3.0 could give users greater control over their data and online identity, as well as enabling them to participate more directly in online communities and economies.
- **Disruption of centralized platforms**: Web 3.0 could disrupt the dominance of large, centralized platforms, such as Google, Facebook, and Amazon, by enabling decentralized alternatives that prioritize user privacy and autonomy.
- **New business models**: Web 3.0 could enable new business models that are more transparent, fair, and community-driven, such as decentralized marketplaces, social networks, and content platforms.
- **Enhanced security and privacy**: Web 3.0 could improve the security and privacy of online transactions and interactions, reducing the risk of fraud and cyber-attacks.

Web 3.0 is a new vision for the internet that intends to address the shortcomings of Web 2.0 by building a more decentralized, user-centric, and privacy-focused web. While it is still in the early stages of development, there are already many projects and initiatives underway to create a more decentralized and user-centric web, and it is likely that we will see significant progress in the coming years.

6 Integration of Metaverse, NFTs, and Web 3.0

The combination of Metaverse, NFTs, and Web 3.0 has the potential to transform how we engage with digital content and with one another. In this section, we will look at how these three technologies can work together to build a more immersive, decentralized, and user-centric digital ecosystem.

- **Digital Ownership**: NFTs can be used in the Metaverse to represent unique digital assets, providing people ownership and control over their virtual property. Virtual real estate, in-game objects, and other digital collectibles may be included. The use of blockchain technology ensures that these assets cannot be replicated or destroyed, and NFTs can give proof of ownership and authenticity.
- **Decentralized Economy**: The Metaverse has the potential to build a new decentralized economy in which users can buy, sell, and exchange digital assets with cryptocurrencies and other digital currencies. This would increase autonomy and

transparency while decreasing the authority of centralized platforms. NFTs can play a key role in this economy by enabling the creation and exchange of unique digital assets.
- **Interoperability**: Web 3.0 could enable data and applications to be shared across different Metaverse platforms, promoting innovation and reducing barriers to entry. This interoperability could allow for the creation of new services and experiences that are not possible with siloed platforms.
- **Immersive Experiences**: The integration of Metaverse and Web 3.0 could create more immersive and personalized experiences, using AI and machine learning algorithms to understand user preferences and provide personalized content and services. NFTs can enhance these experiences by providing unique and valuable digital assets that can be used in the Metaverse.
- **Virtual Communities**: The Metaverse could create new virtual communities, where users can socialize, collaborate, and engage in a variety of activities, such as gaming, shopping, and education. Decentralized protocols can be used to power these communities, giving users more control over their data and interactions.

In conclusion, the integration of Metaverse, NFTs, and Web 3.0 can create a new digital landscape that is more immersive, decentralized, and user-centric. The use of blockchain technology and decentralized protocols can provide greater transparency, security, and autonomy for users, while NFTs can enable the creation and exchange of unique digital assets. As these technologies advance, we may anticipate new and novel use cases that will change the way people engage with one another and with digital material.

6.1 *Business Applications of Metaverse, NFTs, and Web 3.0*

The combination of Metaverse, NFTs, and Web 3.0 can open up several exciting business opportunities. Here are some potential business applications.

- **Virtual Real Estate**: As the Metaverse becomes more popular, there will be a high demand for virtual real estate. NFTs can be used to represent ownership of virtual land and buildings within the Metaverse, and Web 3.0 technology can ensure that the ownership is transparent, verifiable, and tamper-resistant.
- **Virtual Goods and Services**: NFTs can also be used to represent ownership of virtual goods and services, such as virtual clothing, accessories, and even virtual experiences. This can create a new market for businesses to sell their products in the Metaverse, and Web 3.0 technology can ensure that ownership and transactions are transparent and secure.
- **Digital Identity and Reputation**: Web 3.0 technology can enable the creation of decentralized digital identities that are verifiable and tamper-resistant. NFTs can be used to represent these identities, and businesses can use them to build trust and reputation in the Metaverse.

- **Gaming**: NFTs have already been used in gaming to represent in-game items and currencies. With the Metaverse, businesses can create immersive gaming experiences that take place within the virtual world and use NFTs to represent ownership of in-game items and assets.
- **Advertising and Marketing**: The Metaverse provides a new platform for businesses to advertise and market their products and services [21]. NFTs can be used to create unique, limited-edition digital assets that can be used in marketing campaigns, and Web 3.0 technology can ensure that ownership and authenticity are verifiable.
- **Virtual Advertising**: As the Metaverse becomes more popular and immersive, businesses can use it as a platform to advertise their products and services to a global audience. By creating engaging virtual experiences and offering exclusive NFTs as rewards, companies can attract more customers and create stronger brand loyalty.
- **E-commerce**: With the help of NFTs and Web 3.0, businesses can offer their customers unique, personalized experiences while shopping in the Metaverse. Customers can use NFTs to purchase virtual goods, while smart contracts can ensure secure and transparent transactions. This can potentially open up new revenue streams for businesses.
- **Virtual Events**: With the pandemic forcing many in-person events to go virtual, businesses can leverage the Metaverse to create unique, engaging virtual events. Attendees can use NFTs to gain access to these events, receive exclusive rewards, and create memories that they can cherish.
- **Education**: One business application of the Metaverse in education is the creation of virtual classrooms or campuses, which could reduce the cost of maintaining physical classrooms and campuses, reach a wider audience, and provide a more engaging and interactive learning experience for students [22]. This could also provide a business opportunity for companies that specialize in creating virtual environments and educational technology.
- **Tourism**: One potential business application of the Metaverse in the tourism industry is to offer immersive and interactive experiences that simulate the real-world travel and hospitality offerings, enabling customers to explore and engage with virtual destinations, hotels, and attractions without physically moving. This can help travel companies to attract and engage with customers, gather valuable feedback, and enhance their offline offerings based on customers' preferences and expectations. The Metaverse can also provide a new platform for marketing and branding efforts, targeting both individual customers and travel professionals [23].
- **Digital Identity**: In a world where our digital presence is becoming increasingly important, businesses can use the Metaverse and NFTs to create and manage digital identities for their customers. These digital identities can be used to verify personal information, authenticate transactions, and provide a personalized experience across different platforms.

The combination of Metaverse, NFTs, and Web 3.0 has the potential to revolutionize the way businesses operate in the digital world and create new opportunities for growth and innovation.

7 Current Trends and Future Directions

New technologies, such as Web 3.0 and Metaverse, are being developed that have the potential to change the banking industry and improve the way banks provide services to consumers, including retail banking users, investment banking users, and corporates. The emergence of technologies such as blockchain, NFTs, and smart contracts provides new capabilities like NFT-based securities or bonds. Meanwhile, Metaverse provides an immersive experience in a virtual environment that is always growing because to platform updates, research, and Al modules included into Metaverse avatars. Bloomberg Intelligence predicts that the Metaverse market will be worth 800 USD billion by 2024, and financial players are expected to contribute 50 USD billion to the market capitalization by 2026, according to Technavio. Metaverse banking is a step above net banking, offering customers a personalized user experience and data visualization. Major banks worldwide have already started investing in this space, and Union Bank of India recently announced the launch of its Metaverse Virtual Lounge, providing customers with a unique banking experience where they can explore the lounge, learn about bank deposits, loans, government welfare schemes, and digital initiatives, and have a real-world experience.

The Metaverse is often misunderstood as being only related to computer gaming and social media or sometimes dismissed as an overhyped rebranding of virtual reality (VR) and augmented reality (AR). However, the Metaverse is the convergence of rapid and significant technical and sociological developments that include avatars representing us and duplicating many objects around us, such as medical imaging equipment, with potential applications in various fields such as professional training, education, supply chains, and real estate marketing, among others. The growth in telemedicine during the COVID-19 pandemic indicates how quickly elements of the Metaverse can gain traction, and recent perspectives on medical applications of the Metaverse support this indication. Earlier platforms allowed users to choose Avatars, occupations, and social lives and perform real-life activities in virtual spaces. However, with the development of artificial intelligence (AI), deep learning, extended reality technology, content-related services, payment options including digital currencies and cryptocurrencies, and better immersive capabilities, the Metaverse has evolved into an interoperable persistent network of shared virtual environments where people can interact synchronously through their avatars with other agents and objects. As Kim describes it, the Metaverse after these developments is “an interoperated persistent network of shared virtual environments where people can interact synchronously through their avatars with other agents and objects”.

References

1. Lee LH, Braud T, Zhou P, Wang L, Xu D, Lin Z, Kumar A, Bermejo C, Hui P (2021) Allone needs to know about metaverse: a complete survey on technological singularity, virtual ecosystem, and research agenda. arXiv preprint arXiv:2110.05352
2. Wang Q, Li R,Wang Q, Chen S (2021) Non-fungible token (NFT): overview, evaluation, opportunities and challenges. arXiv preprint arXiv:2105.07447
3. Park A, Kietzmann J, Pitt L, Dabirian A (2022) The evolution of nonfungible tokens: complexity and novelty of NFT use-cases. IT Professional 24(1):9–14
4. Kshetri N (2022) Web 3.0 and the metaverse shaping organizations' brand andproduct strategies. IT Professional 24(02):11–15
5. Monaco S, Sacchi G (2023) Travelling the metaverse: potential benefits and main challenges for tourism sectors and research applications. Sustainability 15(4):3348
6. Wang Y, Su Z, Zhang N, Xing R, Liu D, Luan TH, ShenX.(2022) Asurvey on metaverse: fundamentals, security, and privacy. IEEE Commun Surv Tutor
7. Chen Z, Wu J, Gan W, Qi Z (2022) Metaverse security and privacy: an overview. arXiv preprint arXiv:2211.14948
8. Hutson J, Hutson P (2023) Museums & the metaverse: emerging technologies to promote inclusivity and engagement
9. Fernandez CB, Hui P (2022) Life, the Metaverse and everything: An overview of privacy, ethics, and governance in Metaverse. In: 2022 IEEE 42nd international conference on distributed computing systems workshops (ICDCSW). IEEE, pp 272–277
10. Belk R, Humayun M, Brouard M (2022) Money, possessions, and ownershipin the metaverse: NFTs, cryptocurrencies, Web3 and wild markets. J Bus Res 153:198–205
11. Gupta AK (2022) Web 3.0 and its reflections on the future of e-learning. Retrieved from Int J Res Appl Sci Eng Technol: https://www.ijraset.com/research-paper/web-3point-0-and-its-reflections-on-the-future-of-e-learning
12. Taherdoost H (2023) Non-fungible tokens (NFT): a systematic review. Information 14(1):26
13. Cheong BC (2022) Avatars in the metaverse: potential legal issues and remedies. Int Cybersecur Law Rev, pp 1–28
14. Mystakidis S (2022) Metaverse. Encyclopedia 2(1):486–497
15. Gonserkewitz P, Karger E, Jagals M (2022) Non-fungible tokens: use cases of NFTs and future research agenda. Risk Gov Control Fin Mark Inst 12:8–18
16. Chalmers D, Fisch C, Matthews R, Quinn W, Recker J (2022) Beyondthe bubble: Will NFTs and digital proof of ownership empower creative industry entrepreneurs? J Bus Ventur Insights 17:e00309
17. Gan W, Ye Z, Wan S, Yu PS (2023) Web 3.0: the future of internet. arXiv preprint arXiv:2304.06032
18. Murray A, Kim D, Combs J (2022) The promise of a better internet: Whatis web 3.0 and what are we building?. Available at SSRN
19. Hackl C, Lueth D, Di Bartolo T (2022) Navigating the metaverse: a guideto limitless possibilities in a Web 3.0 world. John Wiley & Sons
20. Momtaz PP (2022) Some very simple economics of web3 and the metaverse. FinTech 1(3):225–234
21. Chohan R, Paschen J (2021) What marketers need to know about non-fungible tokens (NFTs). Bus Horiz 66
22. 1016/j.bushor.2021.12.004
23. Hwang GJ, Chien SY (2022) Definition, roles, and potential research issues of the metaverse in education: an artificial intelligence perspective. Comput Educ Artif Intell 3:100082

Classification of Human Cancer Using Meta-Verse with Block-Chain Security

Kapil Joshi, Ajay Singh, Anjali Naudiyal, and Sandeep Sharma

Abstract In today's era, where medical science has made significant progress, the digital world has also contributed to the creation of such a world through its efforts. A headset is used to create such an environment, known as meta-verse, where everything appears to be real but is actually a product of the imagination. A cell is a unit of human body. Controlled growth and division of cells occur in the human body. Normal cells' ages are eventually replaced by new ones. Cancer cells do not obey the laws that healthy cells do, and thus they begin to grow uncontrollably. The physical examination, laboratory tests, imaging tests, and biopsies are some methods used to diagnose cancer. There are several types of imaging tests, including bone scans, computerized tomography (CT), magnetic resonance imaging (MRI), Positron emission tomography (PET) scans, ultrasounds, and X-rays. Three-dimensional (3D) images of cancer can be detected everywhere on the human body using these four technologies: CT, MRI, SPECT, and PET. Using technologies like augmented reality (AR), virtual reality (VR), and block-chain security, a virtual environment can be created to diagnose cancer, Meta-verse help with cancer diagnosis. The chapter concludes that if cancer is detected using meta-verse, it will be detected early and the patient's treatment will be expedited.

Keywords Meta-verse · Block-chain security · AR · VR · CT · Cancer · Digital twins

K. Joshi · S. Sharma
Uttaranchal Institute of Technology, Uttaranchal University, Dehradun 24007, India

A. Singh · A. Naudiyal (✉)
School of Applied and Life Sciences, Uttaranchal University, Dehradun 24007, India
e-mail: naudiyalanjali@gmail.com

G. Chhabra and K. Kaushik (eds.), *Understanding the Metaverse*, Blockchain Technologies, https://doi.org/10.1007/978-981-97-2278-5_4

1 Introduction

A computer-generated world called the meta-verse is made up of the two terms meta, which means numerous, and verse, which means universe. In his 1992 book Snow Crash, Neil Stephenson coined the phrase for the first time [1], the novel's definition of the meta-verse by Stephenson is a large virtual environment in which individuals communicate via digital avatars while coexisting with the real world. Many environments that appear real are created in meta-verse. A three-dimensional virtual environment is created in the meta-verse using augmented reality, virtual reality, mixed reality, high-speed networks, edge computing, and hyperledgers (or block-chain), these are advanced AI technologies. This technology is essential for the development of the meta-verse.

In order to deliver high-quality care at an affordable price, ongoing efforts are made in the fields of software and computers, where new dimensions are being produced. With the aid of technology, new paths have also been continuously developed in the field of therapy. The meta-verse, a virtual dimension of an acting shape, is one example of this.

There is still much work to be done on its advancement; it is a developed service that can be transformed into an advanced technological form that gives the patient and the doctor a realistic experience. Additionally, there can be a lot of efforts made to improve health services with the aid of AR, VR, robotic quantum computing, and other technologies in devices. Medical treatment is primarily focused on the patient due to its effect of solving the patient's problems by creating an unfavorable situation, the settlement of the obligations between the patient and the doctor, and the computer acting as a link between the two so that both sections receive the information as required. Based on the data provided by matter, it is feasible. Information about last year's weapons is simple to collect and analyses. This technology is an appealing and realistic-looking source that conveys all the information to the user with the aid of the programmed the reality show. The user is motivated by this programmed in such a way that everything that is happening in front of him and that he can feel it too. The technology utilized in the shape of block speed, which uses cryptography to protect the customer's herbal information, is comparable to an inherent security that can guard against health security issues. Block-chain technology is comparable to an inherit security that can guard against health security threats. It is a form of block-speed technology, in which the customer's herbal product is protected by cryptography technology. In many domains, the use of the meta-verse is particularly beneficial for finding difficult solutions.

As "the successor to the Internet connectivity," corporations like Google, Microsoft, and Meta have recently invested significantly in constructing the meta-verse. Future service ecosystems in all fields, including health care, education, entertainment, e-commerce, and smart industries, will be transformed by the meta-verse in ways that the Internet was unable to [2].

The utilization of wearable sensors and regeneration are both made possible by the immersive hybrid glasses, which also enable other immersive sensory experiences [3]. The Microsoft HoloLens 2 is a mixed reality headset that the company produces. It uses Windows 10 to operate holographically. Four visible light cameras for tracking, two infrared cameras for eye tracking, a time-of-flight sensor, and an 18 MP camera for depth are all included in the HoloLens 2 device. With a specially designed Snap Dragon 850 processor, HoloLens has a long battery life. It connects to Bluetooth via a USB port. WiFi is also included in the Halo lens. HoloLens culture contains eye-tracking sensors that enable users to check in without utilizing contrast for a more natural dialogue. As HoloLens 2's tracking capabilities advance, mid-air gesture displays to control objects become possible. Ninth more immersive for the user in its simulated environment Mixed Reality experience scaled up to 720 pixel quality. Sensors recognise palms, wrists, and hands. Microsoft developed the remote assistance reality program Dynamics 365 to offer better customer support. Different HoloLens locations One HoloLens 2 Android applications allow for remote and video sharing. On-site personnel can send information to faraway experts using the HoloLens device. The range of Information needed for live streams of films, circuits, etc.

The following two aspects have greatly aided meta-verse's popularity:

1. First, the problem with COVID-19 is the initial factor in meta-verse's popularity. The patient or human being required a virtual environment in the case that it was impossible for people to physically meet, which was made feasible by the meta-verse.
2. Second, using VR/AR technology, the 5G/6G communication system immerses users both physically and aesthetically in a virtual environment. Everybody in the meta-verse, also known as avatar, has a digital identity that is generated with the aid of VR. These "avatars" adopt the appearance and characteristics of their users' natural counterparts [4].

Regarding the dietary habits of the current generation, it is possible that the number of cancer cases may rise in the future, while medical science is working hard to slow this expansion. The patient is able to trust that their data is secure by having their data physically inspected in the meta-verse and then having that data put in the block-chain. It will also result in a drop in the number of deaths per year. Cancer is an incurable disease made up of many different types of cells. These cells divide uncontrollably and spread to all regions of the body in human cells, where they continue to develop until they outgrow the human body. Old cells are damaged and replaced by new ones during cell division. Damaged cells begin to develop and assume the shape of tumors when the cell division mechanism is disrupted. The tumor might or might not be malignant. According to the metastasis process, a cancerous tumor is also known as a malignant tumor, while a benign tumor is referred to as benign. A cancerous tumor affects nearby tissues, produces new tumors, and spreads to distant locations throughout the entire body. Table 1 depicts the number of major cases related to cancer in the whole world in the year 2020, and also Fig. 1 shows the graph related to this data. Table 2 differentiates the cancerous or normal cells. With the help of this difference, stages can be determined.

Table 1 No. of major cancer cases worldwide in 2020

S. No.	Types	No. of cases	Death	Percentage of death (%)
1	Lung	2.21	1.80	81
2	Stomach	1.09	0.76	69
3	Colon and rectum	1.93	0.91	47
4	Breast	2.26	0.68	30
5	Prostate	1.41	0.31	22
6	Skin (non-melanoma)	1.20	0.15	12

Fig. 1 Major cancerous cases in worldwide

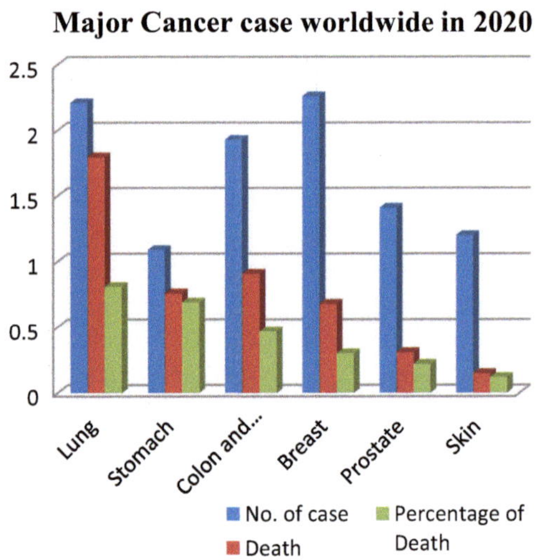

Table 2 Difference between normal cells and cancer cells

S. no.	Parameter	Regular cells	Cancerous cells
1	Growth	When there is enough unchecked growth, stop	Rapid expansion
2	Communication	Heed cellular messages from other cells	Take no action when other cells provide you with signals
3	Cell repair/death	Damaged or ageing cells are replaced or repaired	Cells are not changed or fixed

1.1 Types of Cancer-Causing Genes

DNA, the fundamental physical and functional component of inheritance, makes up genes. The function of genomes is to produce molecules known as proteins. Genes come in a variety of sizes. Cancer is a very deadly condition that is frequently inherited. Cancer is a genetic disease. It is caused by changes in genes that regulate the way cells grow and multiply. Oncogenes, which act as positive growth regulators, and tumor suppressor genes, which act as negative growth regulators, are two types of these cancer genes. Some genes are discussed below.

1.1.1 Proto-Oncogenes

A normal cell's growth or division is regulated by proto-oncogenes. The genes can turn into cancer-causing genes when they are more active than usual, enabling the growth and survival of cells. These genes quickly produce mutations that lead to the development of malignant cells.

1.1.2 Tumor Suppressor Genes

Genes do not assist the tumor in cell division or cell dissemination. A cell that is capable of uncontrolled division is said to be altering.

1.1.3 DNA Repair Genes

These genes repair harmed DNA. The presence of other gene mutations and chromosomal abnormalities, such as chromosomal duplicates and deletions of genomic segments, are frequently found in cells with abnormalities in these genes. Table 3 depicts the genes through which cancer cells are developed.

1.2 Stages Related to Cancer

Stages have been categorized into four groups based on the cell type that produces malignant tumors.

1.2.1 Carcinoma

Carcinoma is the most common type of cancer, accounting for 80–90% of all cancer diagnoses. Between the inside and outside surfaces, a line where epithelial cells

Table 3 Types of cancer-causing genes

S. no.	Type of cancer	Genes
1	Female breast cancer	ATM, BARD1, BRCA1, BRCA2, BRIP1, CHEK2, CDH1, NF1, PALB2, PTEN, RAD51C, RAD51D, STK11, TP53
2	Male breast cancer	BRCA1, BRCA2, CHEK2, PALB2
3	Intestinal cancer	APC, EPCAM, MLH1, MSH2, MSH6, PMS2, CHEK2, PTEN, STK11, TP53, MUTYH
4	Cancer of the uterus	EPCAM, MLH1, MSH2, MSH6, PMS2, PTEN, STK11
5	Primary peritoneal, ovarian, and fallopian tube cancer	ATM, BRCA1, BRCA2, BRIP1, EPCAM, MLH1, MSH2, MSH6, PALB2, RAD51C, RAD51D
6	Stomach cancer	APC, CDH1, STK11, EPCAM, MLH1, MSH2, MSH6,
7	Melanoma	BAP1 (especially uveal melanoma), BRCA2 CDK4, CDKN2A, PTEN, TP53
8	Pancreatic cancer	ATM, BRCA1, BRCA2, CDKN2A, MLH1, PALB2, STK11, TP53
9	Prostate cancer	ATM, BRCA1, BRCA2, CHEK2, HOXB13, PALB2,

develop. Epithelial cells resemble commons when viewed via a microscope. Cancer cells have the ability to separate from tumours and metastasis (move to other places in your body). After 65 or older, the risk of carcinoma increases. Carcinomas are very rare in children.

Adenocarcinoma

Adenocarcinoma is a type of cancer that develops in mucus- or fluid-producing epithelial cells. Occasionally, glandular tissues are referred to as epithelial tissues. Breast, colon, and prostate adenocarcinomas are the most common types of cancer. Renal cell carcinoma (RCC) and hepatocellular carcinoma are the most common form of Adenocarcinoma.

Basal Cell Carcinoma

The form of cancer starts in the epidermis, the skin's outer layer, specifically in the lower or basal layer. This cancer appears as a white, waxy lump or a brown, scaly patch on sun-exposed areas, such as the face and neck. When detected early, most basal cell carcinomas (BCCs) can be treated and cured.

Squamous Cell Carcinoma

Squamous cell carcinoma is a type of cancer that develops in squamous cells, which are epithelial cells found just below the skin's outer layer. Numerous other organs, such as the stomach, intestines, lungs, bladder, and kidneys, are lined by squamous cells. Squamous cells appear flat under a microscope, similar to fish scales. Epidermoid carcinomas are another name for squamous cell carcinomas.

Transitional Cell Carcinoma

A type of epithelial tissue termed transitional epithelium, or urothelium, is where transitional cell carcinoma, a cancer, develops. The linings of the bladder, ureters, kidneys' renal pelvis, and a few other organs are made up of this tissue, which is composed of numerous layers of ectoderm cells that can develop bigger and smaller. Transitional cell carcinomas include a few malignancies of the bladder, ureters, and kidneys.

1.2.2 Sarcoma

Sarcomas are tumors that develop in the muscle, fat, blood vessels, lymphatic vessels, and fibrous tissue that make up soft tissues and bone. The National Cancer Institute estimates that each year in the USA, 12,000 new cases of soft tissue sarcomas and 3000 new cases of bone sarcomas are diagnosed. It is unclear what causes sarcoma exactly. Sarcomas can, like many other malignancies, be caused by DNA mutations that disrupt the genes that control cell growth. These mutations can be inherited from one's parents or developed throughout one's lifetime.

1.2.3 Leukemia

This type of cancer which begins when blood is formed in the bone marrow. They discard healthy blood cells as a result of this modification in the procedure, etc. There are various forms of leukaemia. Some leukaemia types are more prevalent in youngsters. Most cases of other types of leukaemia are in adults. Usually, leukaemia affects white blood cells. Your white blood cells are effective infection-fighting agents; they typically grow and divide in an organized manner as required by your body. But in leukaemia patients, the bone marrow makes an excessive number of aberrant, dysfunctional white blood cells. Depending on the leukemia's nature and other variables, treatment for leukaemia may be difficult. However, there are methods and tools that can aid in the success of your treatment. There are four common forms of leukaemia, which are categorized according to the type of blood cell.

Table 4 Stages related to cancer

S. no.	Stages	Descriptions
1	I	The cancer has not spread to the lymph nodes or other tissues; it is contained in a small location
2	II	Despite expanding, the cancer has not yet spread
3	III	The cancer is now more advanced and may have spread to nearby lymph nodes or other tissues
4	IV	Other body parts or organs have been affected by the cancer's spread. This stage of the disease is also known as metastatic or advanced cancer

Lymphoma

White blood cells offer protection from debt because, in lymphoma, cancer starts in lymphocytes. Hodgkin's lymphoma and non-Hodgkin's lymphoma are the two main kinds of lymphoma. Chemotherapy, drugs, radiation therapy, and very infrequently stem cell transplants are all possible forms of treatment.

Melanoma

Melanocytes can produce melanin and proliferate in the skin and eyes, a condition known as melanoma. There are different types of cancer that starts in the skin. Melanoma can occur anywhere on the skin. Unusual moles, exposure to sunlight, and health history can affect the risk of melanoma. Treatment may involve surgery, radiation, medication or in some cases, chemotherapy.

Brain and Spinal Cord Tumors

Tumours of the brain and spinal cord are collections of aberrant cells that have proliferated uncontrollably in the brain or spinal cord. It's crucial to distinguish between benign (non-cancerous) tumours and malignant tumours (cancers) in the majority of other body areas. Benign tumours in other sections of the body generally never pose a threat to life because they do not invade surrounding tissues or spread to distant locations. The ability of malignant tumours to spread throughout the body is one of the key factors contributing to their danger. A large percentage of brain tumours can spread via the brain tissue, but they seldom do so to other regions of the body. Even so-called benign brain tumours have the potential to cause catastrophic or even life-threatening damage when they advance and obliterate healthy brain tissue. Because of this, medical professionals frequently refer to brain tumours rather than brain cancers. Table 4 depicts the stages related to cancer.

2 Technologies Related to Meta-Verse

The software and computer industries' ongoing efforts are leading to the development of new dimensions. The development of innovative methods in the field of therapy with the aid of technology has allowed for the provision of high-quality care at affordable prices. One of these is the meta-verse, a brand-new kind of digital dimension that still needs a lot of work to be done on it. With the aid of AI, VRL, robots, quantum computing, and other technologies in devices, it is evolving into an advanced technology that offers a realistic experience to both the patient and the doctor as well as further enhances healthcare services. Various technologies related to meta-verse are discussed in this section. Different types of technologies are as follows:

2.1 Digital Twins

Working technologically and recreating actual artefacts in the present day is a particular kind of digital work. Sensors are used to represent realism and assist in converting unrealistic depictions into realistic ones. These sensors operate in accordance with the data systems of any given situation, giving the appearance of reality to the unrealistic depictions.

2.2 Augmented Reality

It is a system that connects actual surroundings and virtual items in a way that makes them appear to coexist side by side in the real world [5]. The physical environment is an immersive feature of the physical world that effectively integrates various sensory modalities, including the visual, auditory which means sense of listening, hap-tic which means sense of feel and olfactory which means sense of scent. These sensory channels could potentially be used to propagate the AR. It gathers information about a situation occurring in reality and applies it to the actual world so that it appears to be happening there. It can build a home that reflects the real world using the information, and it can also utilize the information to construct an interactive object which is dependent on reality because the data we use to create our instructional materials is based on real-world events. It produces a very lifelike, tangible environment that can be extensively employed in the medical industry. Many research projects use digital twin technology to build an accurate representation of the human body that corresponds to the physical data. This has also led to the creation of blood flow models and head sensor models in the field of medicine. When we input the data from a real patient into the model, it creates a virtual representation of the patient and displays a diagnosis recommendation. The computer offers its best result and solution as a result of the high quality on which the data collecting was based. Due to the vast amount of data, a computer is used, which produces high-quality data that is logical. Blood flow models and head sensor models have also been developed in the medical area, allowing for the creation of virtual models using data from actual patients while still allowing for message transmission. Due to the enormous amount of data, the assistance of a computer is taken, ensuring that accurate and practical information is shown to the user with the help of this sensor. It indicates the outcome of its process and diagnosis because the data collection has been done on the basis of high quality; on that basis, a computer delivers its highest result and solution. The healthcare landscape leads to the reality in which methods like artificial neural networks are used because of the quality of both data and sensors, which play a significant role in creating the residential component of the medical scenario.

2.3 *Virtual Reality*

With the use of virtual reality technology, variables can be transferred, such as the ability to interact with items while seeing virtual reality. A virtual reality device contains a small screen on the front of the earpiece in addition to a head-mounted display. The virtual reality headset offers high-quality content. When the user's eyes are divided between the display and the input tracker, a 3D effect is created. Input from hap-tic technology, which provides additional sensory with audio and visual feedback, is used in VR to build virtual environments.

VR and AR differ from one another in this regard: With the aid of still photos, audio, and films created by computer-generated inputs, the camera can immediately observe the real world, whereas VR elaborates the AR through the imagination [6, 7].

2.4 *Mixed Reality*

The user can interact with actual objects like AR and virtual objects like VR in an environment that is created by MR that combines both technologies [8, 9]. A demonstration that holds both physical and virtual elements together. It is conceivable to say that virtual things are combined with the actual surroundings to provide the user a sense of both real and virtual objects. With mixed reality, it is conceivable to play an electronic game while also snatching a container of water from the real-world environment and striking a gaming character with it (MR) [10].

2.5 Extended Reality

It is impossible to envision the meta-verse without AR VR MR because it turns it into a tangible platform, an unreal reality, and it gives users an immersive experience while interacting with the meta-verse. Despite the fact that many sophisticated surgical simulation systems have been developed, Jang used the AR device for 3D visualization of Mayo cardiac scarring based on doctors' understanding of the benefits of various 3D visualization techniques [11]. Figure 2 depicts the role of three realities in meta-verse.

The illustration above demonstrates how users of the real world construct made-up places using virtual reality in the meta-verse. A hybrid space is produced when mixed reality and augmented reality technologies are used together. It is understood that an extended reality exists in the meta-verse and may be evolving. Virtual reality and augmented reality technologies are modified in the meta-verse to simulate a virtual presence. Holographic communication is a method that lets consumers experience discomfort away from their surroundings. By integrating this technology with 3D capturing, hologram production, translation, and 3D projection made feasible with the aid of, holographic communication enables the ability to physically examine 3D digital pictures [12]. On the other hand, it provides an experience that is identical to what occurs in the actual world while allowing users to travel into an imaginary and comprehensive environment in the meta-verse. A virtual environment that is connected to the Internet has become increasingly necessary for healthcare globally as a result of the COVID-19 epidemic. Extended reality gadgets, which are now highly common, can revolutionize healthcare by enabling medical students to learn about real-world scenarios quickly and primarily in traditional classroom settings [13].

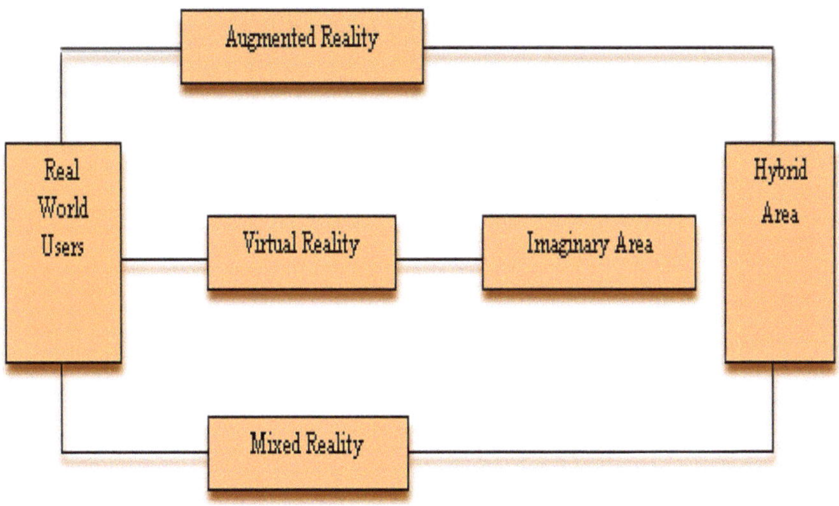

Fig. 2 Diagrammatic representation of three realities in meta-verse

With the aid of consumer headsets that require sensors to view the world, such as in games, extended reality, remote learning encourages design implementation, which necessitates scurrying through 5G in a particular environment. 5G streaming disturbs the workplace. The potential of computers and data warehouses to reduce medical errors and enhance physician abilities, together with immersive streaming's ability to create a dynamic environment, are reasons why people commonly utilise gadgets. When compared to human training tools, immersive learning systems training are extremely beneficial for enabling the growth and dissemination of practitioners' skills.

2.6 *Block-Chain Security*

The transaction database copy is kept digitally separate from the block-chain because it has been fully distributed throughout the network. The chain of transactions is kept up to date by the block-chain, which also records any new transactions in the accounts of all participants. Once Id has been inserted, Id is created. Block-chain is a technological kind in which many administrators employ DLT. Block-chain is a specific sort of DLT that is used to create transaction records for translations. What digital signatures cannot accomplish, it may be argued, is making changes to the agricultural chain transparent to all parties involved. It will be necessary for the hacker to copy and modify the selection of all the blocks if he wants to disable the block-chain system. Through non-fungible tokens or cryptocurrencies, one can acquire decentralized digital asset ownership. Block plays a significant role in China during the next two years, where issues with security, privacy, and data transparency may be readily resolved by utilizing centralized data. Globally speaking, block-chain technology will produce a decentralized, platform-agnostic digital source in the meta-verse. Block-chain technology makes it simple for users to access the online world from their primary institution. Figure 3 depicts the technical diagram of meta-verse.

The technical illustration at top demonstrates how a technology is used in meta-verse to produce a virtual environment that resembles the actual world.

By engaging with the network, the user in the network's environment is turned into a digital avatar with the use of digital twins. On the other hand, the real-world environment is also transformed into a virtual scene by the use of digital twins, as shown in Fig. 3. This occurs information about the transformation parameter and model is kept in the cloud by both the digital avatar and the virtual scene. By utilizing block-chain technology, the cloud environment is produced. Additionally, the cloud contains data about users and the outside world. All of these scenarios generate virtual space, and XR technology enables interaction between users and that area.

With the use of block-chain technology in meta-verse, storing reliable and accurate data will become simpler. Block-chain technology was China's immutable and transparent features to securely manage and store enormous amounts of data in a matter of year, meta-verse will have more options for patients and many environments to communicate with doctors as well as shorten the time it takes to diagnose and

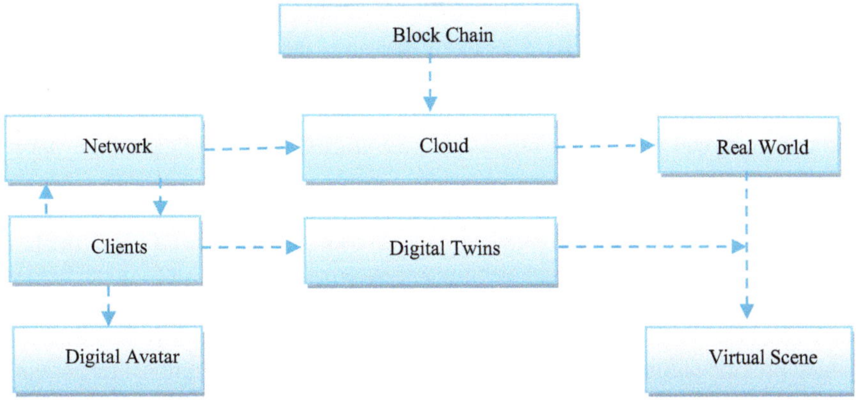

Fig. 3 Technical diagram of meta-verse by using block-chain and digital twins technology

consult a patient. The doctor will receive the patient information stored in the block-chain technology so that doctor will be easy to make sensible decisions. Block-chain technology offers security to stop patient data from being deleted.

2.7 *Artificial Intelligence*

AI has opened up a lot of new possibilities for the healthcare industry. Humans build environments in digital form that are meant to resemble actual objects, can be felt, and can be effectively analyzed. The computer being one of the earliest calculation devices is handy in dealing with vast data, and ML algorithm is an automatically data analyzer that allows us to analyze it even after the complexity of the data. Working in a learning pattern, machine learning algorithm also offers a thorough approach with health prediction, disease analysis, diagnosis, and other useful data is also displayed. We need an accurate system that can produce extremely successful predictions and treatment strategies in order to perform precision medical therapy. All of this knowledge is based on patient data and various individuals' diagnoses.

Polynomial Segmentation Cluster: These are the most significant three issues. The evaluation algorithms used by AI/ML are split into two categories: standard and advanced. In the area of machine learning (ML) in AI, there are static methods, and we can estimate the outcome during training. A significant use of AI/ML in science is the computer's capacity to automatically identify a learning pattern. A trained model can generate a result even for an uncertain input using supervised learning, which forecasts the relationship between the input and the result using a function of mapping for data labels. The main application of MLA is unsupervised learning, which analyses unlabeled data and identifies patterns in it without human interaction. Both trained and untrained semi-supervised learning techniques are utilized in this, and the prediction and size of the result are generated by grouping related data into clusters and labeling them using algorithms. In reinforcement learning, the goal is to create an environment with a series of decisions that can be utilized to resolve challenging and unclear issues. Recently, a subset of ML dubbed deep learning—which incorporates recurrent neural networks, long short-term memory networks, and convolutional neural networks—has been utilized for complex and advanced architects.

Using a network that utilizes multi-layered artificial neural networks, classification regression tasks can be conducted without a lot of performance to get the best results. Consequently, the e-mail technique is automatically employed in the field of health care, and self-oriented results are obtained. The DL architecture is very versatile and can process raw data directly. It additionally has the ability to refine raw data to provide predictions that have been derived from it. Recurrent neural network (RNN) is a prime instance of a well-known deep architecture. The design utilizes legacy processing models acquired by means of time and has been constructed in such a way that it may preserve memory from the prior day. Convolution neural network (CNN) is a network layout that aims to forecast outcomes based on concepts that establish consequences through statistical analysis utilizing linear algebra for identifying intricate patterns. Natural language processing is a rule-based system created by professionals. In this instance, sophisticated ML is used to regulate physical robots in the medical field.

3 Applications of Meta-Verse in Medical Field

3.1 Curative Analysis

By analyzing the patient's direct expression based on technological or oral data, clinicians can determine the physical state of the patient's body when examining the patient's health. Information about patients and doctors is securely exchanged in meta-verse. Doctors can rapidly check patients' CT and MRI scans from a database with the aid of block-chain technology and meta-verse AI.

Figure 4 demonstrates how an individual gets treated in the meta-verse, how their personal information is saved in the block-chain with the use of sensors, how they are transformed into fictional patients, and how doctors can also become imaginary doctors with the aid of extended reality. The patient also receives a converted version of the virtual doctor's information as well as a copy of their image. The virtual doctor receives the physical data that was saved in the unit of the chain via the sensor.

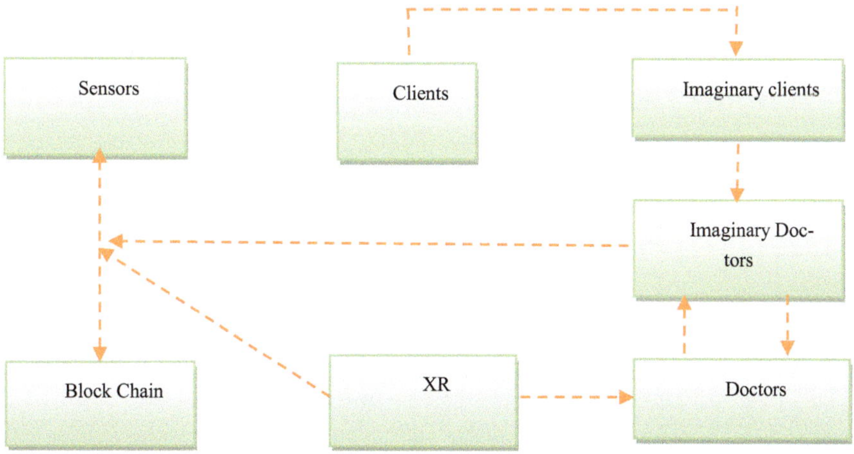

Fig. 4 Diagrammatic representation of curative analysis in meta-verse

3.2 Healthcare Services

Its main objective is to heal the patient. To improve the efficiency of patient care, one must be able to envisage the body for meta-verse body therapy and abstract psychotherapy. In the healthcare industry, professionals are expected to master difficult procedures using a variety of AI tools, as shown by meta-verse. Ontological pictures can be more at ease being unseen in the context of the social world by using virtual space. This capability has allowed virtual space to be used in VR therapy. Meta-verse, between the patient and the doctor, establishes a virtual boundary. The doctor evaluates the patient and offers advice. With the help of augmented reality technology, the tissue that is concealed behind and within the intestine, uterus, gallbladder, and liver can also be accurately navigated by examination [13], and even in spinal surgery, augmented reality technology is utilized to increase accuracy and direction in navigation [14]. In light of the challenging and complex way of life that meta-verse is encouraging, telesurgery lectures using block-chain technology and artificial intelligence techniques are being promoted. The telesurgery system matter bus plays a crucial role in providing complete hardware monitoring and security to the doctor [15]. Figure 5 depicts the diagrammatic representation of healthcare services in meta-verse. The creation of fictional clients and the creation of an illusion were between the fictional therapist and the fictional clients.

The patient's impairment is lessened in the meta-verse, as seen in the above figure. Physical therapist and patient are objects. Physical therapists are transformed into virtual therapists, while patients are transformed into virtual patients. With the aid of a simulated scene extended reality technology, virtual therapist and simulated patients communicate with one another to create a virtual scene. The patient interacts with the simulated patient, and information is transmitted to the orthopedic surgeon with the use of extended reality technology. Any issue can be solved in the meta-verse in a

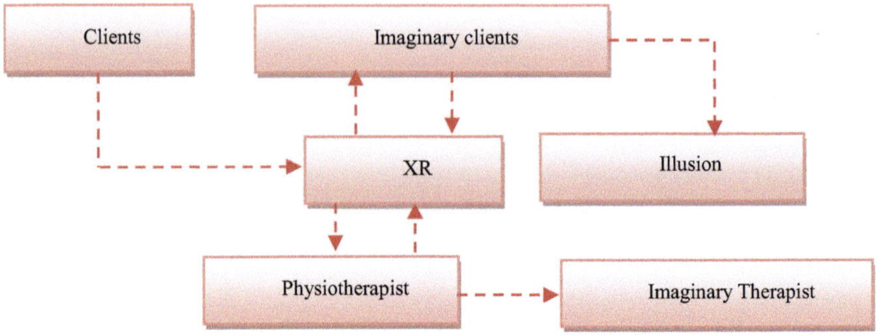

Fig. 5 Diagrammatic representation of Healthcare services in Metaverse

variety of ways, including through reviews and solving of universal healthcare-related problems. A detailed explanation of its characteristics, application capabilities, and challenges has been made. It does excellent work in the area of health care. The health-related solutions developed by meta-verse thus far and more recent additions from the global map have been highlighted. How many a higher medical system be used in different situations.

3.3 Healthcare Redemption

A training method called medical redemption is used to remedy unfavorable effects like accidents, diseases, or their treatments. Digital twins offer a rich graphical perspective in a small area, and the patient may see this action. Various businesses use the meta-verse to deliver various facilities across faraway places. Meta-verse analyses patient data, makes diagnoses, and develops a patient-specific treatment plan. After diagnosing the patient's data, the meta-verse also predicts the patient's health. Meta- verse builds virtual clinics where patients and doctors may connect effectively. Every client in the meta-verse has a unique personal identification number, which is used to securely handle data.

Figure 6 demonstrates how to deliver medical education in the meta-verse. The focus in this illustration is the student. Using extended reality technology, the student engages with a virtual model and relates to surgery block-chain technology is used to create virtual models, and it is also used to create virtual surgeries. This figure's instructional object provides guidance to the pupil. With the aid of the meta-verse, pupils learn quickly. With the aid of block-chain security, the meta-verse offers a computer environment where a real-world counterpart exists.

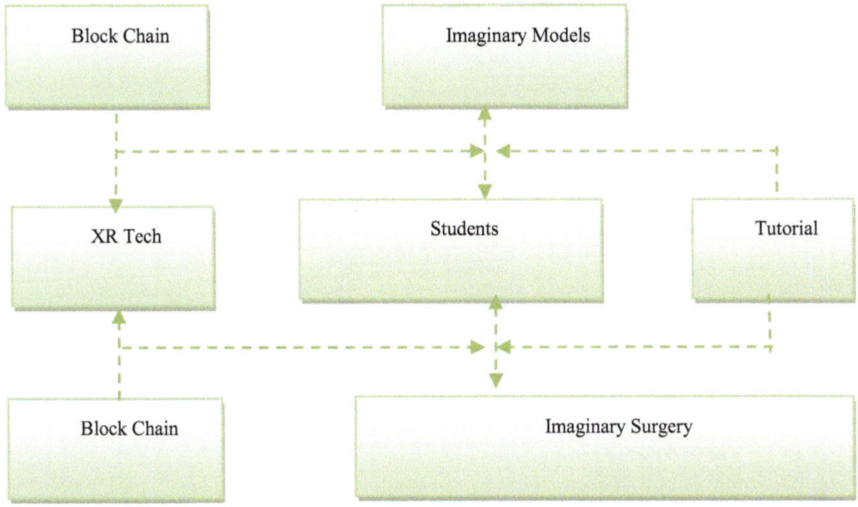

Fig. 6 Diagrammatic representation of health care redemption in meta-verse

3.4 Healthcare Training

Through a communication or sharing platform derived from the meta-verse, teachers and students can carry on a conversation about a particular topic and aid remote learners in processing, medical instruction, and meta-verse communication. A teaching scenario or teaching model based on augmented reality technology will allow students to receive more direct and in-depth medical education. The meta-verse also trains doctors, nurses, and other healthcare professionals in how to communicate with patients or arrange surgeries. According to patient needs, meta-verse offers immersive, interactive, and health discussions. Using a variety of devices and sensors patient information should be tracked, as well as the patient's symptoms.

Figure 7 shows how medical therapy and surgical treatment are carried out in the meta-verse. Psychotherapy is a process in which a patient's digital avatar is produced using block-chain technology and a virtual environment using patient extended reality technology and developed with the aid of block-chain technology, interacts. With the aid of block-chain technology, the therapist and simulation object are changed into a digital avatar and virtual environment. The patient's target organ and area are transferred to and changed into a virtual representation in this surgical treatment environment. With the use of extended reality technologies, surgeons and virtual models communicate online. The technology of meta-verse surgery is quickly gaining importance in the medical sector. Currently, hap-tic gloves, operating rooms, simulator environments, and real surgical procedures are all created using tools for virtual reality headsets. Doctors can score surgeries using augmented reality so they can simply access patient information. The patient can see quicker and simpler hands-free surgery using a 3D virtual model map. The difficulty comes from trying to explain

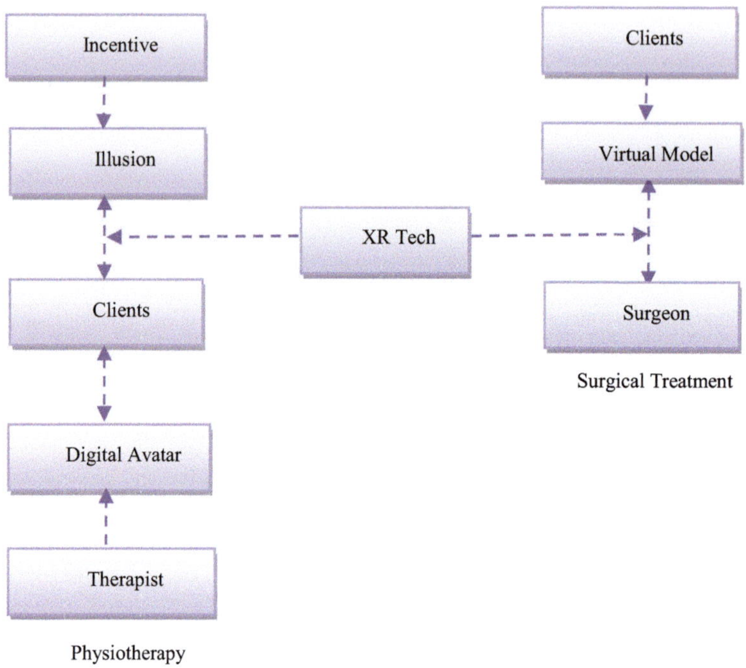

Fig. 7 Diagrammatic representation of health care training in metaverse

the surgery to the professor in these three dimensions. It functions as a sort of virtual environment for post-operative patients to examine how surgery might be carried out and comprehend the potential outcomes [16]. In order to undertake spine surgery over the previous 20 years, doctors had to adapt surgical techniques, which came with dangers. Older work can now be examined in 3D during surgical procedures on the spine thanks to the digital transition, which can make 2D flat displays obsolete [17]. Surgery education in the meta-verse is seen to be crucial for consultation and research on the usage of authorized parliaments, and the database gathered by clients globally that give surgeons and students access to this technology with the help will benefit in meta-verse [18].

4 Challenges and Future

4.1 A Storage Issue with the Information Size

The meta-verse's physical setting reflects the nuanced ways in which users engage with one another and with one another's patients and physicians. Although the meta-verse is tremendously useful, storage problems are caused by the large amounts of big data that are produced quickly. The meta-verse's massive data production during the medical process makes use of distributed and block-chain technology to address this issue. Research is a good course of action [19].

4.2 Problems with Interactive Tools

To produce and use XR, meta-verse needs head-mounted devices like the HoloLens and Pico, which are very heavy for patients and doctors. The employment of several devices by the meta-verse to communicate with its users is crucial, but their security is also crucial because of how lightweight and delicate they are. It is crucial that these devices be lightweight because they can be worn more comfortably if they are and there won't be any issues for the patient.

4.3 Risks with Confidentiality of Information

To protect patient data and medical professionals' diagnostic information, meta-verse needs a block-chain-based digital biometric authentication system. Devices generate a variety of personal information, making it extremely valuable and in need of additional protection, such as block-chain security.

4.4 Ambiguities Related to Expenses for Technology

The advanced technologies that are built into the meta-verse system are difficult to use because of their high cost. Everyone is unable to purchase the sensors that are worn and technologies utilized in the meta-verse because they are so expensive.

4.5 The Equilibrium Between Simulation and the Real World

The meta-verse offers the user an immersive experience, but the user's impression of movement between the inhabited area and the physical environment in the real world may cause issues, leading to VR hallucinations. Because illusion involves more than just the client's inability to distinguish between the delusion and the real world, equilibrium between the two is crucial. After a description of the current state-of-the-art technologies supporting digital and smart healthcare frameworks, the Metaverse's enabling technologies are covered [20].

4.6 Limitations with Interoperability

With the aid of contemporary electronics and technology, patient-assuring devices gather a variety of patient-related data. Due to the increase in data, there are a vast number of devices for interoperability, and difficult hardware software ensures every healthcare service [21]. The maintenance of security processing requirements is another difficult subject.

4.7 Problems of Data Privacy

Large health organizations may share patient information in a variety of ways, with or without their patients' agreement. With the aid of numerous technological gadgets that produce massive amounts of data, it is possible to remotely participate in a simulation of doctors treating a variety of patients in the meta-verse [22]. Communication is used to determine the patients' present status during the procedure, and the information about the patients is shared to minimize the risk of data loss. The patient's therapy will not be finished if the data is lost, and this can also lead to misunderstandings between the patient and the doctor. The goal of fusing the medical images obtained from many sensors is to increase the quality of the diagnostic image modality [23]. Numerical findings show that the recommended secure data-sharing technique outperforms several renowned standard techniques in terms of usual performance, energy usage, and reward [24]. Block-chain improves IoT platform security and privacy possibilities [25]. The area under the curve (AUC) value of the IRMA dataset was high, and 100% accuracy was attained. These findings demonstrate the promise of CNN and other deep learning methods for mammography and histopathology image-based breast cancer diagnosis. This study examined the methods and datasets currently available for the analysis of histology and mammography pictures in order to diagnose breast cancer. According to the results, the IRMA dataset yields 100% accuracy with an AUC score, while the convolutional neural network (CNN) technique on the MIAS database yields a high accuracy of 0.9%. When it comes to segmentation, the U-Net model is demonstrated to be more accurate in MIAS and DDSM when compared to CBIS-DDSM. Random forest and multi-layer perception (MLP) algorithms are used for classification [26].

5 Conclusion

The primary objective is to present the meta-verse in-depth and to illustrate the modifications that its involvement in the field of health care might bring about. By utilizing technologies like electronic and smart health care, the quality of treatment can be further enhanced. With the ability to critically assess the breadth and potential changes that may happen in the future, meta-verse can be utilized as a crucial technology to provide healthcare services, including the diagnosis of diseases, healthcare procedures, etc. Technology of the moment is frequently debated. It is possible to see state-of-the-art health care through the use of a tool called meta-verse. Meta-verse can be used for patient and disease analysis, with results coming back in a shorter period of time. It also offers cutting-edge health care via digital twin and telemedicine. Block-chain was created as a security measure to protect patient data in electronic communications through blocking, which facilitated faster analysis of data by research organizations, hospitals, physicians, and other professionals through encryption algorithms. Everything is quickly understood and assumptions are increased with the aid of meta-verse. With the aid of the meta-verse, numerous assumptions are made. Although the meta-verse is tremendously beneficial in many areas, its use compromises environmental sustainability.

References

1. Duge C (2022) Metaverse: the revolution of the SportsWorld. [Online]. Available: https://www.ispo.com/en/trends/metaverserevolution-sports-world
2. Mozumder MAI, Sheeraz MM, Athar A, Aich S, Kim H-C (2022) Overview: technology roadmap of the future trend of metaverse basedon IoT, blockchain, AI technique, and medical domain metaverse activity. In: Proceedings of 24th international conference on advanced communication technology (ICACT), pp 256–261
3. Riva G, Wiederhold BK (2022) What the metaverse is (really) and why we need to know about it. Cyberpsychol Behav Soc Netw 25(6):355–359
4. Wu T-C, Ho C-T-B (2022) A scoping review of metaverse in emergency medicine. Australas Emerg Care. https://doi.org/10.1016/j.auec.2022.08.002
5. Skarbez R, Smith M, Whitton MC (2021) Revisiting milgram and Kishino's reality-virtuality continuum Front Virtual Reality 2(Art. no. 647997)
6. Bardi J is Marxent's Tech-Obsessed Former Communications Director. (May 2022) Virtual reality De_ned& use cases: 3D cloud by Marx-ent. [Online]. Available: https://www.marxentlabs.com/what-is-virtualreality/
7. (Nov. 2019). The 3 types of virtual reality. [Online]. Available:https://heizenrader.com/the-3-types-of-virtual-reality/
8. Milgram P, Colquhoun H (1999) A taxonomy of real and virtual world display integration. In: Mixed reality, merging real virtual worlds, vol 1,no1999, pp 1–26
9. Learn more about this landscape and the requirements for a computing system that can handle the demands of these new, immersive experiences, 'Virtual Reality vs. Augmented Reality vs. Mixed Reality'. Accessed: 14 May 2022. [Online]. Available: https://www.intel.com/content/www/us/en/tech-tips-and-tricks/virtual-reality-vs-augmentedreality.Html
10. Rohmetra H, Raghunath N, Narang P, Chamola V, Guizani M, Lakkaniga NR (2021) AI-enabled remote monitoring of vital signs for COVID-19: methods, prospects and challenges. In: Computing, pp 1–27
11. Jang J, Tschabrunn CM, Barkagan M et al (2018) Three-dimensional holographic visualization of high-resolution myocardial scar on HoloLens. PloS one 13(10):e0205188
12. He L, Liu K, He Z, Cao L (2023) Three-dimensional holographic communication system for the metaverse. Opt Commun 526(Art. no. 128894)
13. Taylor L, Dyer T, Al-Azzawi M, Smith C, Nzeako O, Shah Z (2022) Extended reality anatomy undergraduate teaching: A literature review on an alternative method of learning. Ann Anatomy-Anatomischer Anzeiger 239(Art. no. 151817)

14. Dhoju R, Alsadoon A, Prasad PWC et al (2021) Augmented reality navigation for liver surgery: an enhanced coherent point drift algorithm based hybrid optimization scheme. Multimedia Tools Appl 80(18):28179
15. Carl B, Bopp M, Sa B et al (2020) Spine surgery supported by augmented reality. Glob Spine J 10(2):41S-55S
16. Bhattacharya P, Obaidat MS, Savaliya D et al (2022) Metaverse assisted telesurgery in healthcare 5.0: an interplay of blockchain and explainable AI. In: 2022 International conference on computer, information and telecommunication systems (CITS). IEEE, pp 1–5
17. Anwer A, Jamil Y, Bilal M (2022) Provision of surgical pre-operative patient counseling services through the metaverse technology. Int J Surg 104(Art. no. 106792)
18. Morimoto T, Kobayashi T, Hirata H, Otani K, Sugimoto M, Tsukamoto M, Yoshihara T, Ueno M, Mawatari M (2022) XR (extended reality: Virtual reality, augmented reality, mixed reality) technology in spine medicine: Status quo and quo vadis. J Clin Med 11(2):470
19. Chapman JR, Wang JC, Wiechert K (2022) Into the spine metaverse: reflections on a future metaspine (uni-)verse. Glob Spine J 12(4):545–547
20. Chengoden R, Victor N, Huynh-The T, Yenduri G, Jhaveri RH, Alazab M, Bhattacharya S, Hegde P, Maddikunta PK, Gadekallu TR (2023) Metaverse for healthcare: a survey on potential applications, challenges and future directions. IEEE Access
21. Shao L, Tang W, Zhang Z, Chen X (2023) Medical metaverse: technologies, applications, challenges and future. J. Mech. Med. Biol. 23(02):2350028
22. Wang Y, Su Z, Zhang N, Xing R, Liu D, Luan TH, Shen X (2022) A survey on metaverse: fundamentals, security, and privacy. arXiv:2203.02662
23. Kumar V, Joshi K, Kumar R, Anandaram H, Bhagat VK, Baloni D (2023) Multi modalities medical image fusion using deep learning and metaverse technology: healthcare 4.0 A futuristic approach. Biomed Pharmacol J 16(4):1949–1959
24. Mishra R, Joshi K, Gangodkar D (2022) Blockchain-enabled secure data sharing scheme in wireless communication. In: 2022 11th International conference on system modeling & advancement in research trends (SMART). IEEE, pp 842–847
25. Saini DKJB, Kumar S, Bhatt A, Gupta R, Joshi K, Siddharth D (2023) Blockchain-based IoT applications, platforms, systems and framework. In: 2023 14th International conference on computing communication and networking technologies (ICCCNT). IEEE, pp 1–6
26. Barthwal A, Joshi K, Kumar A, Memoria M, Kumar R, Mahajan S, Abualigah L (2023) Blockchain and classification of mammograms and histopathology images in breast cancer lesions

Metaverse Emergence with Reference to Education and Teaching–Learning: *Towards Advanced Digital Education*

P. K. Paul

Abstract Within Information and Communication Technology world, Metaverse is an alarming and emerging name. Scientifically Metaverse is a virtual platform with three-dimensional nature and combines with not only virtual reality but also online gaming, and social media from which users can interact each other digitally, and it may be persistent, and to be always active and real-time based and to be effective in proper ecosystem. Metaverse is helpful not only in entertainment but also in minimizing the limitations of geographical boundaries and comes with the valuable services to consumers and users. Metaverse finally dedicated in developing massive virtual universe which are impossible in the real-world context. It comes with beyond gaming and entertainment and offers aspects and features in socializing, education and learning, smooth and effective business opportunities. The education segment is significantly changes in recent past and in traditional education different aspects and systems been included and in digital and online education diverse changes drastically improved traditional way of education. Rapidly evolving technological sphere gives various teaching methods, chalkboard/blackboard to digital smart classes. Metaverse offers higher amount of advancement and virtual world features which is dedicated in promoting digital and online education integrated with effectiveness, entertaining features. This chapter is about Metaverse including its features, characteristics, types with special reference to the impact, potentialities, and possibilities of Metaverse in the education and allied sectors.

Keywords Metaverse · VR and AR · Informatics · Digital education · Digital society · ICT · Higher education · Education technology

P. K. Paul (✉)
Department of CIS, Raiganj University, West Bengal, Raiganj, India
e-mail: pkpaul.infotech@gmail.com

G. Chhabra and K. Kaushik (eds.), *Understanding the Metaverse*, Blockchain Technologies, https://doi.org/10.1007/978-981-97-2278-5_5

1 Introduction

Metaverse is a promising name in the advanced Information and Communication Technology world, and with potential impact in digital world and societal development. As far as education technology is concerned it has started its journey since 1990s when ICT or IT became started practiced in education, teaching and learning [4, 8]. However in early 2000s and specially during the pandemic the changes of ICT applications in Education is significant and tremendous. The way of living also changed and thus also reflects in future technologies belongs to the education, teaching and research.

Learning engagement is considered as effective and successful when better engagement is happened and this also helps in healthy retention of learned information. Metaverse in education helps in healthy, sophisticated and experienced learning with full virtual support with real-life classroom experience. Today Digital Education and EduTech companies are working for removing physical barriers in education systems and helps in immersive, interactive and engaging learning experiences. And in this regard Metaverse is to be helpful in more effective, entertainment and enriching education and learning. According to the expert, simply extensive applications of Virtual Reality and Augmented Reality in education may be called as Metaverse. And according to a distinguished expert Jon Radoff there are seven (07) layers in Metaverse as depicted in Fig. 1. Though it is a minor definition, in broader sense it is offering the digital environment beyond Virtual Reality and Augmented Reality with better communication, engagement and entertaining experiences remotely. Today society is digitally connected using digital tools, devices and systems [18, 26, 43]. It has become possible due to digitally equipped organizations and enterprises. The remote learning, education, knowledge sharing, telemedicine, healthcare systems, digital business only possible because of infinite enterprises which are works on infinitely distributed systems. Metaverse is helpful in bringing digital education, online learning and work culture as well. With Metaverse the scale of the work also increased and ultimately helps in interacting people, technology transformation, secure round o' clock connectivity with security. Using Metaverse in education it can enhance the personal experiences and better collaborations among the stakeholders and thus helps in bring of 'Consumer-Centric Experience'. As mentioned before, teaching practices involved for better collaboration and understanding here 'Metaverse' and allied technologies empowers better online learning experience integrated with immersive technologies for higher amount of immersive experience for fun-integrated learning and works and lies beyond textbook centric education systems [14, 16]. In many countries government and different ministries are working in implementing Metaverse into the education systems for offering modern digital education and many Educational Technology companies are also working in developing a Metaverse-supported educational system development.

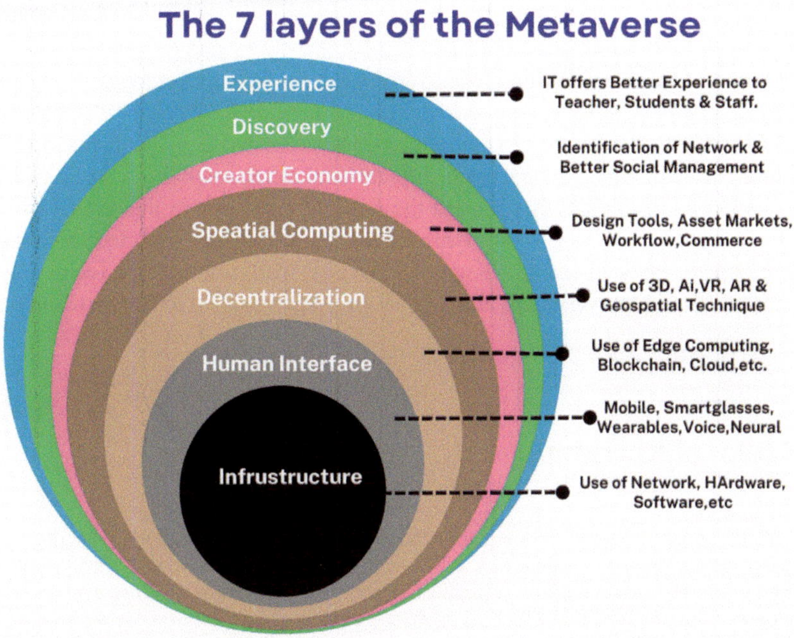

Fig. 1 Seven basic layers of Metaverse

2 Objective of the Work

The core aim of this work entitled 'Metaverse Emergence with Reference to Education and Teaching–Learning: *Towards Advanced Digital Education*' is included but not limited to as follows:

- To know about the basics of Metaverse including its features and characteristics including a brief on historical foundation.
- To gather information and knowledge about the concept, technological background and development of the Metaverse in contemporary scenario.
- To know about the basic and proper advantages of the Metaverse in Education, Teaching and Research specially in Digital Educational Systems.
- To learn about the roles, specific areas where Metaverse applications are existing and emerging and possible.
- To find out the challenges, issues and concern in Metaverse applications and integration in Education Systems specially in Digital Education.

3 Methods

The core aim of this work entitled 'Metaverse Emergence with Reference to Education and Teaching–Learning: *Towards Advanced Digital Education'* is a conceptual work and interdisciplinary in nature and combine with the subject spectrum of Information Technology, Management, Educational Sciences and Social Sciences. Digitalization has significantly changed the education sector and therefore various primary sources, secondary sources have been analyzed and reported. Journals, research papers, book chapters, dissertation related to the Education Technology, Digital Education, Metaverse and allied areas have been studied and reported in this work.

4 Review of Literature

Researchers on diverse areas of Information Science and Technology, Computing and Engineering, Education and Teaching Sciences, Management Sciences have done their work related to the Metaverse in different perspective and here this section of this chapter such work have been discusses based on the importance and focused work chronologically.

Williamson [42] Worked related to E-Learning and specially on Digital education and found several aspects of Digital Education including foundation, features, basic and emerging technologies applicable to the advanced Digital Education practice. Researcher also finds out the basic and possible governance of Digital Education including data visualization, predictive analytics, etc. Digital Education governance is important practice for proper development of the E-Learning and ICT applications in Education practice. There are lot of emerging technologies applicable in Digital Education development and in this context Metaverse is alarming, prominent and impactful. Abdel-Basset et al. [1] of his/her concerned research has emphasized on Internet of Things in promotion of the Digital Education. Among the Digital Education technologies 'Internet of Things' or IoT is considered as valuable and worthy deal. And it is also required in designing and developing smart education environment. Here in Metaverse development and practice IoT is also productive and supportive. Researcher here focused on Supportive framework in the decision-making process toward a Digital and Smart Educational practice. Therefore from the basic to advanced aspects of IoT has also been discussed scientifically.

Suzuki et al. [39], in the concerned research work have discussed on several aspects on Metaverse and focused on Virtual Experiments. The work also been emphasized on the applications to Collaborative Projects for better, smart and sophisticated Digital Educational practices. Therefore here proper framework and its significance also been reported and justified. The group of researchers have studied PubMed database with the word 'Metaverse' and portion of Virtual Reality and Augmented Reality has been described significantly. Kang [15] proposed a potential framework of the Metaverse and have discussed several aspects viz. concept and origin of this futuristic technology. Further researcher also focused on hardware, networking and virtual platforms including tools and standards applicable to the Metaverse. Aspects of the payment, content, services and other technological systems, etc. in relation to the Metaverse also been observed and described in this work. Kye et al. [19] worked on the Metaverse specially on the Educational applications in the context of possibilities and further have analyzed and reported various limitations. Researcher focused on teachers role in regard to Metaverse uses and right product development. They have focused on developing a proper system for avoiding misuse of the students data.

Alfaisal et al. [3], have done a systematic review on the Metaverse systems and specially focused on the educational aspects and development. This research was focused on systematic review of Metaverse with a strong contrast on Information Science. In this work total 41 research papers have been selected and analyzed related to the topic and here Technology Acceptance Model has been adopted. Balica et al. [5], have worked on Metaverse specially on applications as well as on going and using technologies. They have also find-out the infrastructure required for the Metaverse specially the predictive algorithms including real-time customer data analytics. Researchers also synthesized about the virtual navigation tools. Prior finding were analyzed in this work and reported systematically. Proquest, Scopus and Web of Science database was analyzed and total 86 papers met the criteria and thereafter analyzed in their work properly with future prediction. Chen [7] have worked on Metaverse and not only discusses about the potentialities but also different issues were discussed in relation to the Metaverse technology in education. Aspects related to the development of technology development, interaction challenges and issues, content designing and production, problem and issues of game addiction including privacy, and other ethical aspects been developed.

Fitria and Simbolon [11], they worked on Metaverse and analyzed a review of work and explained about the possibility of metaverse in the areas of education. Here they analyzed and reported not only opportunity but also various kind of possible and future threats in brief. Various negative aspects of Metaverse also been analyzed and explained in this work. For conducting this work simple Google search they used with the term 'Metaverse'. In a different move researchers also explained about the universities which have been started initiative on Metaverse to the educational delivery and management. Hwang and Chien [14] also worked on Metaverse and analyzed and reported aspects of different intelligent tutors, intelligent peers required in such system in the context of Artificial Intelligence. Here they have analyzed how Metaverse is not a single technology rather it is combination of different technologies and a concept of futuristic world. Park and Kim [28] also analyzed various aspects of Metaverse and specially found the major components useful for the same. They have mentioned about the hardware, software as well as contents. Further in this work major three approaches have been analyzed and reported, viz. user interaction, implementation as well as application.

Singh et al. [36], Simply but briefly explained about the Metaverse in the context of teaching and learning and education as a whole. Different digital aspects of the Metaverse been analyzed and reported in their work with a strong emphasis on digital resources, interactive features, barriers in the context of technology development have analyzed and reported scientifically. Researchers also discussed about the technological and financial limitations in this work Sriram [37]. This is a very foundation work been done on metaverse describing various aspects on Metaverse including foundation, features and characteristics. Researchers also analyzed and reported on issues, technologies and future prospect on Metaverse in general [34].

Stanoevska-Slabeva [38] in the work entitled 'Opportunities and challenges of metaverse for education: a literature review' various aspects related to the Metaverse been described. Researcher explained how Metaverse is the future of internet technologies and described this is as a post reality universe. Researcher focused on various technologies that are responsible in developing Metaverse Systems and among these important are artificial intelligence, blockchain, Cloud Computing, Internet of Things. In the research work, role of 5G and 6G networks has also been emphasized strongly. Further researcher expressed that technologies and applications of the AR, VR, MR, 3D, and convergence of all these technologies played an important role for solid Metaverse applications in education. Zhang et al. [44], A comprehensive work been done by the group of researchers on Metaverse and there lot of aspects and topics been discussed including Definition and meaning of Metaverse, aspects of framework and features also been well described by the researcher. Further, various aspects of potential applications including current and future challenges, and other future scope and research potentialities been described. Ahuja et al. [2], Group of researchers have focused on Metaverse with a focus on Artificial Intelligence, Machine Learning, and Deep Learning in the context of Education specially in the medical education, as well as integrative health. Researchers explained that Metaverse is not a new concept and early 2000 when virtual 3D game was emerged and popularized that period is called as second life beginning. Apart from the potentialities and importance of the Metaverse in Medical Education, here researchers also able in finding few issues and challenges and reported accordingly.

5 Metaverse: An Overview with Subject Specific Treatment of Metaverse

Metaverse is responsible for bringing new educational environment and experiences specially in online education, and also for the add-on traditional educational modes. Virtual and real-world educational development is effectively possible using Metaverse technologies. Real-time, effective, entertaining education without any kind of geographical boundaries become possible with support of Metaverse and allied systems (Refer Fig. 2 for major element of Metaverse). The wearable devices can move to the wireless with various digital forms of items such as intelligent NPCs, Virtual Learning Resources, etc. and thus rich and fantastic experiences are become possible with this technological adaptation and implementation [9, 29, 35]. Metaverse is a virtual world, where different Information Technologies play a leading role in fulfillment of the digitalization and digital world. In such a world, billion on people can be connected each other using Cloud, 3D, Virtual Reality and Augmented Reality tools and systems. This information technology is a futuristic concept using internet in which human experiences can be higher and entertaining and which is not possible in real world or physical world. Apart from AR and VR glasses high-bandwidth internet, interoperability standard are highly important. It is important to note that the concept of Metaverse not completely new in the year 1992 an author Neal Stephenson first conceptualize about Metaverse; an internet and technology-based world in his novel 'Snow Crash'. Different organizations worldwide working and impacting on Metaverse and very recently Facebook changed with its whole corporate system with a new name called 'Meta' and plans to invest billions of USD in next coming years. Similarly other tech giant like Google, Microsoft, Nvidia, IBM, Apple, Qualcomm are also investing money, effort and technologies into the Metaverse [10, 17, 40].

An emerging research and consultancy firm McKinsey and Company expects to reach USD 5 billion within 2023 as far as Metaverse is concerned. Among different companies Mesh of Microsoft is working hard for Metaverse applications, viz. virtual operating rooms. Different allied technologies helpful in fulfillment of the concepts of Metaverse such as:

- Virtual Reality,
- Augmented Reality,

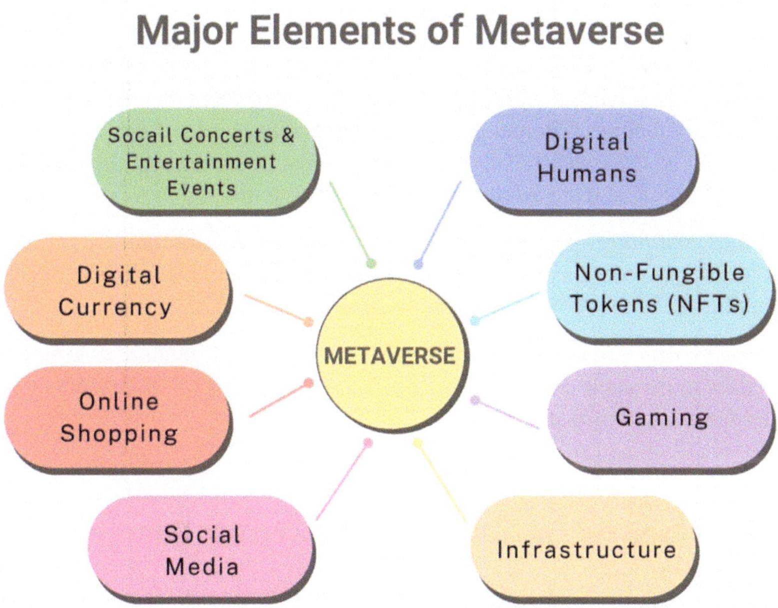

Fig. 2 Most common layers of Metaverse

- Blockchain Technology,
- AI and Machine Learning,
- Spatial Technologies,
- Edge Computing,
- Internet of Things,
- Big Data Technologies and so on.

Due to the impact of Metaverse, it has applications and potential contribution in some of the fields and sectors as mentioned below.

5.1 Different Subject and Different Metaverse Treatments

In History and Heritage Management—Using Metaverse, anyone can easily able to understand and learn History and Heritage Management. Like any kind of previous or old story can be learn using Digital media and students can feel 'like-real'. A student can learn about the Black Revolution or French Revolution using Metaverse-related systems in Virtual Reality way.

In Geography and Environmental Management—As far as Geography or Environmental Science Students are concerned students can learn about different aspects of Geography using AR and VR using Metaverse concept. Here Solar System, Water Systems, Soil Systems are possible to learn using Digital tools in virtual context and real-life experience can be gathered effectively.

In Basic and Applied Science—Metaverse is impactful and can be effectively useful in different Basic Science facilities and laboratory management for creation of virtual setup, anatomy feeling, different physical and life science-related topics and experiments. Some of the concepts like mass and gravity, Newton's weight, etc. can be easily understandable using Metaverse-supported systems.

In Art and Sculpture—As Far as Art and Sculpture is concerned Metaverse is useful to give a sense and feel to the students and learners regarding various art- and sculpture-related products virtually in digital manner.

In Language and Literature—Regarding Language and Literature is concerned, Metaverse can be nicely adopted for clearing any kind of novel or concept of story digitally for giving original feeling.

6 Benefits and Advantages of the Metaverse in Education

As far as Metaverse-related benefits (also refer Fig. 3) and advantages are concerned it offers wide range of advantages and features viz.

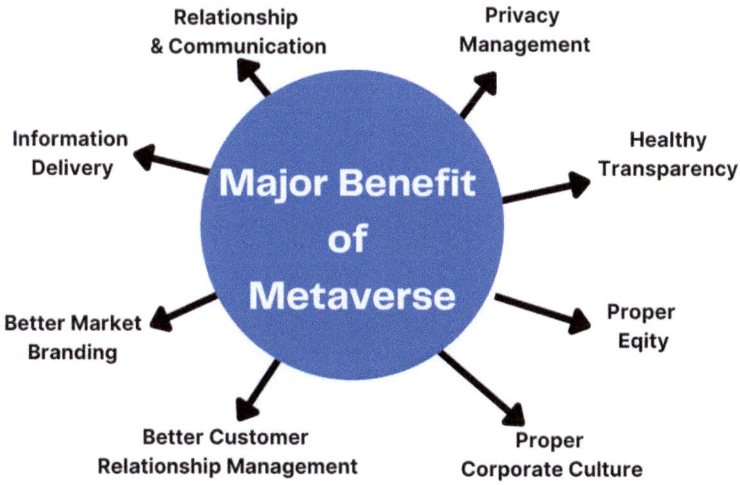

Fig. 3 Metaverse and its impactful beneficiary areas

6.1 Amazing Real-Life Learning Experiences

Metaverse offers educational and teaching–learning activities entertaining and amazing using 3D objects and seems like original and physical classroom. The historical tour become effectively possible using Metaverse in which student can experienced historical periods, feelings of the periods, aspects, etc. Easily one can enter into any of the age or civilization using latest technologies which are supportive to the Metaverse systems or concept.

6.2 Gamification and Enhanced Learning

Gamification is the combination of the game and education (skilling) for entertaining educational systems. Metaverse support modern educational systems with full of enjoy and entertainment way, here education integrates with fun. The concept and growth of Gamification is noticeable and Metaverse empower such systems effectively with various allied tools and technologies. Therefore juniors can learn with fun and can complete the task on-time.

6.3 Hands-On and Real-Life Practice

Metaverse further allows digitalization of traditional education systems and helps in better and sophisticated educational practices which are difficult to bring in real-world setting. Understanding of complex theories become easy with Metaverse. Students hands-on practices, real-life experience is possible virtually whenever possible in a safe environment. Various studies like understanding complex surgeries, explosive experiments is possible using Metaverse-supported systems [12, 13].

6.4 Life-Like Ecosystem Enhancement

Metaverse as supported by latest Information Technological tools and systems, therefore it not directly able in traditional classroom environment to a greater extend, but in Digital and Online Education System Metaverse is effective, supportive and modern. Life-like ecosystem becomes effectively designed and offered using various allied technologies. Teachers, professors and educators can teach the concept and theories including practical using Metaverse ecosystems easily.

6.5 Enhanced and Advanced Learning Speed

Experts believe that Metaverse can be helpful in enhancing the speed of the teaching–learning process than traditional mode of education. As Metaverse uses 3D and Virtual Reality with Augmented Reality, therefore higher amount of speed can be countered in this mechanism. 3D Simulations, IoT enable devices are fruitful in effective and speedy education system. A study of PwC shows that Virtual Reality enables courses are effective to under four times than general traditional classroom systems as far as some group of employees are concerned.

7 Metaverse in Education Systems: The Wider Applications

Metaverse is helpful not only to the students or trainees, researchers but also to the corporate learners, adult and mature students and it is possible due to the virtual classroom systems supported by cloud and AR, also Virtual Reality based systems. Here for proper and immersive educational and learning experiences Metaverse is impactful and to be widely used in coming information age, as mentioned below and depicts in Fig. 4.

7.1 In Enhanced Virtual 3D Classrooms Development

The growing online and digital educational programs, centers, bodies, units and colleges are able in giving new age in educational delivery and people now have

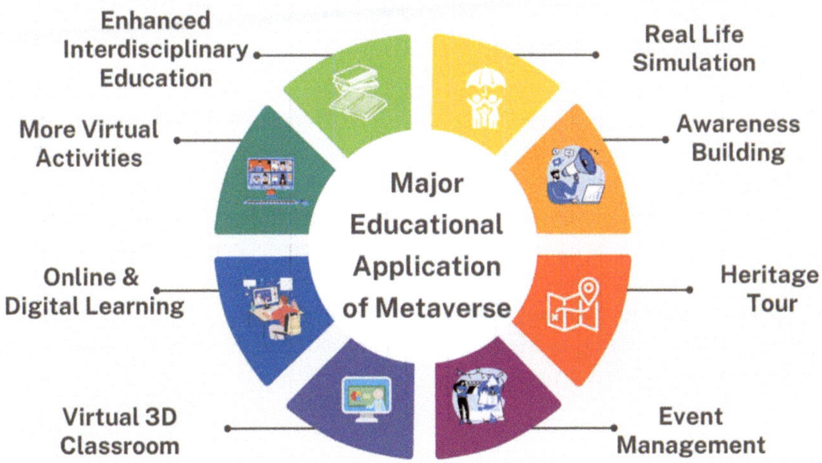

Fig. 4 Major educational applications of Metaverse

choice between online, face to face, and blended. Here is the requirement and urgency of Metaverse in order to move forward new age 3D virtual classroom environment as per their time and location convenience. Metaverse-supported teaching tools and educational methods are able in offering new age systems and development in modern teaching–learning mechanism. It is a fact that Face to Face class or traditional class room environment has its own significance and this is can't be replaced or possible to replicate, though digital and online education powered by 3D can help in minimizing such distances to some extent. With Metaverse-supported systems, healthy learning experience is effectively possible.

7.2 In Promotion of Digital Learning

Metaverse is completely able in designing and developing smart classes not only in higher education and degrees but also in school education. Students and teachers both can enhance existing digital learning experience using Metaverse-supported systems powered by other additional tools and technologies such as cloud computing, big data, Internet of Things (IoT) [20, 41]. It provides wider space in digital content creation and delivery with real-life examples to achieve proper or fixed goals.

7.3 *In Sophisticated and Digitalized Virtual Campus Activities and Virtual Tour*

As far as extra-curricular activities are concerned Metaverse is important in various educational context. Here sports and digital culture- and art-related experiences can be gathered and offered using Metaverse-supported systems. Different fun and entertaining activities such as music or mathematics or physical science practice possible in virtual environment. Students using Metaverse systems can participate in virtual tours, learn in virtual environment therefore in such cases where physical or face to face is going to difficult can adopt this concept of mechanism [21, 22]. Virtual world tour is effectively possible in Metaverse in a second, and therefore, it destroys the general boundaries as much as possible.

7.4 *In Promotion of Interdisciplinary Learning*

This is the age of interdisciplinary science and technologies; and merger of humanities with science or humanities with management or commerce. Research, skilling and degrees are developing and increasing in the field of interdisciplinary knowledge world. With Metaverse-supported systems one can easily teach the core of mathematics logic or formula and at the same time history or cultural studies. Therefore Metaverse is promoting such educational systems and models effectively [24, 25]. Holistic learning experience and real-life applications of various theories become possible using Metaverse.

7.5 *In Enhancing and Bringing Awareness*

In education general, and also in corporate education enhancing and keeping awareness at par with contemporary scenario is important, and thus aim fulfillment is effectively possible with Metaverse. Awareness of different issues can be created digitally in promotion of digital society or digital humanities using Metaverse. Aspects like pollution, weather and climate, gender studies, journalism, democracy, famine is effectively possible with Metaverse-based systems. Therefore using Metaverse students and learners not only gains theoretical knowledge but also able in gain emotions.

7.6 *Events and People Collaborate Engagement-Remotely and Effective Teaching*

Using Metaverse one can easily able in gaining knowledge anywhere anytime with proper digital experiences and with 3D or Virtual Reality kind of exposure and therefore worldwide this system has an unique way for better collaboration with the experts, among the experts, students and experts. Educators can hold knowledge session, seminar and conferences with eminent and expert speakers and students can learn anywhere anytime and experienced their self. Cost-effective academic, research and professional events can be developed and exposed for expansion of the knowledge, wider fest and exhibitions, and so on [23, 27, 30]. Therefore in collaborative learning Metaverse is helping in enhancing good teaching skills, healthy opportunities in experiments, enhancing teaching professionalism, developing social interaction in the contemporary and real-life situation. In improving problem solving skills, Metaverse is therefore worthy, timely and impactful for the teachers, learners and students of different circles.

7.7 *Interactive Curriculum and Flexible Learning*

Reading and understanding a concept is little difficult than understanding the same using digital technologies like Metaverse, as it has support of AR, VR and 3D-related subtechnologies with other technologies like Internet of Things (IoT), Big Data, Cloud Computing, etc. Using Metaverse one can understand and learn solar ecosystem in virtual walk or platform. Here learners can feel and imagine about the orbits of the planets easily. Further, here step by step rocket launching is possible to understand easily using 3D setup. Different modern applied curricula, interactive curricula, teaching–learning process can be easily understand using Metaverse-supported systems. Concepts like medical surgeries, astrophysics, geology, pharmacy and even anthropology can be easily conceptualized with Metaverse-supported systems.

7.8 *In Easy and Quick Information Finding*

Today we are living in internet and digital age, and people are having wide sense of knowledge in internet searching. But still Metaverse-supported systems can more effective in easy and quick information finding and through this individualized learning path is possible easily and effectively. Metaverse as connected with AI and Machine Learning therefore a clear understanding and thoughts clarification possible perfectly.

7.9 In Developing and Enhancing Practical Learning

Days are gone when only theoretical knowledge considered as enough, today in the age of information and digital technologies knowledge without applications or practical become meaningless. Therefore Metaverse offers good concepts, and clarification in practical applied learning in online and digital environment. Different mathematical and statistical formula, physics problem can be easily understandable using Metaverse-supported systems for a complete holistic learning [6, 31].

7.10 In Managing Time and Locations of the Stakeholders

Metaverse is also applicable in time management than traditional educational systems or digital or online education. As traditional teaching–learning process requires both the teachers and students in a same platform, therefore it depends on time and location bound. Though the digital education or online education gives the flexibility in location and timing for remote based learning but more flexibility of learning is effectively possible using Metaverse [32, 45]. Smart wearable devices are supported by high internet connectivity gives wider opportunities in education, i.e., 'Flexible Education' powered by synchronous and asynchronous learning environments.

7.11 In Finding Learner Identity

Identity of the learners considered as impactful and worthy and here in the context of learners identity 'Metaverse' is worthy, and real identification is possible using appropriate tools. Further it is responsible in creating more advanced features, and therefore supports in joyful and creative sense effectively.

7.12 Interaction of Learners, Students and Trainees

In the traditional and conventional educational systems, students basically interact with the teachers and peer in-person. In digital education also it has screen for knowledge material display but Metaverse-supported systems works with the wider scale and its own style of interactive digital platform that offers real-time interaction, practical experience, socialization. In conventional educational systems, there are important concerns in practical learning and education, and sometimes screen-based learning may be boring, whereas Metaverse-supported systems empowers realistic scenario of education and learning experiences.

7.13 *Learning Resources and Activities and Assessment*

In traditional educational systems and learning, there are concern in static environment and lectures, books, videos are sharable in such formats whereas Metaverse based education may empowers participation based education. Thus sharing of learning resources become easy, effective and advanced with realistic approaches. Further in non-Metaverse education getting feedback is difficult whereas Metaverse-supported system easily feedback can be gathered and here formative and summative assessment also to be easy and effective. Thus comprehensive assessment become possible using Metaverse-based systems.

8 Enhancement of Education Powered by Metaverse: A Present and Future Trend

Metaverse is emerging and applicable in diverse areas of education and teaching–learning, educational management and development. Applications of the Metaverse and allied technologies give the following and at the same time requires mentioned set of practices, viz.

8.1 *Demand of the High-Speed Communication and Networks*

Establishing Metaverse requires advanced and robust wireless communication systems and also advanced high-speed networks like 5G or more advanced network systems without such strong network implementation of Metaverse is not possible. Further, high-speed network is required for offering various intelligent services for the teaching and educational management. Steadiness, low latency regarding transmission of data, proper and immediate feedback is effectively possible in Metaverse if it is supported by the high-speed and intelligent network [33].

8.2 *Higher Degrees of Computing Technologies Involvement*

As an intelligent, modern and future technologies Metaverse is depends on various technologies such as Internet of Things, Cloud Computing, Big Data, Edge Computing and also Distributed Computing. And this is required for various purposes, viz. processing, computing, storing and transmit, interchange of the data, and this will help in real-world classroom environment. Sharing, storing of data of the students become easy with various supportive allied technologies. Information of the learners including records, learning resources become easy to manage effectively using allied technologies of Metaverse.

8.3 Analytical Technologies and Its Support

Various kinds of analytical technologies and their support are highly important in proper Metaverse utilization in higher education. Among the analytical technologies important are Artificial Intelligence, Text Mining, Machine Learning and Deep Learning. AI and Robotics in Metaverse offers intelligent NPC tutoring, intelligent NPC and proper educational services for various activities such as simulation, decision making, etc. Ultimately analytical technologies and subtechnologies helps in collecting, measuring, tracing, analyzing various types of educational and learning-related data of the students. As a whole such allied technologies are helpful in educational resource management, and the same can be fruitful for the teachers and students, both the circles.

8.4 Advancing Modeling and Rendering Technologies

Metaverse depends on various modeling techniques and technologies which are connected with the Virtual Reality and Augmented Reality, and many other rendering technologies dedicated in creating 3D world for better and simulated scenario and context. Though Virtual Reality and Metaverse is not the same but a higher amount of portion of Metaverse is closely connected with the Virtual reality, Different kind of virtual items such as sketch up, unity, digital twins, holography are require various modeling and simulations. Educational contents, educational models, clarification of any age are highly depends on Metaverse. There are lot of difficulties in clearing concepts in traditional educational systems can be effective and easy to understand using proper modeling and rendering technologies.

8.5 Interaction Technologies

Conventional internet is not biased with the embodied or multimodal interaction, whereas Metaverse is different. There are technologies, viz. Virtual Reality, sensor, tracking (real-time), XR, Analytics and IoT is required highly in conceptualizing and developing Metaverse-related items and products. Such technologies are mainly deployed for the following as far as proper Metaverse development is concerned:

- Manipulations.
- Navigations.

- Collaborations.
- Proper Sensory feedback systems.

Using interaction technologies students or knowledge seekers are able in mobilizing the bodies for various exploratory learning experiences for offering better collaboration, interaction, socialization and so on. Metaverse is able in finding and offering its educational experiences only upon support on interaction technologies [7].

8.6 Proper Advancement and Authentication Technologies

Metaverse is highly depends on proper advanced and authentic technologies, and many experts and scientist expressed the same. According to expert proper blockchain is highly important and required for offering transparent, open and decentralized including sustainable ecosystems. Blockchain is highly important in order to gather learners' data and at the same time for proper tracing and avoiding different fraud and plagiarism-related issues. Therefore proper technological infrastructure regarding Metaverse in education must be employed for real-life educational experience and content delivery to the wide range of knowledge seekers.

8.7 Requirement of the Smart Wearable Device

Different kind of smart and wearable devices are highly required in Metaverse development viz. non-see-through and optical-see-through. Here headsets, HMD (head-mounted displays), smart glasses are highly required. As smart wearable devices are highly connected with the hardware therefore its necessity is mandatory in Metaverse educational product development and experiencing.

8.8 Advanced Way of Avatar or Digital Character

Metaverse highly depends on Digital Character called 'Avatar' and this character may be teachers and learners. Real-time tracking technology, simulation technologies are highly required in order to prepare and develop avatar or digital character. Here teachers as per their requirement can customize the features such as changes of shape, skin color, gender, dress or gender. Similar to the teachers, here learner can also change the features as per the requirement. In such system facial expression, gesture of the digital character is displayed like physical appearances. Using sensor, controllers and also real-time tracking an interactive, joyful, entertaining moment and experiences are able to design effectively.

8.9 Non-player Character (NPC)

Metaverse is highly technology centric and supported by various Artificial Intelligence Systems, VR and AR Systems. It has Avatar or Play Character already learned, as far as other character is concerned NPC or Non Player Character is considered as important. Here intelligent NPC teachers, Intelligent NPC Student and Supporting NPC peers are there in order to support in educational process whenever required by the Player Character-Teachers or Player Character-Student. Regarding simulation, decision making, supporting of the arbitration such NPC are highly required. Tutoring, discussion, regarding practicing skillsets Metaverse depends on NPC.

8.10 Healthy and Sophisticated Learning Scene

There are different realistic learning scenes highly important in Metaverse, and it is possible to achieve by the rendering technologies, viz. digital twins, XR, VR, 3D, etc. to identification of the material, color, ornaments, etc., advanced technologies are highly required in order to offer healthy and sophisticated learning experiences.

8.11 In Learning Resource

In Metaverse world, learning resources considered as worthy and important and here various supporting tools and technologies considered as prime for real-life experience development compared to the physical world. Using VR and 3D and allied technologies, learning resources can be shared as the multimodal nature and here learners can learn effectively with fun, knowledge with ability in editing, creating learning resources of their own. An example of such system is Roblox, which is allows players for the creation of the works in virtual context. Another example ZEPPTO also allows users to make AR fashion ready, and the same can be applied in educational context effectively, whenever required.

8.12 For Learning Logging

As Metaverse is connected with different stakeholders therefore various allied technologies helps in keeping record of the students, learners, teachers, associates, etc. and here every objects of the teachers and learners is possible to track from the logging. Real-time status of information, historical information, is possible to collect and analyze as per the requirement. Ultimately, both learners as well as teachers not only able in review but also further decision making regarding learning process and also for the completion of the conduct upon personal experiences.

8.13 For Proper Learning Analysis

Regarding proper learning analysis as well Metaverse is worthy. Various technologies such as advanced intelligent computing, RDBMS, AI and Big Data helps in collecting huge amount of data and also proper analyzing of the data. And from this massive data trend pattern, effectiveness, use pattern are possible to identified using proper tools. Furthermore, visual and personalized analysis reports are effectively possible with Metaverse-supported systems.

8.14 For Learning Authentication

Metaverse is digital and existing system also Digital but Metaverse is more open, shared and decentralized in nature. In Metaverse-supported system information of the learners can be stored effectively and secured. Privacy can be managed in such systems, further here virtual works of the users can be shared to others. Here blockchain and other allied technologies, viz. Non-Fungible Token are highly useful in authentication and also be traced and therefore the systems of Metaverse become sage, persistent, and also secured.

9 Future Potentials

Therefore Metaverse is dedicated in enabling both the teachers and students advanced, systematic, entertaining-educational contents in diverse situation. And the same can be achieved similar to the physical mode of studies. Asynchronous and face-to-face learning is effectively possible using Metaverse-supported system. Though it is worthy to note that a proper Metaverse development requires some of the things as a prerequisite, including it concern about some of the issues as mentioned in Fig. 5.

Different kind of learning activities may grow further in future for the academic activities such as lecture, individual work, group panel as well as collaborative work. Incredible experience is possible using this latest technology for wide range of subjects, viz.

- Basic Science,
- Applied Science,
- Fine Arts and Sculpture,

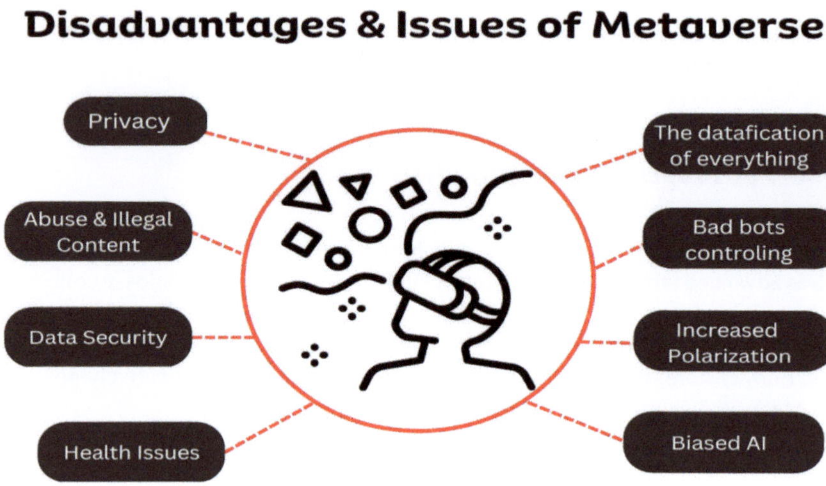

Fig. 5 Major issues and concern in Metaverse

- History and Heritage Management,
- Tourism and Transportation Sciences,
- Archives,
- Public Health and Bio Diversity,
- Life Science,
- Core Medicine,
- Engineering-Civil,
- Engineering-Mechanical,
- Nano-Technology,
- Geology and Earth Sciences,
- Environmental Sciences, etc.

Various allied Information Technology components also going to be worthy in managing, exploring and expanding Metaverse system at par futuristic goal. Among the subtechnologies few important are Cloud Computing, Big Data, Internet of Things (IoT), Blockchain Technologies, Edge Computing, etc., and all these become impactful in future in order to complete the expectation and goal of the Metaverse world.

10 Concluding Remarks

Today we are living in Digital Society; this is such a society where different tools and devices are inter-connected and able in sharing of signal and data for different sectors. Further in this society different emerging Information Technology components also played a leading and vital role. A virtual world may be created using Metaverse like real-world. The education can be more updated, modern and skill focus using Metaverse. And such futuristic system is not only to the development of the Digital Education but also helpful in traditional onsite educational systems. Today top brands are working on better Educational Technology development including Metaverse applications for a healthy, advanced and easiest way of teaching–learning process. There are some issues related to the Metaverse, viz. proper funding, technology transformation, technology integration, availability of skilled manpower, teacher-students digital and working skills. As far as challenges are concerned reputation and identity Metaverse, community and network challenges, ownership and property challenges, time and space, currency and payment considered as vital for the growth of Metaverse-in general, and also in educational applications specially in school and Higher Educational sector.

References

1. Abdel-Basset M, Manogaran G, Mohamed M, Rushdy E (2019) Internet of things in smart education environment: supportive framework in the decision-making process. Concurr Comput: Pract Exp 31(10):e4515
2. Ahuja AS, Polascik BW, Doddapaneni D, Byrnes ES, Sridhar J (2023) The digital metaverse: applications in artificial intelligence, medical education, and integrative health. Integr Med Res 12(1):100917
3. Alfaisal R, Hashim H, Azizan UH (2022) Metaverse system adoption in education: a systematic literature review. J Comput Educ 1–45
4. Almahasees Z, Mohsen K, Amin MO (2021) Faculty's and students' perceptions of online learning during COVID-19. In: Frontiers in education, vol 6. Frontiers Media SA, p 638470
5. Balica RŞ, Majerová J, Cuţitoi AC (2022) Metaverse applications, technologies, and infrastructure: predictive algorithms, real-time customer data analytics, and virtual navigation tools. Linguist Philoso Invest 21:219–235
6. Bhardwaj A, Kaushik K (2023) Metaverse or Metaworst with cybersecurity attacks. IT Prof 25(3):54–60
7. Chen Z (2022) Exploring the application scenarios and issues facing metaverse technology in education. Interact Learn Environ 1–13
8. Contreras GS, González AH, Fernández MIS, Martínez CB, Cepa J, Escobar Z (2022) The importance of the application of the metaverse in education. Mod Appl Sci 16(3):1–34
9. Dahan NA, Al-Razgan M, Al-Laith A, Alsoufi MA, Al-Asaly MS, Alfakih T (2022) Metaverse framework: a case study on E-learning environment (ELEM). Electronics 11(10):1616
10. De Gagne JC, Randall PS, Rushton S, Park HK, Cho E, Yamane SS, Jung D (2022) The use of metaverse in nursing education: an umbrella review. Nurse Educ 10–1097
11. Fitria TN, Simbolon NE (2022) Possibility of metaverse in education: opportunity and threat. SOSMANIORA: Jurnal Ilmu Sosial dan Humaniora 1(3):365–375
12. Han D (2022) Exploration for educational application of metaverse: focusing on implication for use in English education. Robot AI Ethics 7:10–21
13. Huh S (2022) Application of computer-based testing in the Korean medical licensing examination, the emergence of the metaverse in medical education, journal metrics and statistics, and appreciation to reviewers and volunteers. J Educ Eval Health Prof 19
14. Hwang GJ, Chien SY (2022) Definition, roles, and potential research issues of the metaverse in education: an artificial intelligence perspective. Comput Educ: Artific Intell 3:100082
15. Kang YM (2021) Metaverse framework and building block. J Korea Instit Inf Commun Eng 25(9):1263–1266

16. Kaushik K, Ahuja NJ, Nayyar A (2021) Blended learning tools and practices: a comprehensive analysis
17. Khalil A, Haqdad A, Younas S (2023) Educational metaverse for teaching and learning in higher education of Pakistan. J Positive School Psychol 1183–1198
18. Kumar A, Krishnamurthi R, Bhatia S, Kaushik K, Ahuja NJ, Nayyar A, Masud M (2021) Blended learning tools and practices: a comprehensive analysis. IEEE Access 9:85151–85197
19. Kye B, Han N, Kim E, Park Y, Jo S (2021) Educational applications of metaverse: possibilities and limitations. J Educ Eval Health Prof 18
20. Lee H, Woo D, Yu S (2022) Virtual reality metaverse system supplementing remote education methods: based on aircraft maintenance simulation. Appl Sci 12(5):2667
21. López-Belmonte J, Pozo-Sánchez S, Carmona-Serrano N, Moreno-Guerrero AJ (2022) Flipped learning and E-learning as training models focused on the metaverse. Emerg Sci J 6:188–198
22. Marini A, Nafisah S, Sekaringtyas T, Safitri D, Lestari I, Suntari Y, Sudrajat A, Iskandar R (2022) Mobile augmented reality learning media with metaverse to improve student learning outcomes in science class. Int J Interact Mobile Technol 16(7)
23. Mohamed ES, Naqishbandi TA, Veronese G (2023) Metaverse!: possible potential opportunities and trends in e-healthcare and education. Int J E-Adoption (IJEA) 15(2):1–21
24. Mughal MY, Andleeb N, Khurram AFA, Ali MY, Aslam MS, Saleem MN (2022) Perceptions of teaching-learning force about metaverse for education: a qualitative study. J Positive School Psychol 6(9):1738–1745
25. Mustafa B (2022) Analyzing education based on metaverse technology. Technium Soc Sci J 32:278
26. Ng DTK (2022) What is the metaverse? Definitions, technologies and the community of inquiry. Australas J Educ Technol 38(4):190–205
27. Ortiz Anacona JD, Rojas Millán EE, Cano Gómez CA (2022) Application of metaverse and virtual reality in education. Metaverse 3(2):13
28. Park SM, Kim YG (2022) A metaverse: taxonomy, components, applications, and open challenges. IEEE access 10:4209–4251
29. Pau PK, Sridevi KV, Ghosh M, Lama A (2012) Education technology: the transparent knowledge delivery through QPN and cloud computing. IJSD-An Int J 12(2):455–462
30. Paul PK (2021) Digital education: from the discipline to academic opportunities and possible academic innovations-international context and indian strategies. In: Digital education for the 21st century: technologies and protocols, vol 255
31. Paul PK, Bhuimali A, Kalishankar T, Aithal PS, Rajesh R (2018) Digital education and learning: the growing trend in academic and business spaces—an international overview. Int J Recent Res Sci Eng Technol 6(5):11–18
32. Paul PK et al (2014) Education technology: emphasizing edunxt knowledge transformation systems of Sikkim Manipal University (SMU), Gangtok, Sikkim, India. Int J Embed Syst Comput Eng 4(2):109–113
33. Paul, P.K., K L Dangwal, Asok Kumar Garg.: Education Technology and Sophisticated Knowledge Delivery. Techno-Learn-International Journal of Education Technology. Vol. 2, No. 2, pp. 169–175 (2012)
34. Şentürk MF, Aydin ZG, Aydin MA (2022) A study on metaverse and its applications in education. El-Cezeri 9(4):1424–1430
35. Şeyma EŞİN, Özdemir E (2022) The metaverse in mathematics education: the opinions of secondary school mathematics teachers. J Educ Technol Online Learn 5(4):1041–1060
36. Singh J, Malhotra M, Sharma N (2022) Metaverse in education: an overview. Appl Metalytics Meas Custom Exp Metaverse 135–142
37. Sriram GK (2022) A comprehensive survey on metaverse. Int Res J Modern Eng Technol 4(2):772–775
38. Stanoevska-Slabeva K (2022) Opportunities and challenges of metaverse for education: a literature review. EDULEARN22 Proc 10401–10410
39. Suzuki SN, Kanematsu H, Barry DM, Ogawa N, Yajima K, Nakahira KT, Shirai T, Kawaguchi M, Kobayashi T, Yoshitake M (2020) Virtual experiments in metaverse and their applications to collaborative projects: the framework and its significance. Procedia Comput Sci 176:2125–2132

40. Tlili A, Huang R, Shehata B, Liu D, Zhao J, Metwally AHS, Wang H, Denden M, Bozkurt A, Lee LH, Burgos D (2022) Is metaverse in education a blessing or a curse: a combined content and bibliometric analysis. Smart Learn Environ 9(1):1–31
41. Wang M, Yu H, Bell Z, Chu X (2022) Constructing an Edu-metaverse ecosystem: a new and innovative framework. IEEE Trans Learn Technol 15(6):685–696
42. Williamson B (2016) Digital education governance: data visualization, predictive analytics, and 'real-time' policy instruments. J Educ Policy 31(2):123–141
43. Yu JE (2022) Exploration of educational possibilities by four metaverse types in physical education. Technologies 10(5):104
44. Zhang X, Chen Y, Hu L, Wang Y (2022) The metaverse in education: definition, framework, features, potential applications, challenges, and future research topics. Front Psychol 13
45. Zhong J, Zheng Y (2022) Empowering future education: learning in the Edu-metaverse. In: 2022 international symposium on educational technology (ISET). IEEE, pp 292–295

The Metaverse in Healthcare: Emerging Technologies, Challenges, and Research Directions

K. Aditya Shastry and S. G. Mohan

Abstract The Metaverse, a virtual shared space, is a promising avenue for transforming healthcare delivery by providing new and innovative solutions for patients, providers, and researchers. This paper presents a comprehensive examination of the existing state of medical care in the Metaverse and the technological advancements driving its development. The paper highlights the potential benefits of Metaverse healthcare, including the ability to provide remote healthcare services, increase access to healthcare, and offer immersive and engaging patient experiences. It also examines the challenges faced in implementing Metaverse solutions in healthcare. The protection of patient privacy and the prevention of illegal access to sensitive information are two of the utmost important concerns which ought to be resolved. Technical feasibility is another important consideration, as the progress and integration of Metaverse technologies into existing healthcare systems may require significant infrastructure and resource investments. Regulatory compliance is also a challenge, as Metaverse healthcare solutions must adhere to existing healthcare regulations and standards. Additionally, the work reviews future research perspectives in Metaverse healthcare. One important area of research is the development of immersive and interactive patient experiences, that could aid patient engagement and enhance the overall healthcare experience. The integration of "artificial intelligence" and "machine learning" is another promising opportunity for future research, as these technologies could be utilized to analyze patient data, provide personalized treatment plans, and improve healthcare outcomes. Finally, the paper explores new revenue models to support sustainability in Metaverse healthcare, as the costs of developing and maintaining these solutions can be significant.

Keywords Metaverse · Healthcare · Technologies · Challenges

K. Aditya Shastry (✉) · S. G. Mohan
Nitte Meenakshi Institute of Technology, Bengaluru 560064, India
e-mail: adityashastry.k@nmit.ac.in

S. G. Mohan
e-mail: mohan.sg@nmit.ac.in

G. Chhabra and K. Kaushik (eds.), *Understanding the Metaverse*, Blockchain Technologies, https://doi.org/10.1007/978-981-97-2278-5_6

1 Introduction

The Metaverse, a virtual shared space, represents a new frontier in healthcare, offering the promise to revolutionize the manner in which medical facilities are distributed and experienced. The use of Metaverse technologies has the potential to provide innovative solutions for patients, healthcare providers, and researchers by offering new and improved methods of communication, collaboration, and data management [1]. In this paper, we explore the current state of healthcare in the Metaverse, examine the technological advancements that are driving its development, and highlight the challenges faced in implementing Metaverse solutions in healthcare. In addition, we engage in the examination of prospective avenues for future study in Metaverse healthcare, including the development of immersive and interactive patient experiences, the grouping of "Artificial Intelligence (AI)" and "machine learning (ML)", and the investigation of new revenue models to support sustainability. This paper provides a complete overview of the current developments in Metaverse healthcare and the opportunities and challenges that lie ahead.

The Metaverse represents a platform in which anyone can communicate among themselves and digital objects. The incorporation of AI technologies into Metaverse healthcare systems has started up a novel frontier in the delivery of healthcare services. AI-empowered Metaverse medical care techniques and applications provide opportunities for delivering personalized, efficient, and effective healthcare services to patients, while easing medical care expenses and improving patient outcomes [2].

Some potential applications of AI in Metaverse healthcare include virtual diagnosis and treatment, telemedicine, virtual rehabilitation and therapy, and personalized medicine. Virtual diagnosis and treatment allow healthcare providers to diagnose and treat patients in real-time, without the requirement for physical visits. Telemedicine enables patients to access healthcare services from remote locations, reducing the requirement for travel and increasing access to care. Virtual rehabilitation and therapy provide patients with immersive and interactive therapy experiences that enhance their recovery and promote well-being. Personalized medicine leverages AI and big data analytics to personalize medical treatment and management plans for individual patients, based on their unique characteristics and health histories [3].

AI-empowered Metaverse healthcare systems have the ability to transform the manner in which the medical services are presented, rendering them more available, competent, and effective. Nevertheless, there are difficulties that must be addressed, such as privacy and security concerns, regulatory hurdles, and the need for standardized and interoperable systems. Nonetheless, the potential benefits of AI in Metaverse healthcare systems make it an exciting and rapidly evolving field, with substantial promise for changing healthcare delivery in the forthcoming years [4].

Figure 1 shows the four axes for implementing the Metaverse in the medical domain.

The Metaverse is a simulated environment that is generated by computer software and designed to simulate real-life environments and experiences. In medical field, the Metaverse could be utilized to give patients and healthcare providers with immersive

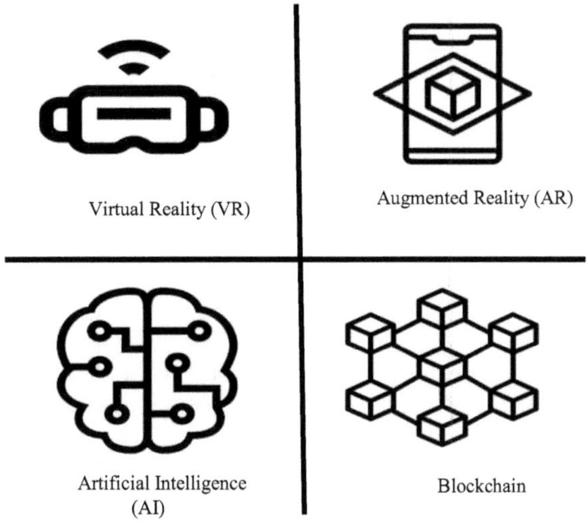

Fig. 1 Four axes of the Metaverse implementation in healthcare

and collaborative digital experiences that support the delivery of healthcare services. The four axes for implementing the Metaverse in the medical domain are described below.

"Virtual Reality (VR)" establishes a fabricated world that is comparable to the real world or an imagined world. This technology creates a wholly immersive and collaborating involvement for consumers, permitting them to communicate with simulated objects and conditions as if they were real. In healthcare, VR has the promise to transform the way medicinal training is provided for improving patient treatments. One of the primary advantages of VR in medicine is its capability to simulate surgical procedures. Surgeons can use VR to practice and prepare for complex surgical procedures, reducing the risk of complications during the actual surgery. Additionally, VR could be employed to provide training and education to medical professionals, permitting them to understand about complex medical concepts in a more interactive and engaging way. Medical scholars may also use VR to explore anatomy and physiology in simulated circumstances, enhancing their understanding and retention of the material.

Additional application of VR in healthcare is telemedicine consultations. Telemedicine refers to the remote delivery of healthcare services using communication expertise such as video conferencing. VR could be employed to provide an engaging telemedicine experience, allowing patients to feel more connected to their healthcare provider despite being in different locations. VR could also be utilized to simulate home environments and provide remote monitoring of patients with tailored care and improved patient outcomes. In summary, VR provides an immersive and interactive experience that can simulate real-life environments and permit consumers to interact with virtual objects and environments as if they were real. In healthcare, VR has the promise to transform medical education and training, simulate surgical procedures, and improve telemedicine consultations, leading to better patient outcomes.

Augmented Reality (AR) superimposes numeral data onto the physical realm, creating an enhanced view of the environment. AR applications can be accessed through devices such as smartphones, tablets, and smart glasses, and they use the device's camera to capture images of the real world. The digital information is then overlaid onto the images, creating a composite view that combines the real world with virtual objects. AR provides real-time information, guidance, and support during medical procedures. For instance, a surgeon can wear AR glasses that overlay a patient's anatomy onto their field of view, allowing them to see the patient's internal organs and structures in real-time during a procedure. This can help the surgeon to navigate around delicate structures and to perform procedures more accurately and with greater confidence.

AR can visualize medical data and imaging in real-time. For example, a doctor can use AR to overlay medical images onto a patient's body, providing a more intuitive and interactive way to view and understand the patient's condition. This can help to improve communication between doctors and patients and to better explain medical conditions and treatment options. In addition, AR can train medical professionals. AR can provide a safe and controlled environment for medical students to practice procedures and learn about anatomy, without the need for expensive and complex medical simulators. AR can provide on-the-job training for medical professionals, allowing them to receive real-time guidance and support during procedures.

Artificial Intelligence (AI) handles the improvement of intelligent instruments which need human intelligence. AI systems use algorithms, data, and advanced computing power to be trained from knowledge and enhance their implementation over time. AI has a diverse array of capabilities encompassing image identification, audio recognition, the processing of natural languages, and making choices. In the Metaverse, AI supports decision-making, improves accuracy and efficiency, and personalizes treatments. AI can help individuals navigate the virtual world by providing real-time recommendations and suggestions based on their preferences, behavior, and past experiences. For example, an AI assistant assists workers find the most relevant content or products based on their interests and previous interactions. AI can improve accuracy and efficiency in various applications. In the healthcare industry, AI can help doctors and researchers to examine sizable volumes of information and recognize samples besides trends that would be challenging for individuals to identify. AI can also be employed to computerize routine jobs like records entry and processing, freeing up medical professionals to focus on more complicated and critical jobs.

AI can also personalize treatments based on individual patient data, such as their medical history, genetic information, and lifestyle. AI can analyze this data and provide individualized therapy plans that are fitted to the individual's unique demands as well as traits. This aids in enhancing the effectiveness of treatments and lowers the danger of harmful impacts. AI can help to improve decision-making, accuracy, and efficiency while providing individualized results which are tailored to the individual's needs and preferences.

Blockchain is a decentralized ledger system that facilitates the maintenance of safe, translucent, and unalterable records. It consists of a decentralized network of computers that collectively maintain a ledger of transactions, which are secured using cryptographic algorithms. In the blockchain, every block encompasses a collection of interactions, while every block is interconnected with its preceding block, consequently establishing an immutable sequence of events. In the Metaverse, blockchain is utilized to secure and manage patient data and identities, enabling secure and private sharing of information between patients and healthcare providers. The utilization of blockchain technology offers a robust and unalterable method for the storage and dissemination of patient information, ensuring its security and integrity, while concurrently upholding the principles of secrecy and security. Patients can retain control upon their documents and allow access to healthcare providers on a need-to-know basis, without having to rely on intermediaries such as health insurance companies.

Blockchain enhances the precision and trustworthiness of patient data by ensuring that the information is consistent and up-to-date across multiple systems. This could lessen the errors and enhance the effectiveness of healthcare delivery. In addition, blockchain can be used to manage identities and access to healthcare services in a protected and decentralized manner. Patients can maintain their identity and medical history on the blockchain, that could be retrieved by healthcare providers with the patient's consent. This aids in lessening the danger of identity theft and fraud and enhances the precision and efficacy of medical delivery. Blockchain enhances the precision and consistency of patient data, while retaining the privacy and confidentiality of the data, and could aid in reducing the errors and increase the effectiveness of healthcare delivery.

These four axes form the foundation of the Metaverse implementation in healthcare, providing a platform for delivering immersive and interactive digital experiences that support the delivery of healthcare services. By leveraging these technologies, healthcare providers can improve patient engagement and outcomes, increase efficiency and accuracy, and support personalized and effective treatments [5]. The subsequent sections of this work are structured in the subsequent manner. Section 2 discusses some of the existing technologies being used in healthcare. Section 3 provides descriptions of the challenges and research directions faced when implementing Metaverse in the healthcare domain. This is followed by conclusion in Sect. 4.

2 Prevailing Digital and Intelligent Medical Care Enablement Technologies

This section offers a summary of the cutting-edge equipment which is presently being utilized to facilitate digitized and intelligent medical care.

2.1 *Sensors*

Sensors play a fundamental part in enabling digital and smart healthcare. They are used to collect and transmit real-time health data, providing healthcare providers with significant perceptions into patients' health status and facilitating early detection of potential health problems.

- Medical Sensors: Medical sensors are devices which are specifically designed to monitor and measure precise fitness constraints, like heart rate, blood pressure, temperature, and oxygen levels. These sensors can be wearable or implantable within the body, providing continuous monitoring of patients' health status.
- Wearable Sensors: Wearable sensors are incorporated into tools, like fitness trackers, smartwatches, and smart clothing, providing real-time data about a person's health, activity levels, and sleep patterns. Wearable sensors are able to examine patients' health and detect potential issues, such as changes in heart rate, that may indicate the requirement for medical intervention.
- Implantable Sensors: Implantable instruments are miniscule tools that can be implanted within the body, providing continuous monitoring of specific health parameters, like the blood glucose levels in diabetic patients. Implantable sensors provide a distinctive prospect for distant supervising of patients' health status, enabling early detection of potential health problems and reducing the necessity for frequent doctor visits.
- Environmental Sensors: Ecological instruments are mechanisms which monitor ecological features which could influence health, such as air quality, temperature, and humidity. Environmental sensors could be employed to observe and manage indoor atmosphere condition, for example, reducing the risk of respiratory problems and other health issues.

Sensors in digital and smart healthcare play a critical part in enabling live healthiness supervising and early detection of potential health problems. By providing healthcare providers with continuous and accurate health data, sensors help to enhance the precision of diagnoses and treatments and support better patient outcomes. Despite their benefits, there are also challenges that must be addressed, such as privacy and security concerns, data accuracy and reliability, and the need for standardized and interoperable systems. Nevertheless, the potential benefits of sensors in digital and smart healthcare make them an exciting and rapidly evolving field, with considerable promise for changing healthcare delivery in the future [6].

2.2 Extended Reality

"Augmented reality (AR)", "virtual reality (VR)", and "mixed reality (MR)" are all parts of extended reality. These techniques rely on AI, machine vision, and networked devices like smartphones, wearables, and HMDs to function. New technologies are increasing the quality of service delivery across several industries by adding elements like as speech, motions, activity recognition, visual, and haptic feedback. There was once a consensus that XR would only find applications in the entertainment business. An immersive interaction was thought to only improve a player's or viewer's time spent with a gameplay or film. The reality of XR use, however, has been astounding. Its applications are expanding rapidly, from medicine to industry. XR would develop to its maximum capacity in the Metaverse. To achieve this impression of being "there", the Metaverse makes use of both virtual and augmented reality capabilities. Individuals can hop from one digital world to the next in the Metaverse and carry out simulated versions of actions they would normally conduct in the reality [7, 8]. The worldwide necessity XR in medicine has been heightened by the COVID-19 pandemic. The utilization of XR equipment has witnessed a notable surge owing to the escalating demand for distant health inspections and interventions [9].

Using XR and the Metaverse, we can completely alter the way patients are treated. People studying medicine benefit more from hands-on experience than theoretical lectures. The physical planet, however, may be a dangerous place in which even a minor slip-up could have catastrophic results. Metaverse XR technologies will make it possible for future doctors to hone their skills in a 3D simulation as lifelike as the actual stuff. Physicians may gain from the dynamic nature of this Metaverse 3D virtual environment for the purpose of skill development. Through the use of XR, the Metaverse allows surgeons to see tissues, tumors, X-rays, and ultrasound scans in live time and from a variety of perspectives without taking their interest away from the patient [10].

The Metaverse's 3D models of a participant's anatomy help doctors save time and provide better care. With the help of XR technology, clinicians could assist patients through recovery in a Metaverse setting in which they can simulate real-life conditions. The individual could get occupational treatment, therapeutic services, and physiotherapy across the Metaverse using XR and remote-controlled equipment. Medical records are safe and simple to retrieve in the Metaverse, so physicians may look at them with XR gadgets and have online conferences with patients to formulate appropriate prescriptions [11].

While XR-enabled Metaverses help provide better medical care in a range of places, substantial hurdles exist in putting them into practice. Due to the nature of XR-related technologies, there exists a risk of a security breach in the Metaverse, that might exhibit significant concerns for patients' heath, confidentiality, and even the physician's professional standing. It could be quite expensive to create and deploy XR tools that improve delivery of healthcare in the Metaverse. Nonetheless, not all patients can purchase or benefit from this cutting-edge equipment, which complicates doctor–patient interactions [12].

2.3 Blockchain

A blockchain refers to a decentralized record containing a comprehensive account of all the occurrences that have transpired within the public blockchain. Whenever a fresh event occurs on the blockchain, a copy of it is uploaded to all users' accounting records [13]. Every node in the network comprises many transactions. Blockchain is often referred as disseminated archive technology since it is managed by a number of different entities (DLT). A distributed ledger is a technology (DLT) that uses a checksum to store operations permanently. Due to this, it is obvious that any change to even a single block in a chain will be detected instantaneously. To get into a blockchain network, intruders ought to vary each chunk within the chain on all of the scattered types of the network. Decentralized resources can be issued, owned, and used via the use of quasi assets and bitcoins on the blockchain. Considering the many issues with confidentiality, security, and data transparency that come with centralized system, the notion of the Metaverse simply cannot exist without blockchain. The blockchain will transform the Metaverse into a decentralized digital source, making it cross-platform and universally accessible [14].

The blockchain-based Metaverse provides decentralized accessibility to any kind of virtual world. By utilizing blockchain technology, the Metaverse would be able to more easily acquire genuine, high-quality data. The blockchain's preservation and accessibility would guarantee the safekeeping of the Metaverse's massive data. Improvements in accessibility to medical care for patients are anticipated as an outcome of Metaverse implementation. With additional alternatives and a more welcoming atmosphere, patient–doctor communication will be facilitated in the Metaverse. As a consequence, patients can get their diagnoses and consultations quicker. With the help of the Metaverse, which is powered by blockchain technology, physicians will have access to more accurate patient information and be able to make more informed choices. The immutability of the blockchain avoids third parties from attaining unauthorized changes to the ledger [15].

The healthcare Metaverse provided by the blockchain is not without its limitations. The high price tag and extensive infrastructure needs provide obstacles to widespread adoption of this method. Individuals are at risk since this technique is unrestrained and uncontrolled. Because of the technology's intricacy, it may be challenging for consumers or patients to adopt it. The blockchain-enabled Metaverse could be painfully sluggish [16] owing to the need to store all information at each and every node connected to the network. Since this equipment is so complicated and requires a lot of power, it is out of reach for smaller medical centers.

2.4 *Artificial Intelligence*

"Artificial intelligence" (AI) refers to the study, creation, and administration of machines which can train to act independently of their creators and complete human-directed activities and decisions [17]. The usage of AI will improve the Metaverse's foundation by bolstering the simulated worlds' internal systems and providing a more immersive 3D experience. The value of services and the entire Metaverse network will increase thanks to advancements in AI technology [18].

Groundbreaking new methods, such as augmented reality (XR) and big data coupled with AI in software and hardware, are being implemented in the field of healthcare to improve the efficiency of pharmaceutical products, lower the cost of treatment, boost medical processes, and enhance access to healthcare services [19]. Medical professionals and patients likewise could gain from the holistic education, comprehension, and exchange of individuals' medical problems and health information made possible by the Metaverse. Artificial intelligence is being used to examine and evaluate patients' medical files. With the use of AI, the Metaverse provides doctors with access to high-quality 3D photos and images of patients in need of treatment. Physicians could benefit greatly from the features provided by AI, which could also aid in patient prioritization, error reduction when evaluating electronic health information, and improved diagnostic precision [20].

Large amounts of health information and patient records hinder clinicians' ability to keep up with rapidly evolving medical science and deliver individualized care. Artificial intelligence (AI) systems in the Metaverse could quickly analyze electronic health records and physiological evidence obtained by medical units and experts to provide instant and reliable suggestions to clinicians. Drug development, medical prognosis, and disaster relief are all domains where the Metaverse and AI could work together to benefit humanity. Despite its ability to provide additional insight and speed up exchanges with health records for both patients and practitioners, an AI-enabled Metaverse may present considerable threats to patient confidentiality and ethical concerns and even produce clinical blunders, that could deceive the physicians in administering therapy. Embracing the AI-enabled Metaverse would be difficult due to the lack of rationale for outcomes [21].

2.5 *Internet of Things*

The term "Internet of Things" (IoT) encompasses the vast array of interconnected gadgets which are presently linked to the World Wide Web and engaged in the interchange of knowledge. IoT has become increasingly pervasive due to the widespread adoption of cellular connections and the development of affordable computer processors. This has enabled the connection of a wide range of devices, ranging from cellphones to complex interplanetary processes, to the IoT. Because of the integration of sensors and the capability to engage in interaction with each other, such gadgets possess the capability to exchange live information autonomously, eliminating the necessity for operator supervision. The integration of the IoT with the Metaverse is presumed to facilitate the convergence of both the virtual and real realms, leading to the formation of a more sophisticated and adaptable ecosystem [22].

When it pertains to healthcare, the most common usage of the IoT-enabled Metaverse is for remote health monitoring. Patients that are unable to visit a medical facility or hospital in person can nevertheless get their critical healthcare information (like pulse rate, hypertension, fever, and much more) instantly gathered by IoT gadgets and presented in the Metaverse's simulation world. Due to this, consumers no longer need to go out of their way for seeing their doctors or get their important medical files. Clinicians would be able to execute intricate surgeries which would be challenging for human hands alone by sending miniature consumer nanotechnology within the body system and analyzing operations in a simulated three-dimensional environment facilitated by the Metaverse. Furthermore, miniature IoT systems or nanomachines used in robotic systems might lessen the amount of cuts significantly. Because of this, the procedure is less harmful to the person and recovery time is shortened. The combination of the Metaverse with the Internet of Things could also aid in the treatment of long-term illnesses by facilitating better relaxation, medicine reorders, and emergency notifications [23].

Concerns about privacy and security of information are of significance in the medical industry. As IoT-enabled gadgets do not really conform to communication standards and frameworks in live time, privacy and security problems would be a worry of the IoT-enabled Metaverse. It would also be challenging to combine IoMT gadgets with Metaverse. It will become challenging to combine information for critical insights and evaluation in the IoT-enabled Metaverse because of the lack of consistency in information and communication methods. The expense of these tools is yet another barrier to implementing IoT-enabled Metaverse applications in hospital environments [24].

2.6 Digital Twin

A digital twin is a computer-generated model that simulates the behavior of a real-world product or procedure in live time. The growth of digital twins could be credited to the evolution of commercial engineering and manufacturing methods. The term "digital twin" is used to describe a digital clone of a real or analogue good, service, or process. A digital twin is an electronic replica of a real item. This could range from a piece of equipment to a healthcare gadget, or something as huge as a city. Conversely, the Metaverse concept depicts a digital space where each and every individual interacts in the constant manner as they would in the tangible realm. There is a digital twin for each real mechanism in the Metaverse, making digital twins essential to the development of the Metaverse [25].

By creating a digital twin of the complete infirmary in the Metaverse, we could evaluate the needs with regard to organizational processes, personnel, and healthcare techniques. Such virtual models in the Metaverse could be useful in addressing issues like space shortages, the spread of disease, the appointment of medical staff, and the accessibility of surgical units. The use of digital twins offered by Metaverse could enhance patient care, reduce expenses, and boost staff efficiency. In the healthcare world, which is both complex and politically fraught, this is of paramount importance. Utilizing digital twins, hospitals could be completely digitized in the Metaverse, providing a risk-free patient experience. Through the usage of digital twins, the Metaverse would also facilitate the development of patient-specific prosthetics. Digital twin-powered Metaverse could also aid to conduct cardiac and brain doctors in simulation exercises of surgeries before performing sophisticated real-world operations [26].

The issue of constructing a digital counterpart of every physical item demands the production of a vast collection of digital twins together with enormous processing capabilities. The highest challenging task is to build a tissue or living thing digital copy in the Metaverse. A significant problem for the healthcare sector's digital twin-enabled Metaverse is the creation of a clone of a human that has all of their bodies and systems functioning in synchrony with one another in live time [27].

2.7 *Quantum Computing*

"Quantum computing" is a computational paradigm that leverages the principles of quantum physics processes like symmetry, resonance, and interaction to perform its operations. Quantum computers are tools that can perform calculations using particle physics. They can solve a wide range of calculation problems. Crucial decisions may be made with the aid of quantum computing more quickly than with traditional computing. Since then, quantum computing has been extremely prevalent in a wide scope of industries, like the financial sector and the pharmaceutical industry. Alternatively, the goal of the Metaverse is to develop a virtual environment that is highly realistic; therefore, it plans to combine many diverse businesses into one unified ecosystem. The goal is to design a user experience that combines elements of online communities, video games, and computer simulations into a world that is a close facsimile of our own. Particularly, the pandemic's effect on medical advice has led to the rise of digital healthcare. The high-quality visuals and realism of the Metaverse will be put to use in medical consultations with patients. With this function of the Metaverse, patients and doctors will be able to interact often in this virtual space. These virtual worlds require massive amounts of computing power to create. The Metaverse influences the establishment of a realistic, dynamic medical digital environment [28].

Overcoming the massive obstacles of privacy, higher processing capacity, and data breaches would be feasible with the help of the Metaverse facilitated by quantum computing. Quantum computing will safeguard the Metaverse by facilitating the growth of quantum-enabled encryption methods. As the scope of the Metaverse increases, quantum-resistant protection will become a necessity for all operations and transactions in the medical sector. Incorporating quantum computing into the Metaverse would greatly increase the speed and accuracy of healthcare systems. Health-related Metaverse applications rely heavily on computational and simulated data. Computing is thus enhanced by quantum computing, as is the whole environment for both patients and physicians. Using quantum randomization, programmers may help ensure that rules they have set up prohibit residents and processes from tampering with healthcare applications in the Metaverse [29].

Although quantum computers have demonstrated potential, they are yet to attain the industry standard. Existing Metaverse-enabling systems and methodologies are built on binary systems, while quantum bits are used in quantum computing. As a result, this makes it much harder to use quantum computing in the Metaverse. The high-energy requirements of quantum computing are an issue that needs to be fixed before its true capacity can be realized in the framework of the Metaverse. Metaverse encryption and protections would fail if quantum dominance was to escape into the hands of hackers [30].

2.8 Human–Computer Communication

The domain of research identified as "human–computer interaction" (HCI) investigates how people and machines interrelate while designing new forms of computing technology. Written or demonstration-centered control is no longer the central pattern in HCI. The expansion of the Metaverse will be facilitated by human–computer collaboration, simulated realism, enhanced realism, and forthcoming content production and collaboration technologies. HCI technology representing wearable consumer head-mounted displays will power all graphical interactions within the Metaverse (HMD). In the Metaverse, these HMDs will be important in facilitating interaction between users and their environments. In the Metaverse, these HCI tools will enhance the sensations of users. Haptic wearable devices are another form of HCI technology which would permit operators to sense and interrelate with the Metaverse via contact, sight, and taste. These devices will likewise permit users to collaborate with robotics in a Metaverse setting, regardless of their physical location [31].

The HCI-enabled Metaverse will incorporate XR technologies such as hologram construction, lenticular reproduction, XR incorporation, and XR interconnectivity. There are numerous methods in which medicine might profit from the HCI-aided Metaverse. By enhancing the interaction and absorption offered by the HCI-enabled Metaverse, medicinal training for learners would be facilitated, research results would be disseminated, consultations would be enhanced, and precise, rated treatment and diagnosis would be made available. Incorporating HCI into the Metaverse will enhance the value of care provided to users in a diversity of situations, comprising but not restricted to: preventing disease, telehealthcare, medical exam, prognosis, therapeutic interventions, rehabilitative services, management of chronic conditions, in-home care, and first aid. Doctors in other locations can participate in the operations with the assistance of robots in the Metaverse's realistic setting [32].

There are many ways in which HCI technologies aid the Metaverse, yet these developments likewise come with their fair share of difficulties. The true capacity of the Metaverse cannot be realized in recent years, however, because contemporary HCI technologies including head-mounted screens and tactile wear gadgets remain in their infancy. The lack of standardization among these HCI tools makes it difficult to incorporate them into the Metaverse. Another factor which sticks in the way of widespread adoption of these technologies is the price tag [33].

There are many ways in which HCI technologies aid the Metaverse, yet these advancements also come with their fair share of difficulties. The true capacity of the Metaverse cannot be realized for quite some years, however, because contemporary HCI technologies including head-mounted screens and tactile wear gadgets remain in their infancy. The lack of standardization among these HCI tools makes it difficult to incorporate them into the Metaverse. One more thing that would stand in the way of widespread adoption of these technologies is the price tag [33].

Table 1 shows the outline of the enabling technologies of the Metaverse for healthcare.

3 Challenges and Future Research Directions

We discuss the new challenges and future research directions facing the use of Metaverse, AI, and Data Science in smart health and intelligent healthcare systems.

3.1 Challenge-1 Explainable Systems

Explainable Systems refer to AI systems that are transparent and can provide understandable reasons for their decisions and predictions. It is a significant view of AI development and deployment, especially in sensitive domains such as healthcare, finance, and criminal justice, where the decisions made by AI systems can have significant impact on individuals and society. Key challenges in developing Explainable Systems include [34]:

- Balancing Explanation with Performance: Ensuring that the system's ability to provide explanations does not forfeit precision or quickness in its operation.
- Human Understanding: Ensuring that the explanations delivered by the system are easy to understand and interpret by humans.
- Trust and Transparency: Building trust and transparency in AI systems, particularly when dealing with sensitive data.
- Bias and Fairness: Addressing bias and fairness concerns in the training information and models used by AI systems.

Table 1 Summary of the Metaverse technologies in healthcare

Technology	Description
Virtual reality (VR) and augmented reality (AR)	Technology that creates simulated environments and enhances real-world experiences, enabling immersive patient experiences and remote consultations
5G networking	High-speed, low-latency communication tools which empowers seamless, real-time data transfer and enables VR/AR applications
Cloud computing	Tools which allow scalable, flexible, and cost-effective data storage and management, enabling real-time data analysis and collaboration
Artificial intelligence (AI)	Machinery which aids machines to learn from data and make predictions, enabling predictive healthcare and more accurate and efficient clinical decision-making
Blockchain	Technology that enables secure, decentralized data storage and management, enhancing data privacy and security
Natural language processing (NLP)	Technology that enables machines to understand and process human language, permitting more successful interaction linking patients and healthcare providers
Internet of Things (IoT)	Technology that enables the interconnectedness of devices, enabling real-time data collection and analysis for improved patient outcomes
3D printing	Technology that enables the creation of real entities from digital prototypes, enabling tailored therapeutic mechanisms and prosthetics

- Integration with Complex Models: Ensuring that explanations can be provided for complex AI models like deep learning networks.
- Regulation and Standardization: The requirement for consistent regulations and standards to guarantee that Explainable Systems are developed and deployed in an ethical and responsible manner.

Research directions on explainable AI are as follows [35]:

- Interpretable Machine Learning: Building "machine learning" representations that are transparent and can provide understandable explanations for their decisions.
- Explainable Decision-Making: Studying the processes involved in decision-making by AI systems and developing methods to make these decisions more transparent.
- Human–AI Interaction: Developing user-friendly interfaces that enable humans to interact with AI systems and understand the reasoning behind their decisions.

- Post hoc Explanation: Developing methods to generate explanations for AI models after they have been trained and deployed, especially for black-box models such as deep neural networks.
- Explainable Predictive Models: Building extrapolative representations which are transparent and can provide understandable explanations for their predictions.
- Bias and Fairness: Developing methods to detect and mitigate bias in AI models and ensuring that AI systems are fair and equitable.
- Explanation Generation: Developing methods to automatically generate explanations for AI models, such as rule-based systems, decision trees, and case-based reasoning.
- Integration with Human Expertise: Exploring ways to integrate human expertise and knowledge into AI systems to enhance their transparency and explainability.

3.2 Challenge-2: Connecting Patients with Healthcare Facilities

The complicated interaction involving patients and healthcare professionals is yet an additional concern which continually occurs in a healthcare center. Although artificial intelligence (AI) and data science solutions could reduce patients' out-of-pocket costs and long queues [36], patients might still be unsatisfied for a number of reasons, including treatment outcomes, misunderstandings, and poor communication. A medical provider's time is best spent treating patients, not mediating disputes over patients' interests, results, or trust in doctors.

Research path in this challenge is as follows: Smart software could assist in explaining some clinical signs, causes, treatment alternatives, and therapy procedures to people who lack professional understanding on diagnosis and treatment. There exist numerous prospective applications for virtual worlds like Metaverse in the healthcare industry, including the visualization of institutional layouts and the simulation of surgeries. Metaverse, AI, and data science could also be employed to improve and streamline telehealth and follow-up treatment. Work on health programs that employ AI to personalize and provide healthcare information to patients is also crucial [37].

3.3 Challenge-3: Gaining Confidence of Patients

Patients are less likely to put their faith in AI systems and medical robotics because they lack human feelings like compassion and empathy [37]. This is why patients put their faith in their physicians rather of the more precise AI technologies [38]. Consequently, medical practitioners have an additional obstacle while utilizing AI: gaining patients' confidence. Additionally, the advancement of health monitoring system depends on the accuracy of treatments and therapies being improved, the number of incorrect diagnostics being reduced, and the establishment of trustworthy connections between humans and AI.

Research path in this challenge is as follows: We should study how to give artificial intelligences human-like emotions like kindness and understanding. While genuine empathy and compassion may be out of reach for AI in the near future, they can be emulated. Artificial intelligence with "emotional" intelligence is valuable in the medical field. Nonetheless, one must be careful not to overstep the line into immoral or dishonest imitation [39].

3.4 Challenge-4: Data Security and Privacy

Machine learning (ML) using a massive dataset is a fundamental aspect of several AI technologies. However, to construct a repository of such data, it might be essential to gather a vast amount of sensitive information, making data security and privacy a significant concern. Meaningful and freely given permission, as well as privacy and information security, must be respected and controlled. Unfortunately, attacks and hacks on these repositories are not uncommon, and there is a great deal of sensitive information at risk. The sophistication and level of assault and espionage will only grow in the future, making it even more critical to protect individual's data.

The medical industry has unique data security requirements that necessitate cutting-edge protection methods. Sensitive patient data must be safeguarded from unauthorized access or disclosure. AI and data science could help strengthen online privacy and safeguard sensitive information. The usage of these tools in healthcare must be carefully planned and implemented to prevent security breaches and data misuse.

Research in this area must focus on ongoing development in the medical industry regarding data security and privacy. The study of these fields could help create new methods and technologies to protect sensitive patient data. Smart health cannot progress without addressing these issues.

In summary, data security and privacy are crucial concerns in the medical industry, especially regarding the usage of AI and data science. To prevent security breaches and data misuse, cutting-edge protection methods must be implemented. Research in this area is ongoing and essential to guarantee the safe usage of these tools in healthcare. Protecting sensitive patient data is critical to the future of smart health.

3.5 Challenge-5: Ethical Issues

The ethical usage of information in AI and data science is critical to confirming that these tools are used for legitimate purposes and do not violate individuals' privacy or other rights. It is equally important that AI methods are proposed to exhibit moral behavior and follow ethical standards. The preservation of AI material, assignment of duty, and oversight of such systems all require new legislation and regulations.

In the medical industry, regulations for the application of AI and data science should take a page from HIPAA, which sets standards for the confidentiality and safety of medical information. There is an immediate need for rules, laws, and regulations concerning the moral usage of AI. Without these guidelines, there is a danger that AI systems may be used for nefarious purposes, such as discriminating against certain groups or violating individuals' privacy.

Research in this area must focus on the development of responsible AI and the application of AI to the task of teaching AI experts about moral questions. This is a critical field of study because progress in the Metaverse, AI, and data science cannot be made without also making progress on the moral front. Most in the AI field place a premium on the principle of responsible AI use, and there is an urgent need for ethical guidelines and standards to guide the development and use of these technologies.

The ethical use of data in AI and data science must also address issues of bias and fairness. AI systems can only be considered ethical if they are designed to avoid reinforcing existing biases or perpetuating unfairness in society. For example, an AI system that is used to make decisions about employment or loan applications must be designed to prevent distinguishing alongside specific clusters based on race, gender, or other factors.

Table 2 shows the different challenges facing the use of Metaverse, AI, and Data Science in smart health and intelligent healthcare systems and the ways to resolved them.

These proposed solutions aim to address the new challenges arising from the integration of the Metaverse, AI, and data science in smart health and intelligent healthcare systems. However, it is essential to recognize that the field is continually evolving, and ongoing research and collaboration will be necessary to tackle these challenges effectively.

Some future research directions of implementing Metaverse in healthcare are as follows:

- Virtual Health Consultations: Research will aim to refine the use of virtual reality (VR) and augmented reality (AR) technologies for remote healthcare consultations. This includes improving the fidelity of virtual environments, enhancing user interfaces for both patients and healthcare providers, and ensuring the safety and confidentiality of these interactions.
- Telemedicine Integration: The Metaverse can be integrated with telemedicine platforms to provide a more immersive and interactive healthcare experience. Forthcoming investigation would investigate practices to seamlessly blend Metaverse elements with telemedicine tools to improve patient engagement and outcomes.
- Medical Education and Training: The Metaverse offers a realistic and safe environment for medical education and training. Research will focus on developing immersive simulations, surgical training modules, and medical classrooms within the Metaverse. This permits health experts to hone their skills and knowledge in a risk-free setting.

Table 2 Challenges of Metaverse in healthcare systems with solutions

Challenges	Proposed solutions
Data privacy and security	• Implement robust encryption and access controls for sensitive healthcare data • Ensure compliance with data protection regulations (e.g., HIPAA, GDPR) • Continuously monitor and update security measures to address emerging threats
Ethical and legal concerns	• Build apparent moral recommendations for AI and data science in healthcare • Ensure transparency and responsibility in algorithmic decision-making • Cooperate with legal experts to navigate complex regulatory landscapes
Bias and fairness in AI	• Implement bias detection and mitigation techniques in AI algorithms • Continuously audit and retrain AI models to reduce bias • Promote diversity in AI development teams to address cultural bias
Integration and interoperability	• Develop standardized data formats and APIs for seamless data sharing • Invest in interoperable healthcare systems and technologies • Promote industry-wide collaboration to establish common standards
Data quality and reliability	• Invest in data validation and cleansing processes • Ensure real-time data quality checks and corrections • Foster a culture of data stewardship among healthcare professionals
Resource constraints and scalability	• Optimize AI algorithms for resource-constrained environments • Explore cloud-based solutions for scalability • Evaluate the cost-effectiveness of AI implementations
User education and training	• Provide comprehensive training to healthcare professionals on AI and data science tools • Develop user-friendly interfaces for AI-driven healthcare systems • Promote ongoing education to keep users updated on advancements
Ethical AI decision-making	• Implement explainable AI models to provide transparency in decision-making • Allow users to challenge AI-generated recommendations • Ensure human oversight in critical healthcare decisions

(continued)

Table 2 (continued)

Challenges	Proposed solutions
Data ownership and consent	• Establish clear data ownership rights for patients • Implement consent management systems for data sharing • Train patients regarding their information usage and empower them to make informed choices
Healthcare disparities	• Analyze AI-generated insights to identify and address healthcare disparities • Develop targeted interventions for underserved populations • Collaborate with healthcare providers and communities to bridge gaps

- Patient Education and Rehabilitation: Virtual reality environments can be used to educate patients about their medical conditions and treatment options. Research will investigate how to create engaging and informative experiences that empower patients to take control of their health. Additionally, VR-based rehabilitation programs could assist people improve from injuries or surgeries more effectively.
- Psychological Condition and Therapy: The Metaverse can play a significant role in mental health treatment. Research will explore the use of VR environments for exposure therapy, mindfulness exercises, and virtual support groups. Additionally, AI-driven virtual therapists could become a reality in the Metaverse.
- Health Data Visualization: With the increasing amount of health data generated, there will be a need for innovative ways to visualize and interpret this data. Researchers will work on creating data-rich Metaverse environments that allow healthcare professionals to explore and analyze complex medical data in real-time.
- Remote Monitoring and IoT Integration: The Metaverse can integrate with the IoT devices to enable remote monitoring of patients' vital signs and health metrics. Research will focus on developing secure and reliable connections between the Metaverse and IoT devices, and methods for presenting this data in immersive ways.
- Healthcare Gamification: Gamification techniques can be employed within the Metaverse to motivate patients to adhere to treatment plans, exercise routines, and healthy lifestyles. Research will delve into the design of gamified healthcare experiences that are both enjoyable and effective.
- Ethical and Regulatory Considerations: As the Metaverse becomes more integrated with healthcare, there will be a need for comprehensive ethical guidelines and regulations. Researchers will explore the ethical implications of using immersive technologies in healthcare and work on frameworks to ensure patient safety, privacy, and consent.
- Accessibility and Inclusivity: Ensuring that the Metaverse is available to all, involving people with incapacities and individuals in underserved communities, would be a critical research focus. This includes developing adaptive technologies and addressing issues related to the digital divide.

In summary, the ethical usage of information in AI and data science is critical to safeguarding that these machineries are used for legitimate purposes and do not violate individuals' privacy or other rights. The advancement of responsible AI must be a priority, and there must be a focus on AI ethics to guide the development and utilization of these technologies. Regulations and guidelines for the ethical use of AI and data science in the medical industry must also be developed to ensure that these technologies are used to improve patient outcomes and not harm them. The integration of the Metaverse into healthcare represents a groundbreaking frontier for research and innovation. As technology continues to evolve, interdisciplinary collaboration between healthcare professionals, technologists, ethicists, and regulators will be essential to comprehend the complete potential of the Metaverse in improving healthcare outcomes and patient experiences.

4 Conclusion

In conclusion, the Metaverse presents a promising avenue for transforming healthcare delivery, with its potential to provide remote healthcare services, increase access to healthcare, and offer immersive and engaging patient capabilities. While the development and integration of Metaverse technologies into existing healthcare systems require significant infrastructure and resource investments, the advantages of Metaverse healthcare are substantial. However, several challenges, including data secrecy and confidentiality, technical feasibility, and regulatory compliance, must be addressed to ensure successful implementation. Future research directions in Metaverse healthcare include the development of immersive and interactive patient experiences and the integration of AI and ML to improve healthcare outcomes. New revenue models are also necessary to support the sustainability of Metaverse healthcare solutions. In this chapter, we covered the introduction of Metaverse in healthcare followed by relevant technologies in intelligent healthcare management. The challenges and future research directions of implementing Metaverse in healthcare were also discussed. Overall, the Metaverse has the potential to revolutionize healthcare delivery and improve patient outcomes, but it is essential to address the challenges and invest in its development to ensure its success in the future.

References

1. Park S-M, Kim Y-G (2022) A Metaverse: taxonomy, components, applications, and open challenges. IEEE Access 10:4209–4251. https://doi.org/10.1109/access.2021.3140175
2. Musamih A, Yaqoob I, Salah K, Jayaraman R, Al-Hammadi Y, Khalifa University, Omar M (2022) Metaverse in healthcare: applications, challenges, and future directions. IEEE Consum Electron Mag. https://doi.org/10.1109/MCE.2022.3223522
3. Wang G, Badal A, Jia X et al (2022) Development of Metaverse for intelligent healthcare. Nat Mach Intell 4:922–929. https://doi.org/10.1038/s42256-022-00549-6

4. Yang Y, Siau K, Xie W, Sun Y (2022) Smart health: intelligent healthcare systems in the Metaverse, artificial intelligence, and data science era. J Organ End User Comput 34:1–14. https://doi.org/10.4018/JOEUC.308814
5. Petrigna L, Musumeci G (2022) The Metaverse: a new challenge for the healthcare system: a scoping review. J Funct Morphol Kinesiol 7(3):63. https://doi.org/10.3390/jfmk7030063
6. Dwivedi YK et al (2022) Metaverse beyond the hype: multidisciplinary perspectives on emerging challenges, opportunities, and agenda for research, practice and policy. Int J Inf Manag 66. https://doi.org/10.1016/j.ijinfomgt.2022.102542
7. Mehta V, Devraj HC, Banerjee P (2018) Applications of augmented reality in emerging health diagnostics: a survey. In: International conference on automation and computational engineering (ICACE), Greater Noida, India, pp 45–51. https://doi.org/10.1109/ICACE.2018.8687114
8. Musamih A, Yaqoob I, Salah K, Jayaraman R, Al-Hammadi Y, Omar M, Ellahham S (2022) Metaverse in healthcare: applications, challenges, and future directions. IEEE Consum Electron Mag 1–13. https://doi.org/10.1109/MCE.2022.3223522
9. Tai Y, Gao B, Li Q, Yu Z, Zhu C, Chang V (2021) Trustworthy and intelligent COVID-19 diagnostic IoMT through XR and deep-learning-based clinic data access. In: IEEE Internet of Things J 8(21):15965–15976. https://doi.org/10.1109/JIOT.2021.3055804
10. Guan J, Irizawa J, Morris A (2022) Extended reality and internet of things for hyper-connected Metaverse environments. In: IEEE conference on virtual reality and 3D user interfaces abstracts and workshops (VRW), pp 163–168. https://doi.org/10.1109/VRW55335.2022.00043
11. Schlichting MS, Füchter SK, Schlichting MS, Alexander K (2022) Metaverse: virtual and augmented reality presence. In: International symposium on measurement and control in robotics (ISMCR), pp 1–6. https://doi.org/10.1109/ISMCR56534.2022.9950565
12. Bale A, Ghorpade N, Hashim M, Vaishnav J, Almaspoor Z (2022) A comprehensive study on Metaverse and its impacts on humans. Adv Human-Comput Interact 1–11. https://doi.org/10.1155/2022/3247060
13. Alabdulatif A, Khalil I, Saidur Rahman M (2022) Security of blockchain and AI-empowered smart healthcare: application-based analysis. Appl Sci 12(21). https://doi.org/10.3390/app122111039
14. Ali S, Abdullah, Armand TPT, Athar A, Hussain A, Ali M, Yaseen M, Joo M-I, Kim H-C (2023) Metaverse in healthcare integrated with explainable AI and blockchain: enabling immersiveness, ensuring trust, and providing patient data security. Sensors 23(2). https://doi.org/10.3390/s23020565
15. Badruddoja S, Dantu R, He Y, Thompson M, Salau A, Upadhyay K (2022) Trusted AI with blockchain to empower Metaverse. In: 4th International conference on blockchain computing and applications (BCCA), San Antonio, TX, USA, pp 237–244. https://doi.org/10.1109/BCCA55292.2022.9922027
16. Bansal G, Rajgopal K, Chamola V, Xiong Z, Niyato D (2022) Healthcare in Metaverse: a survey on current Metaverse applications in healthcare. IEEE Access 10:119914–119946. https://doi.org/10.1109/ACCESS.2022.3219845
17. Yang Y, Siau K, Xie W, Sun Y (2022) Smart health intelligent healthcare systems in the Metaverse, artificial intelligence, and data science era. J Organ End User Comput 34(1):1–14. https://doi.org/10.4018/JOEUC.308814
18. Huynh-The T, Pham Q-V, Pham X-Q, Nguyen TT, Han Z, Kim D-S (2023) Artificial intelligence for the Metaverse: a survey, engineering applications of artificial intelligence, vol 117, Part A, pp 105581, ISSN 0952-1976, https://doi.org/10.1016/j.engappai.2022.105581
19. Garavand A, Aslani N (2022) Metaverse phenomenon and its impact on health: a scoping review. Inf Med Unlock 32:101029. https://doi.org/10.1016/j.imu.2022.101029
20. Petrigna L, Musumeci G (2022) The Metaverse: a new challenge for the healthcare system: a scoping review. J Funct Morphol Kinesiol 7(3):63. https://doi.org/10.3390/jfmk7030063. PMID: 36135421; PMCID: PMC9501644

21. Dilibal C, Tur Y (2022) Implementation of developed esantem smart healthcare system in Metaverse. In: 2022 international symposium on multidisciplinary studies and innovative technologies (ISMSIT), Ankara, Turkey, pp 1027–1031. https://doi.org/10.1109/ISMSIT56059.2022.9932849
22. Lee CW (2022) Application of Metaverse service to healthcare industry: a strategic perspective. Int J Environ Res Public Health 19(20):13038. https://doi.org/10.3390/ijerph192013038
23. Mozumder MAI, Sheeraz MM, Athar A, Aich S, Kim H-C (2022) Overview: technology roadmap of the future trend of Metaverse based on IoT, blockchain, AI technique, and medical domain Metaverse activity. In: 2022 24th international conference on advanced communication technology (ICACT), PyeongChang Kwangwoon_Do, Korea, Republic of, 2022, pp 256–261. https://doi.org/10.23919/ICACT53585.2022.9728808
24. Yang D, Zhou J, Chen R, Song Y, Song Z, Zhang X, Wang Q, Wang K, Zhou C, Sun J, Zhang L, Bai L, Wang Y, Wang X, Lu Y, Xin H, Powell CA, Thüemmler C, Chavannes NH, Chen W, Wu L, Bai C (2022) Expert consensus on the Metaverse in medicine, Clinical eHealth, vol 5, pp 1–9. ISSN 2588-9141, https://doi.org/10.1016/j.ceh.2022.02.001
25. Aloqaily M, Bouachir O, Karray F, Ridhawi IA, Saddik AE (2022) Integrating digital twin and advanced intelligent technologies to realize the Metaverse. IEEE Consum Electron Mag. https://doi.org/10.1109/MCE.2022.3212570
26. Jamshidi MB, Ebadpour M, Moghani MM (2022) Cancer digital twins in Metaverse. In: 2022 20th International conference on mechatronics—Mechatronika (ME), Pilsen, Czech Republic, pp 1–6. https://doi.org/10.1109/ME54704.2022.9983328
27. Song Y-T, Qin J (2022) Metaverse and personal healthcare. Procedia Comput Sci 210:189–197. ISSN 1877-0509, https://doi.org/10.1016/j.procs.2022.10.136
28. Ganapathy K (2022) Metaverse and healthcare: a clinician's perspective. Apollo Med 19:256–261
29. Kumar A, Bhushan B, Shriti S, Nand P (2022) Quantum computing for health care: a review on implementation trends and recent advances. In: Kumar R, Sharma R, Pattnaik PK (eds) Multimedia technologies in the internet of things environment, volume 3. Studies in Big Data, vol 108. Springer, Singapore. https://doi.org/10.1007/978-981-19-0924-5_2
30. Solenov D, Brieler J, Scherrer JF (2018) The potential of quantum computing and machine learning to advance clinical research and change the practice of medicine. Mo Med 115(5):463–467. PMID: 30385997; PMCID: PMC6205278
31. Wang Y, Siau KL, Wang L (2022) Metaverse and human-computer interaction: a technology framework for 3D virtual worlds. In: Chen JYC, Fragomeni G, Degen H, Ntoa S (eds) HCI international 2022—late breaking papers: interacting with eXtended reality and artificial intelligence. HCII 2022. Lecture notes in computer science, vol 13518. Springer, Cham. https://doi.org/10.1007/978-3-031-21707-4_16
32. Xi N, Chen J, Gama F et al (2022) The challenges of entering the Metaverse: an experiment on the effect of extended reality on workload. Inf Syst Front. https://doi.org/10.1007/s10796-022-10244-x
33. Ruiz Mejia JM, Rawat DB (2022) Recent advances in a medical domain Metaverse: status, challenges, and perspective. In: 2022 thirteenth international conference on ubiquitous and future networks (ICUFN), Barcelona, Spain, 2022, pp 357–362. https://doi.org/10.1109/ICUFN55119.2022.9829645
34. Ahmadi Marzaleh M, Peyravi M, Shaygani F (2022) A revolution in health: opportunities and challenges of the Metaverse. EXCLI J 21:791–792. https://doi.org/10.17179/excli2022-5017. PMID: 35949490; PMCID: PMC9360475
35. Sun M, Xie L, Liu Y, Li K, Jiang B, Lu Y, Yang Y, Yu H, Song Y, Bai C, Yang D (2022) The Metaverse in current digital medicine. Clin eHealth 5:52–57. ISSN 2588-9141, https://doi.org/10.1016/j.ceh.2022.07.002
36. Chengoden R, Victor N, Huynh-The T, Yenduri G, Jhaveri R, Alazab M, Bhattacharya S, Hegde P, Reddy P, Gadekallu T (2016) Metaverse for healthcare: a survey on potential applications, challenges and future directions. IEEE Access 4

37. Siau K, Wang W (2018) Building trust in artificial intelligence, machine learning, and robotics. Cutter Bus Technol J 31:47–53
38. Bedué P, Fritzsche A (2021) Can we trust AI? An empirical investigation of trust requirements and guide to successful AI adoption. J Enterprise Inf Manag 35. https://doi.org/10.1108/JEIM-06-2020-0233
39. von Eschenbach WJ (2021) Transparency and the black box problem: why we do not trust AI. Philos Technol 34:1607–1622. https://doi.org/10.1007/s13347-021-00477-0

Transforming the Metaverse: Overcoming Challenges and Shaping the Future

S. C. Vetrivel and K. C. Sowmiya

Abstract The concept of a metaverse, an immersive virtual reality space where users can interact with each other and digital objects, has gained significant attention in recent years. This chapter explores the challenges and future aspects associated with the development and implementation of a metaverse. The challenges include technological limitations, such as scalability, network infrastructure, and computational power required to support a seamless metaverse experience. Additionally, ensuring user privacy, security, and data protection within the metaverse poses significant challenges. Social and ethical considerations, including issues related to virtual economies, governance, and the impact on real-world interactions, are also discussed. Furthermore, the chapter examines the potential future aspects of a metaverse, including its applications in various fields such as gaming, education, healthcare, and business. It highlights the potential for enhanced collaboration, creativity, and entertainment, as well as the transformative impact on industries and society as a whole. The Chapter concludes by acknowledging the need for interdisciplinary research, technological advancements, and collaborative efforts to address the challenges and unlock the full potential of a metaverse.

Keywords Metaverse · Transformation · Challenges · Shaping · Future

S. C. Vetrivel (✉)
Department of Management Studies, Kongu Engineering College, Perundurai 638060, India
e-mail: scvetrivel@gmail.com

K. C. Sowmiya
Research Department of Physics, Sri Vasavi College, Erode 638316, India

G. Chhabra and K. Kaushik (eds.), *Understanding the Metaverse*, Blockchain Technologies, https://doi.org/10.1007/978-981-97-2278-5_7

1 Introduction

1.1 Definition and Concept of a Metaverse

The metaverse refers to a virtual reality space where users can interact with a computer-generated environment and other users in real-time. It is often described as a collective virtual shared space that incorporates aspects of augmented reality, virtual reality, and the Internet. In the metaverse, users can explore, create, communicate, and engage in various activities, blurring the lines between the physical and digital worlds [1]. The concept of the metaverse has gained significant attention in recent years as advancements in technology have brought us closer to realizing this immersive virtual environment. The metaverse aims to create a seamless, interconnected digital realm that goes beyond traditional video games or virtual reality experiences. It envisions a fully interactive and immersive space where users can access a vast array of virtual worlds, each with its own rules, environments, and possibilities. In the metaverse, users can embody digital avatars that represent their virtual identities, allowing them to navigate and interact with the virtual space and other users. This interaction can take various forms, including socializing, conducting business transactions, attending virtual events, participating in games and simulations, exploring new landscapes, or pursuing creative endeavors. The metaverse holds the potential to revolutionize numerous industries, including entertainment, education, commerce, healthcare, and social interactions [2]. It can enable new forms of collaboration, learning, and entertainment, transcending geographical and physical limitations. However, realizing the full potential of the metaverse involves addressing several challenges, such as technological infrastructure, privacy and security concerns, standardization, accessibility, and ensuring equitable participation. Overall, the concept of the metaverse represents a vision of a fully immersive and interconnected virtual space that can transform how we live, work, and interact in the digital age. It is an ongoing area of research and development, with various stakeholders working toward creating a more inclusive, accessible, and comp.

1.2 Brief Overview of the Current State of the Metaverse

The term "metaverse" refers to a virtual reality space where people can interact with a computer-generated environment and other users in real-time. As of my knowledge cutoff in September 2021, the concept of the metaverse was still evolving, and various companies and platforms were exploring different aspects of it. However, I can provide a brief overview based on the trends and developments up to that point. At that time, the metaverse was primarily being pursued by tech giants, including Facebook (now Meta), Google, Microsoft, and Epic Games, as well as smaller startups and blockchain-based platforms. These companies were investing heavily in virtual reality (VR) and augmented reality (AR) technologies, creating immersive experiences, and building platforms to facilitate social interactions [3, 4]. Virtual reality headsets, such as Oculus Rift and HTC Vive, were gaining popularity and becoming more accessible. They allowed users to enter virtual worlds and interact with others through avatars. Social VR platforms like VRChat, Rec Room, and AltspaceVR emerged, enabling people to connect, socialize, and participate in various activities together. Augmented reality also played a role in the metaverse, with platforms like Snapchat and Instagram incorporating AR filters and effects. These applications allowed users to overlay virtual elements onto the real world through their smartphones [5]. Blockchain technology and cryptocurrencies were starting to be integrated into the metaverse. Non-fungible tokens (NFTs) gained attention, enabling the ownership and trading of unique digital assets. Some projects aimed to create decentralized virtual worlds where users could own and trade virtual land, items, and creations. However, please note that the metaverse is a rapidly evolving concept, and new developments might have occurred since my knowledge cutoff. I recommend exploring the latest news and updates from relevant sources to get a more accurate and up-to-date understanding of the current state of the metaverse.

2 Technological Challenges

2.1 Scalability and Performance Issues

The concept of the metaverse, a virtual reality-based collective shared space, is still evolving, and as such, there are several scalability and performance issues associated with its development. Here are some common challenges:

- Infrastructure Scalability: Building a metaverse requires robust and scalable infrastructure capable of handling a massive influx of users and data. Scaling up the network, servers, and storage systems to accommodate a potentially vast user base can be a significant challenge.
- Network Bandwidth: The metaverse relies heavily on network connectivity to deliver real-time interactions and immersive experiences [6]. The demand for high-speed, low-latency connections may strain existing network infrastructure, leading to performance issues such as lag, latency, and dropped connections.
- Processing Power: The computational requirements of rendering complex virtual environments in real-time can be substantial. As the metaverse grows in complexity, ensuring that users' devices have sufficient processing power to handle the rendering demands becomes crucial [7]. This can pose a challenge, especially for users with lower-end hardware or limited access to computing resources.
- Data Storage and Retrieval: The metaverse generates and processes vast amounts of data, including user interactions, virtual assets, and environmental data. Efficiently storing, indexing, and retrieving this data to provide seamless experiences

and enable dynamic interactions is a significant challenge, particularly as the volume of data increases.

- Interoperability and Standardization: The metaverse aims to connect various virtual worlds, platforms, and technologies seamlessly. However, achieving interoperability and standardization across different systems, software, and hardware configurations can be complex. Incompatibilities between platforms or virtual worlds may hinder performance and scalability, requiring significant effort to resolve.
- User Density and Concurrency: The metaverse's success relies on a large number of concurrent users engaging with the virtual environment simultaneously [8]. Ensuring optimal performance and responsiveness with a high user density can be challenging. Balancing the load across servers and maintaining consistent user experiences can strain system resources.
- Security and Privacy: With a metaverse consisting of multiple interconnected platforms and user-generated content, security and privacy concerns become critical. Protecting user data, preventing unauthorized access, and mitigating risks such as virtual theft, fraud, or harassment are significant challenges in maintaining a scalable and trustworthy metaverse.

Addressing these scalability and performance issues requires ongoing research, technological advancements, and collaboration among various stakeholders, including developers, infrastructure providers, and standards organizations. As the metaverse continues to evolve, innovative solutions will be necessary to overcome these challenges and create a robust and accessible virtual reality experience for users.

2.2 Interoperability and Standards

Interoperability and standards play a crucial role in shaping the metaverse, a digital realm where people can interact, explore, and create across various virtual environments. Interoperability refers to the seamless exchange of information and assets between different metaverse platforms and experiences. It involves establishing common protocols, data formats, and technical specifications that enable users to move fluidly between virtual worlds, regardless of the underlying technologies or platforms [9, 10]. By embracing interoperability, the metaverse can break down barriers and foster a unified ecosystem that transcends individual platforms, allowing users to carry their identities, virtual possessions, and experiences with them wherever they go. Standards, on the other hand, define a set of guidelines and best practices for the development and operation of the metaverse. These standards ensure consistency, compatibility, and safety across virtual environments, creating a more cohesive and user-friendly experience. They cover various aspects, such as user interfaces, avatars, virtual currencies, privacy, security, content creation, and virtual object interactions. With well-defined standards, developers can build metaverse applications and experiences that seamlessly integrate with each other, encouraging collaboration, innovation, and widespread adoption. The establishment of interoperability and standards in the metaverse has numerous benefits [11]. It allows users to access a vast array of virtual experiences without being limited to a single platform or walled garden. It promotes competition and innovation among developers, as they can leverage existing standards and interoperability frameworks to build upon and enhance each other's work. Interoperability and standards also encourage the growth of a vibrant and diverse metaverse community, fostering social connections and enabling new forms of collaboration, education, commerce, and entertainment.

However, achieving interoperability and establishing standards in the metaverse is a complex and ongoing process. It requires collaboration and consensus among industry stakeholders, including technology companies, developers, content creators, and regulatory bodies. Open standards, community-driven initiatives, and transparent governance structures can facilitate the development and adoption of interoperability frameworks and standards that benefit the entire metaverse ecosystem. By working together to address technical, legal, and ethical challenges, we can shape a metaverse that is open, inclusive, and built on interoperability and standards to unlock its full potential.

2.3 *Data Storage and Processing Requirements*

The metaverse, a virtual reality space that encompasses interconnected digital worlds, brings with it immense data storage and processing requirements. As users immerse themselves in this vast and dynamic environment, the need for storing and managing vast amounts of data becomes crucial. The metaverse demands colossal storage capacity to accommodate various forms of content, including high-resolution graphics, 3D models, audio, and video files. With millions of users interacting simultaneously, the data generated from their actions, movements, and interactions within the metaverse creates a continuous stream of information that must be captured, processed, and stored in real-time [12, 13]. Furthermore, processing this data is equally critical for a seamless metaverse experience. The metaverse relies on complex algorithms, artificial intelligence, and machine learning techniques to render lifelike graphics, simulate physics, manage user interactions, and create dynamic environments. Real-time processing capabilities are required to handle the massive influx of data, enabling the metaverse to respond instantaneously to user inputs and interactions. Processing power is crucial for tasks such as rendering high-definition graphics, running physics simulations, supporting realistic audio and visual effects, and facilitating natural language processing for communication within the metaverse. To meet these demanding storage and processing requirements, the infrastructure supporting the metaverse must be highly scalable, flexible, and robust. Cloud-based solutions with vast storage capacities, distributed computing capabilities, and edge computing resources become essential to handle the massive volumes of data and ensure low-latency experiences for users across the globe [14]. Advancements in hardware, such as high-performance GPUs and specialized processors, are crucial for accelerating data processing tasks and delivering immersive virtual reality experiences within the metaverse. The metaverse necessitates significant data storage and processing capabilities to handle the vast amounts of content generated by users and provide seamless, real-time experiences. The ability to store, process, and analyze this data efficiently and effectively is crucial for creating a dynamic and immersive virtual reality space that engages and captivates its users.

2.4 Network Infrastructure and Bandwidth Limitations

The metaverse, a virtual reality space where users can interact with a computer-generated environment and other users, has gained significant attention in recent years. To realize the full potential of the metaverse, a robust network infrastructure is crucial. The network infrastructure forms the backbone of the metaverse, facilitating seamless communication, data transfer, and real-time interactions between users across different platforms and locations. However, the metaverse presents significant challenges in terms of bandwidth limitations. As the metaverse relies heavily on immersive and high-resolution content, the demand for bandwidth-intensive data such as video, audio, and 3D graphics is immense [15]. This places a strain on existing network infrastructure, which may struggle to provide the required bandwidth to support a large number of concurrent users. In order to overcome these limitations, advancements in networking technologies and infrastructure are necessary. The development of faster and more reliable networks, such as 5G and beyond, along with improvements in data compression and optimization techniques, will be essential to ensure a smooth and immersive metaverse experience for all users. Additionally, collaboration between network providers, content creators, and technology companies will be crucial to address the bandwidth challenges and create a scalable and efficient network infrastructure that can support the metaverse's growth.

2.5 Security and Privacy Concerns

As the concept of the metaverse evolves and gains more prominence, there are several security and privacy concerns that arise. Here are some of the key concerns associated with the metaverse:

- Data Privacy: In the metaverse, users are likely to generate and share vast amounts of personal data, including their preferences, behaviors, and even biometric information. The collection and use of this data raise concerns about privacy, as it could be exploited or misused by malicious actors or corporations [16]. Proper data protection measures and transparent privacy policies are necessary to safeguard user information.

- Identity Theft and Fraud: With a virtual presence in the metaverse, individuals may be at risk of identity theft and fraud. Cybercriminals could impersonate users or create fake accounts, leading to financial losses or reputational damage. Robust identity verification processes and strong authentication mechanisms are crucial to mitigate these risks.
- Cyberattacks and Malware: The interconnected nature of the metaverse opens up opportunities for cyberattacks. Malicious actors could exploit vulnerabilities in virtual platforms, applications, or devices to launch attacks such as Distributed Denial of Service (DDoS) attacks, phishing attempts, or the spread of malware. Regular security audits, effective intrusion detection systems, and prompt patching of software vulnerabilities are essential to protect against such threats.
- Virtual Asset Theft: In the metaverse, users can acquire and trade virtual assets, such as digital currencies, virtual real estate, or unique digital items. However, the ownership and security of these assets can be compromised by hackers or scams. Strong encryption, decentralized systems, and smart contract audits can help enhance the security of virtual assets and prevent theft.
- Virtual Harassment and Abuse: Just like in the real world, the metaverse can also be a space where individuals experience harassment, bullying, or other forms of abuse. Proper moderation systems, community guidelines, and reporting mechanisms are necessary to address such issues and ensure a safe environment for all users.
- Lack of Regulation and Legal Frameworks: The metaverse is a relatively new concept, and regulations and legal frameworks surrounding it are still evolving. The absence of clear guidelines can create uncertainties and challenges in addressing security and privacy concerns effectively. Collaboration between policymakers, technology companies, and user communities is crucial to develop appropriate regulations that balance innovation and protection.

3 Social Challenges

3.1 Digital Divide and Accessibility

The digital divide refers to the gap between individuals, communities, and countries in terms of access to and use of digital technologies. It encompasses disparities in internet connectivity, hardware and software resources, digital literacy, and the ability to effectively participate in the digital world [17]. The digital divide can have significant social, economic, and educational implications, as it can restrict opportunities and hinder the ability of certain groups to fully engage in the digital age. When it comes to the accessibility of the metaverse, which refers to a virtual reality space where users can interact with a computer-generated environment and other users, the digital divide becomes relevant [18]. While the metaverse holds the potential for new forms of social interaction, entertainment, education, and commerce, its accessibility can be limited by existing inequalities in digital access.

(a) Internet Connectivity: The metaverse relies heavily on internet connectivity for users to access and interact with virtual environments. However, in areas with limited or unreliable Internet infrastructure, accessing the metaverse may be challenging or impossible.
(b) Hardware Requirements: Participating in the metaverse often requires specialized hardware such as virtual reality (VR) headsets, controllers, or high-performance computers. These devices can be expensive, making them inaccessible to individuals or communities with limited financial means.
(c) Digital Literacy: Effective use of the metaverse demands a certain level of digital literacy. Users need to understand how to navigate virtual environments, interact with other users, and utilize the available tools and features. Limited digital literacy can pose a barrier to entry for individuals who are less familiar with technology.
(d) Economic Disparities: The metaverse has the potential to become a significant platform for commerce, entertainment, and employment. However, economic disparities can hinder access to these opportunities. Those who lack financial resources may find it difficult to participate fully or benefit from the economic potential within the metaverse.

Addressing the accessibility challenges of the metaverse and reducing the digital divide requires efforts at various levels. Governments, organizations, and technology companies need to invest in expanding Internet infrastructure, providing affordable access to necessary hardware, promoting digital literacy programs, and ensuring inclusivity in metaverse development [19]. Initiatives focused on bridging the digital divide should aim to make the metaverse accessible to diverse populations, including those with disabilities, to foster an inclusive and equitable digital future.

3.2 *Social Inequality and Inclusivity*

While the metaverse is still a developing concept and its exact form and implementation are subject to ongoing debates and iterations. Therefore, the social inequality and inclusivity aspects of the metaverse are speculative and can vary depending on how it is designed and utilized. There are some general considerations and potential implications:

- Access and Infrastructure: The metaverse's inclusivity will largely depend on access to the necessary technology and infrastructure [20]. If participation in the metaverse requires expensive equipment or high-speed Internet connections, it may exclude individuals and communities with limited resources, exacerbating existing inequalities.
- Digital Divide: The metaverse could potentially widen the digital divide if marginalized groups, such as those with lower incomes, disabilities, or limited

technological literacy, are unable to access or navigate the virtual space effectively. Efforts should be made to bridge this divide and ensure equitable access to the metaverse.

- Representation and Diversity: Inclusivity in the metaverse should also consider representation and diversity. It is essential to create a space that reflects the diversity of real-world populations in terms of gender, race, ethnicity, culture, and other identities. By doing so, the metaverse can promote inclusivity, understanding, and empathy among its users.
- Economic Disparities: The metaverse may introduce new economic opportunities, such as virtual entrepreneurship, digital asset ownership, and virtual real estate. However, there is a possibility that existing economic disparities could carry over into the virtual space, resulting in further inequality [21]. Efforts should be made to prevent concentration of wealth and power in the hands of a few, while enabling economic participation and upward mobility for all.
- Digital Citizenship and Governance: As the metaverse evolves, it will require frameworks for digital citizenship and governance to address issues such as privacy, security, harassment, and the enforcement of rights and responsibilities. Ensuring that these frameworks are inclusive, transparent, and accountable will be crucial for promoting fairness and protecting the rights of all individuals in the metaverse.
- Education and Skill Development: Inclusivity in the metaverse will also require efforts to provide accessible education and skill development opportunities. Individuals should have the chance to acquire the digital literacy and skills necessary to navigate and contribute to the virtual space effectively.

It is important to note that the metaverse is still a theoretical and evolving concept. Its impact on social inequality and inclusivity will depend on the choices made by developers, policymakers, and communities as it becomes a more tangible reality.

3.3 *Governance and Regulation*

The governance and regulation of the metaverse is a complex and evolving topic, as the metaverse itself is still a relatively new concept and is constantly evolving. As it gains popularity and becomes more integrated into our daily lives, there is a growing need to establish governance and regulation to ensure ethical, legal, and secure use of the metaverse. Here are some key aspects of governance and regulation in the metaverse:

- Interoperability: One of the challenges in the metaverse is ensuring interoperability between different virtual worlds, platforms, and services. Governance efforts may focus on establishing standards and protocols to enable seamless interaction and transfer of assets between different parts of the metaverse.
- Intellectual Property Rights: As users create and trade digital assets within the metaverse, issues related to intellectual property rights arise. Regulations may be

needed to protect creators' rights, prevent copyright infringement, and establish mechanisms for licensing and monetizing virtual assets.

- Privacy and Data Protection: In the metaverse, users may share personal information and engage in various activities. Regulations should address privacy concerns, data protection, and consent mechanisms to ensure that users have control over their data and are protected from unauthorized use or misuse.
- Security and Cybercrime: The metaverse presents new challenges in terms of security and combating cybercrime. Governance efforts may focus on implementing robust security measures, addressing hacking and fraud, and establishing mechanisms to resolve disputes and enforce compliance within the metaverse.
- Financial Transactions and Virtual Economies: With the rise of virtual currencies, in-game economies, and real-world economic activities within the metaverse, regulations are needed to prevent money laundering, fraud, and unfair practices. This may involve regulating virtual currencies, ensuring transparency in transactions, and establishing consumer protection mechanisms.
- Content Moderation and Community Standards: Governance efforts may involve setting community standards, addressing hate speech, harassment, and inappropriate content within the metaverse. This may include mechanisms for reporting and removing offensive content and establishing guidelines for acceptable behavior.
- User Rights and Accessibility: Regulations should ensure that users have equal rights and access to the metaverse, regardless of their abilities, background, or socio-economic status. Efforts may focus on accessibility standards, preventing discrimination, and promoting inclusivity within the virtual environment.
- Government Oversight and International Cooperation: As the metaverse transcends national borders, governance and regulation may require international cooperation and coordination. Governments may need to work together to establish common standards, address jurisdictional issues, and collaborate on cross-border enforcement.

3.4 User Behavior and Online Communities

User behavior and online communities within the metaverse are fascinating and dynamic. In this virtual realm, users interact and engage with each other in ways that transcend the limitations of physical space. The metaverse provides a platform for individuals to create and embody digital personas, fostering a sense of identity and expression. User behavior varies widely, ranging from explorers who immerse themselves in the vast landscapes of the metaverse to social butterflies who thrive in vibrant virtual hangouts. Online communities within the metaverse are formed around shared interests, passions, and activities [22]. These communities transcend geographical boundaries, allowing individuals from around the world to connect, collaborate, and build relationships. Users come together to participate in virtual events, engage in collaborative projects, and even form economic systems within the metaverse. The metaverse encourages user creativity, empowering individuals to build and customize their virtual environments, establish businesses, and even design and sell virtual goods. User behavior and online communities within the metaverse represent the convergence of technology, imagination, and human connection, redefining the way we socialize, create, and interact in the digital age.

3.5 Intellectual Property Rights and Content Ownership

In the Indian context, intellectual property rights (IPRs) play a crucial role in safeguarding the content ownership within the emerging realm of the metaverse. The metaverse refers to a virtual reality space where users can interact with one another and engage in various activities. As this digital environment expands and becomes an integral part of people's lives, the protection of intellectual property becomes paramount. In India, the legal framework governing intellectual property rights encompasses various laws and acts, including the Copyright Act, Trademarks Act, and Patents Act. These laws extend their protection to various forms of creative and innovative content that can be produced within the metaverse. This includes virtual artwork, music, designs, characters, trademarks, and other digital assets [22]. Creators and developers in the metaverse can seek copyright protection for their original works, ensuring exclusive rights over their creations. Copyright law grants them control over reproduction, distribution, adaptation, and public display of their content. Similarly, trademarks can be registered to protect virtual brands and distinctive signs used within the metaverse.

It is important for individuals and businesses operating within the metaverse to understand and respect intellectual property rights. Unauthorized use, reproduction, or distribution of protected content can lead to legal consequences, including claims of infringement and damages. However, as the metaverse blurs the boundaries between reality and virtuality, new challenges arise in the context of content ownership. Determining ownership of user-generated content, collaborations, and derivative works can be complex and require careful consideration. In response to these emerging challenges, Indian policymakers, technology companies, and legal experts are continuously working to adapt and evolve the legal framework to address the unique aspects of the metaverse [23]. They are exploring innovative solutions and considering the application of blockchain technology to establish transparent and immutable records of content ownership. Intellectual property rights and content ownership in the Indian context are of significant importance within the metaverse. Existing intellectual property laws provide a foundation for protection, but as the metaverse continues to evolve, it is essential to adapt and refine these legal frameworks to effectively address the emerging complexities and ensure fair and equitable treatment for creators and users alike.

4 Ethical Challenges

4.1 Digital Identity and Authenticity

Digital identity and the authenticity of the metaverse have become crucial aspects in India's rapidly evolving digital landscape. In an era where virtual reality and augmented reality are merging with our everyday lives, establishing a secure and reliable digital identity has become paramount. The metaverse, a collective virtual shared space, has gained significant traction in India, allowing users to interact, transact, and create in a digitally simulated environment. However, ensuring the authenticity and integrity of digital identities within this metaverse presents unique challenges. To address these challenges, India has been actively working on implementing robust systems for digital identity verification. The Aadhaar program, a biometric-based identification system, has played a pivotal role in establishing a secure digital identity infrastructure [24]. Aadhaar provides individuals with a unique identification number linked to their biometric and demographic information, enabling seamless verification and authentication processes. Integrating Aadhaar with the metaverse platforms can enhance user trust and establish a strong foundation for digital identity management. Furthermore, India has been exploring blockchain technology as a means to enhance the authenticity of digital identities within the metaverse. Blockchain's decentralized nature and immutability offer a promising solution to combat identity fraud and ensure the integrity of user data. By leveraging blockchain, India can create a transparent and tamper-proof system for digital identity verification, enabling users to have greater control over their personal information and preventing unauthorized access.

Moreover, collaboration between the government, private sector, and technology innovators is essential for building a trustworthy metaverse ecosystem. Establishing industry standards, protocols, and best practices for digital identity verification will instill confidence among users and promote wider adoption. Additionally, incorporating multi-factor authentication mechanisms and encryption technologies can further strengthen the security and authenticity of digital identities within the metaverse [25]. India recognizes the significance of digital identity and authenticity in the evolving metaverse landscape. By leveraging technologies such as Aadhaar and blockchain, India aims to establish a secure and reliable framework for digital identity verification. The collaborative efforts of stakeholders will play a crucial role in creating a trusted metaverse environment that empowers users and safeguards their digital identities.

4.2 Virtual Economies and Real-World Implications

Virtual economies are digital systems where users engage in trade, production, and consumption of virtual goods and services within virtual worlds or metaverses. These virtual economies have real-world implications that extend beyond the boundaries of the digital realm. One significant impact is the emergence of a new market for virtual assets and currencies, with real-world value attached to them. People can buy, sell, and exchange virtual goods using real money, creating a thriving marketplace [26]. This has given rise to a new breed of entrepreneurs and businesses that specialize in virtual commerce, such as virtual real estate developers, virtual fashion designers, and virtual currency exchange platforms.

The real-world implications of the metaverse and virtual economies are far-reaching. First, they have opened up new avenues for employment and income generation. People can now earn a living by creating and selling virtual goods, providing services within virtual worlds, or by engaging in virtual currency trading. This has blurred the lines between work and play, as individuals can turn their hobbies and passions into profitable ventures within the metaverse. Additionally, virtual economies have implications for traditional industries. Companies are beginning to recognize the potential of virtual advertising and marketing, as the metaverse offers a captive and highly engaged audience. Brands can now create virtual experiences and engage with consumers in immersive and interactive ways, forging deeper connections and driving real-world sales [27]. Moreover, virtual economies can have an impact on wealth distribution and social dynamics. Just like in the real world, virtual economies can experience economic inequalities, with some individuals amassing vast fortunes while others struggle to make ends meet. This raises questions about fairness, governance, and regulation within the metaverse. Furthermore, the integration of virtual and real-world economies brings forth legal and ethical challenges. Issues such as intellectual property rights, fraud, and virtual asset theft require careful consideration and regulation. Governments and legal systems are still grappling with how to address these matters effectively, ensuring the protection of users' rights and maintaining a fair and secure environment.

4.3 Behavioral and Psychological Impacts

The emergence and widespread adoption of the metaverse have brought forth significant behavioral and psychological impacts on individuals and society. The metaverse, a virtual reality-based collective space, has revolutionized the way people interact, work, learn, and entertain themselves. One notable impact is the alteration of social dynamics and relationships. As people immerse themselves in this digital realm, they often experience an increased sense of anonymity, leading to changes in behavior and self-presentation. This can result in both positive and negative consequences, such as individuals feeling more liberated to express themselves authentically or engaging in toxic behaviors shielded by the virtual environment. Furthermore, the metaverse's immersive nature can blur the boundaries between real and virtual experiences, potentially leading to challenges in distinguishing between the two and affecting mental well-being [28, 29]. Excessive time spent in the metaverse may also contribute to issues such as social isolation, addiction, and neglect of physical health. On the other hand, the metaverse offers new opportunities for collaboration, creativity, and personal growth. It allows individuals to explore different identities, experiment with novel experiences, and connect with like-minded people from across the globe. However, it is crucial to be mindful of the potential psychological impacts, develop guidelines for healthy engagement, and ensure the metaverse promotes inclusivity, well-being, and ethical conduct to foster a positive digital environment for all.

4.4 Surveillance and Data Exploitation

The concept of surveillance and data exploitation in the metaverse refers to the monitoring and gathering of user data within virtual reality (VR), augmented reality (AR), and other immersive digital environments. While the metaverse is a hypothetical space that combines physical and virtual realities, the potential for surveillance and data exploitation exists due to the interconnected nature of these environments and the information users generate while interacting within them.

Here are some aspects to consider regarding surveillance and data exploitation in the metaverse:

I. User Tracking: In the metaverse, user tracking technologies can monitor and record user movements, interactions, and behaviors. This data can be used to analyze user preferences, habits, and interests, enabling companies and advertisers to deliver targeted content and personalized experiences.
II. Data Collection: Various forms of data are collected in the metaverse, including personal information, biometric data, social connections, and transactional records. These data points can be exploited for commercial purposes or even abused if they fall into the wrong hands.
III. Privacy Concerns: The metaverse blurs the boundaries between the physical and digital realms, raising concerns about user privacy [30]. With extensive data collection and tracking, individuals may feel their privacy is compromised, leading to potential backlash and debates about data ownership, consent, and security.
IV. Behavioral Analysis: Analyzing user behavior patterns within the metaverse can provide valuable insights into consumer preferences and decision-making. Companies can leverage this information to optimize their products, services, and marketing strategies, but it also raises ethical questions about the extent of behavioral manipulation and control.
V. Third-Party Access: In a decentralized metaverse, where multiple platforms and services coexist, sharing user data across different applications and platforms becomes more prevalent. This raises concerns about data breaches, data sharing without explicit consent, and the potential for unauthorized access by malicious actors.
VI. Regulation and Governance: The evolving nature of the metaverse necessitates adequate regulation and governance frameworks to protect user privacy and

prevent data exploitation. Governments and regulatory bodies may need to establish guidelines and laws to safeguard user rights and ensure responsible data practices.

It's important to note that the metaverse is still largely in the conceptual stage, and its actual implementation and the associated surveillance and data exploitation practices will depend on the specific technological advancements, policies, and societal norms that emerge over time.

4.5 Ethical Considerations in AI and Virtual Beings

Ethical considerations in AI and virtual beings of the metaverse are crucial to ensure the responsible development and use of these technologies. Here are some key ethical considerations to keep in mind:

- Privacy: AI systems and virtual beings in the metaverse often gather and analyze vast amounts of personal data. Respecting users' privacy and ensuring robust data protection measures is essential to prevent unauthorized access or misuse of sensitive information.
- Bias and Fairness: AI algorithms can inadvertently perpetuate bias, discrimination, and inequality if they are trained on biased datasets or designed with biased objectives. It is important to address and mitigate these biases to ensure fair treatment and equal opportunities for all users within the metaverse.
- Transparency and Explainability: AI systems should be transparent and provide explanations for their actions and decision-making processes. Users should have a clear understanding of how AI and virtual beings operate, including the data they use, the algorithms employed, and the criteria used to make decisions.
- User Consent and Autonomy: Users' consent should be obtained before collecting and utilizing their personal data [31]. They should have control over the information they share and the interactions they have with AI systems and virtual beings. Users should also have the ability to easily opt out or disengage from these technologies if desired.
- Accountability and Liability: Clear accountability frameworks are necessary to assign responsibility when AI systems or virtual beings cause harm or make errors. Developers, operators, and users should have a shared understanding of who is accountable for the actions and consequences of AI technologies within the metaverse.
- Safety and Security: AI systems and virtual beings should be designed with robust safety measures to prevent malicious use or hacking attempts. Ensuring cybersecurity and protecting against potential risks or vulnerabilities is crucial to maintain user trust and prevent harm.
- Emotional and Psychological Impact: Virtual beings in the metaverse can have a significant emotional and psychological impact on users. It is essential to

consider the potential effects of these interactions and take measures to prevent manipulation, addiction, or negative mental health outcomes.
- Human-Like Simulation: As AI systems and virtual beings become more sophisticated, there is a need to clearly distinguish between humans and AI entities. Users should not be deceived into believing they are interacting with real humans when they are actually engaging with AI-powered virtual beings.
- Cultural Sensitivity and Representation: AI systems and virtual beings should be designed to respect diverse cultures, values, and beliefs. Care should be taken to avoid stereotypes, cultural appropriation, or offensive content that may arise from biased data or inadequate representation.
- Long-Term Impact: It is important to consider the long-term societal impact of AI and virtual beings in the metaverse. Potential implications, such as job displacement, economic inequality, or social isolation, should be carefully evaluated and steps taken to mitigate any negative consequences.

5 Economic Challenges

5.1 Business Models and Monetization Strategies

The concept of the metaverse is still evolving, and its business models and monetization strategies are being explored by various companies and entrepreneurs. In the Indian context, similar to the global landscape, there are several potential business models and monetization strategies for the metaverse. Here are a few examples:

- Virtual Real Estate: One prominent business model is the sale and leasing of virtual land or property within the metaverse. Companies can create virtual worlds and offer individuals or businesses the opportunity to purchase or rent virtual spaces for various purposes, such as virtual events, advertising, or setting up virtual storefronts. Revenue can be generated through upfront sales or recurring rental fees.
- Virtual Goods and Services: In the metaverse, businesses can create and sell virtual goods, such as avatars, virtual clothing, accessories, and digital art. They can also provide virtual services like virtual event management, virtual tourism experiences, or virtual consulting services [32]. These goods and services can be monetized through microtransactions, subscriptions, or one-time purchases.
- Advertising and Sponsorships: Companies can monetize the metaverse by incorporating advertising and sponsorships. They can offer opportunities for brands to advertise within virtual worlds or sponsor virtual events, providing exposure to a large user base. Advertising and sponsorship revenue can be generated through partnerships, product placements, or sponsored in-world experiences.
- In-World Currency and Economy: A metaverse can have its own virtual currency that users can earn, purchase, and spend within the virtual world. Businesses can facilitate transactions by creating platforms for buying and selling virtual

goods and services using this virtual currency. Revenue can be generated through transaction fees or by acting as a virtual currency exchange.

- Licensing and Intellectual Property: Intellectual property within the metaverse can be valuable, and businesses can monetize their virtual assets by licensing them to others. This could include licensing virtual land, branded items, or virtual experiences to individuals or other businesses. Licensing agreements can be structured based on upfront fees, royalties, or revenue sharing models.
- Subscription and Membership Models: Businesses can offer premium memberships or subscriptions within the metaverse, providing users with exclusive benefits, content, or virtual experiences. Revenue can be generated through monthly or annual subscription fees, allowing users to access additional features or virtual environments not available to free users.
- Partnerships and Collaborations: Companies can form strategic partnerships or collaborations within the metaverse to leverage each other's strengths and create synergies. This could involve joint virtual events, cross-promotions, or shared virtual experiences. Revenue sharing or mutually beneficial arrangements can be established to monetize these partnerships.

It's worth noting that the metaverse ecosystem is still developing, and these business models and monetization strategies may evolve over time. The specific implementation and success of these strategies would depend on factors such as user adoption, platform capabilities, and the overall metaverse landscape in India.

5.2 *Intellectual Property and Copyright Issues*

- Virtual Assets and Ownership: In the metaverse, users can create and own virtual assets such as virtual land, objects, avatars, and virtual currencies. The issue of ownership and rights over these virtual assets can raise questions regarding intellectual property. Developers and users may assert copyright claims over their creations, which could include virtual designs, 3D models, textures, or scripts.
- Trademark Infringement: The metaverse is a vast digital space where various brands, logos, and trademarks may be used by different individuals or entities. Unauthorized use of trademarks or attempts to create confusion by using similar or identical marks can lead to trademark infringement claims.
- Copyright Infringement: Users in the metaverse can create and share content such as artwork, music, videos, and written works. Copyright infringement can occur when someone reproduces, distributes, displays, or performs copyrighted works without permission from the copyright owner.
- Licensing and Permissions: Content creators and developers may choose to license their creations within the metaverse. Determining the terms and conditions for the use of licensed content, including the scope of use, duration, and royalties, can be complex.

- Virtual Performances and Streaming: The metaverse may provide platforms for virtual performances, concerts, and other entertainment events. Streaming or recording these events without permission from the rights holders could infringe on their intellectual property rights.
- User-generated Content (UGC): Many metaverse platforms allow users to create and share their content. Platforms need to establish clear policies regarding ownership and rights associated with user-generated content to avoid potential IP disputes.
- Patents and Technology: The development of new technologies and functionalities within the metaverse may involve the creation of unique algorithms, software, or hardware innovations. Protecting these inventions through patents can be important for companies operating in the metaverse.

5.3 Market Competition and Consolidation

Market competition in the metaverse has reached new heights as various companies and entities strive to establish their dominance in this immersive digital realm. With the concept of the metaverse gaining widespread recognition and interest, tech giants, gaming companies, social media platforms, and even startups are vigorously vying for their slice of the metaverse pie. The metaverse represents a convergence of virtual reality, augmented reality, and other emerging technologies, offering limitless possibilities for entertainment, communication, commerce, and beyond [33]. As the race intensifies, we witness the consolidation of the metaverse market, where larger players acquire smaller ones, forge strategic partnerships, or develop their proprietary platforms. This consolidation aims to leverage the strengths, resources, and user bases of different entities to create more comprehensive and seamless metaverse experiences. The competition and consolidation in the metaverse market not only drive innovation but also set the stage for the metaverse's future, shaping the way we interact, collaborate, and transact in this interconnected virtual universe.

5.4 Value Creation and Distribution

The metaverse has the potential to create and distribute value in various ways. Here are some aspects of value creation and distribution in the metaverse:

- Digital Assets: In the metaverse, users can create, own, and trade digital assets such as virtual real estate, virtual goods, avatars, and virtual currencies. These assets can have real-world value, and their creation and distribution can generate economic opportunities for individuals and businesses.
- Virtual Commerce: The metaverse provides a platform for virtual commerce, allowing businesses to sell virtual products and services. Users can purchase items

like clothing, accessories, virtual experiences, and digital artwork. This creates new revenue streams and business opportunities.

- Social Interactions: The metaverse enhances social interactions by enabling users to connect with people from all over the world. This can lead to the creation of communities, social networks, and new forms of entertainment. Value is created through the shared experiences, collaboration, and networking opportunities that arise within the metaverse.
- Advertising and Marketing: As the metaverse grows, businesses will have opportunities to advertise and market their products or services to a highly engaged audience. Brands can create immersive experiences, product placements, and interactive campaigns within the metaverse, reaching consumers in innovative ways.
- Virtual Work and Education: The metaverse has the potential to revolutionize remote work and education. Companies can establish virtual offices and meeting spaces, allowing employees to collaborate in a digital environment. Educational institutions can provide immersive learning experiences and access to global resources. This creates value by enabling increased productivity, cost savings, and access to new learning opportunities.
- Intellectual Property and Licensing: In the metaverse, intellectual property rights become significant. Creators and innovators can protect their virtual assets, inventions, and designs. Licensing agreements can be established to monetize and distribute virtual content, resulting in value creation for content creators and licensors.
- Infrastructure and Technology: The development and maintenance of the metaverse require infrastructure and technology solutions [34]. Companies involved in creating virtual reality hardware, software platforms, networking, and server infrastructure can provide value through their products and services.
- The distribution of value in the metaverse can vary depending on the specific platform, business models, and governance structures. A fair and inclusive distribution of value will be crucial for the long-term success and sustainability of the metaverse.

5.5 *Job Displacement and Workforce Transformation*

The emergence of the metaverse has brought about significant job displacement and workforce transformation across various industries. As this virtual reality environment evolves, traditional jobs that relied heavily on physical presence and manual labor are being replaced by digital alternatives. For instance, retail employees are now being replaced by virtual store assistants and automated systems that allow users to browse and purchase products within the metaverse. Similarly, customer service representatives are being substituted by AI-powered chatbots and virtual assistants that can provide instant support and guidance to users. Moreover, the metaverse has also given rise to new job opportunities and skill requirements [35]. Professionals with expertise in virtual reality development, immersive design, 3D modeling, and virtual economy management are in high demand. As the metaverse continues to expand, individuals will need to adapt and acquire new skills to remain relevant in this rapidly evolving digital landscape. Workforce transformation programs and initiatives aimed at upskilling and reskilling individuals are crucial to ensure a smooth transition and to leverage the vast potential of the metaverse in a way that benefits both businesses and workers.

6 Future Aspects and Opportunities

6.1 Advancements in Virtual Reality and Augmented Reality Technologies

The metaverse has witnessed remarkable advancements in virtual reality (VR) and augmented reality (AR) technologies, transforming the way we interact and perceive digital environments. VR has evolved to offer more immersive experiences through improved display resolutions, wider field of view, and enhanced tracking systems. Users can now explore virtual worlds with a heightened sense of presence, engaging in realistic simulations and interactive activities. AR, on the other hand, has made significant progress in blending virtual elements with the real world, allowing for seamless integration of digital information into our physical surroundings [36]. Through advanced computer vision and spatial mapping techniques, AR technologies enable users to overlay virtual objects, data, and interactive elements onto their immediate environment, revolutionizing industries such as gaming, education, healthcare, and communication. These advancements in VR and AR technologies have propelled the metaverse into a new era of limitless possibilities, where people can engage, collaborate, and create in a digitally immersive and interconnected world.

6.2 Cross-Platform Integration and Metaverse Convergence

Cross-platform integration and metaverse convergence are two key concepts that are shaping the future of the metaverse. Cross-platform integration refers to the seamless interoperability and connectivity between different virtual worlds and platforms within the metaverse [37]. It enables users to seamlessly navigate and interact across multiple virtual environments regardless of the underlying technology or platform they are using. This integration fosters a unified and cohesive metaverse experience, allowing users to transcend boundaries and explore diverse virtual realms without limitations.

Metaverse convergence, on the other hand, takes cross-platform integration a step further by bringing together different metaverses into a cohesive and interconnected virtual universe. It aims to break down the silos that exist between individual metaverse platforms and create a shared space where users can seamlessly transition from one metaverse to another. This convergence opens up new possibilities for collaboration, social interaction, and exploration, as users can traverse a vast interconnected metaverse and engage with a diverse community of participants from various virtual worlds.

The integration of these two concepts, cross-platform integration and metaverse convergence, holds immense potential for the future of the metaverse. It paves the way for a truly immersive and interconnected digital reality where users can seamlessly navigate between different metaverse platforms, communicate with users from diverse virtual environments, and engage in a wide range of activities that transcend individual metaverse boundaries. This convergence and integration foster an environment of creativity, collaboration, and shared experiences, enabling the metaverse to become a thriving and dynamic digital ecosystem that transcends the limitations of any single platform [38]. As technology continues to advance, we can expect cross-platform integration and metaverse convergence to play a pivotal role in shaping the future of the metaverse and revolutionizing the way we interact and experience the digital world.

6.3 Education, Healthcare, and Other Potential Applications

The metaverse presents a vast scope and incredible potential in the field of education, revolutionizing traditional learning methods and opening up new frontiers for knowledge acquisition and collaboration. The metaverse refers to a virtual, interconnected space where users can engage in immersive and interactive experiences, transcending physical limitations. In education, the metaverse offers a transformative learning environment where students can explore realistic simulations, virtual laboratories, and historical reconstructions, enhancing their understanding and engagement.

One of the key advantages of the metaverse in education is its ability to break down geographical barriers. Students and educators from around the globe can connect and interact in a shared virtual space, fostering cultural exchange and expanding access to quality education. This opens up opportunities for remote learning, bringing education to underserved areas and marginalized communities. Moreover, the metaverse enables personalized and adaptive learning experiences. With the use of artificial intelligence and virtual reality, educational content can be tailored to individual student needs and learning styles, promoting deeper understanding and retention. Interactive simulations and gamified elements in the metaverse can make learning more engaging, stimulating critical thinking, and problem-solving skills.

Collaboration and social interaction are also enhanced in the metaverse. Students can collaborate on projects, engage in virtual debates, and participate in group activities, fostering teamwork and communication skills. Educators can create immersive virtual classrooms, where students feel an increased sense of presence and community, leading to a more inclusive and participatory learning environment [39]. The metaverse also holds the potential for lifelong learning and professional development. Professionals can engage in virtual workshops, attend conferences, and gain practical experience in a safe and controlled environment. Continuous learning becomes seamless, as the metaverse provides access to a wealth of knowledge resources and experts across various fields. However, it is important to acknowledge that the metaverse in education comes with challenges and considerations. Issues related to accessibility, privacy, and digital literacy need to be addressed to ensure equitable and responsible usage. Additionally, there should be a balance between virtual and physical experiences, recognizing the value of real-world interactions and tangible learning environments. The metaverse presents an expansive frontier for education, offering immersive, personalized, and collaborative learning experiences. With its potential to transcend boundaries and empower learners, the metaverse has the capacity to reshape education, equipping individuals with the skills and knowledge they need to thrive in a rapidly evolving world.

Healthcare is an essential sector that has the potential to benefit greatly from the integration of the metaverse. With the metaverse's immersive and interconnected nature, healthcare professionals can leverage this technology to enhance patient care, medical education, and research. In the metaverse, doctors and specialists can collaborate in virtual environments, sharing knowledge and expertise across geographical boundaries. Virtual reality (VR) and augmented reality (AR) can play a significant role in telemedicine, allowing physicians to provide remote consultations, monitor patients' vital signs, and even perform virtual surgeries. Moreover, patients can access healthcare services conveniently through virtual clinics, receive personalized health education, and engage in virtual support groups [40]. The metaverse can also revolutionize medical training, providing realistic simulations for surgical procedures, emergency situations, and rare medical cases. Additionally, researchers can utilize the metaverse to visualize complex data, simulate clinical trials, and accelerate the discovery of new treatments. By embracing the metaverse, the healthcare industry has the potential to transform the way we deliver and experience healthcare, ultimately improving patient outcomes and accessibility to quality care.

Beyond healthcare, the metaverse holds tremendous potential in various other domains. Education can be revolutionized by immersive virtual classrooms, where students can engage in interactive learning experiences, explore historical events, or visit distant planets. Businesses can utilize the metaverse for virtual conferences, product launches, and collaborative workspaces, enabling global teams to collaborate seamlessly regardless of their physical location. Entertainment industries can create immersive experiences, allowing users to participate in their favorite movies, games, or concerts as active participants rather than passive viewers. Virtual marketplaces within the metaverse can facilitate e-commerce, enabling users to buy and sell virtual and physical goods, creating new economic opportunities. Social interactions can be enhanced, with people connecting in virtual spaces, attending virtual events, or participating in shared experiences, transcending geographical limitations. Overall, the metaverse has the potential to reshape numerous aspects of our lives, offering endless possibilities for communication, collaboration, entertainment, and economic growth.

6.4 Metaverse as a Platform for Creativity and Innovation

The metaverse has emerged as an extraordinary platform for creativity and innovation, revolutionizing the way we interact, create, and collaborate in the digital realm. By seamlessly blending the physical and virtual worlds, the metaverse transcends traditional boundaries and provides a limitless playground for human imagination. In this immersive and interconnected space, individuals can unleash their creativity, experiment with new ideas, and bring their wildest visions to life. Artists, designers, musicians, and creators of all kinds are now equipped with powerful tools and technologies to push the boundaries of what is possible, fostering a renaissance of artistic expression and innovation [25]. Within the metaverse, virtual environments serve as canvases where ideas manifest in breathtaking ways. Architects can design fantastical structures that defy the laws of physics, while fashion designers can dress avatars in innovative and boundary-pushing attire. Musicians can compose and perform groundbreaking compositions, leveraging spatial audio and interactive experiences to immerse audiences in their sonic creations. Artists can experiment with immersive and interactive mediums, pushing the boundaries of traditional art forms and inviting viewers to actively engage with their work. Moreover, the metaverse offers a rich ecosystem for collaborative creation, enabling people from different corners of the world to come together and co-create in real-time. Boundaries of distance, language, and culture are blurred as individuals collaborate on projects, sharing ideas, expertise, and resources instantaneously. This global network of creative minds sparks synergies and cross-pollination of ideas, fueling innovation across disciplines.

In this expansive digital universe, innovation thrives as entrepreneurs and inventors develop groundbreaking applications and technologies. The metaverse serves as a breeding ground for startups, fostering an ecosystem of technological advancements that push the boundaries of what we thought was possible. From virtual reality and augmented reality experiences to advanced artificial intelligence systems, the metaverse becomes a catalyst for cutting-edge innovations that impact industries ranging from entertainment and education to healthcare and beyond. As the metaverse continues to evolve, it holds the potential to democratize creativity and innovation like never before. With accessible tools and platforms, individuals from diverse backgrounds can contribute their unique perspectives and ideas to the global conversation. This inclusivity unlocks a wealth of untapped talent and diverse viewpoints, propelling the metaverse into a hotbed of creativity, innovation, and endless possibilities for the future.

6.5 Collaboration Between Academia, Industry, and Policymakers

The metaverse represents a visionary concept that has captured the attention of academia, industry, and policymakers alike. This digital realm, which combines virtual and augmented reality with immersive technologies, has the potential to revolutionize the way we live, work, and interact. Recognizing the transformative power of the metaverse, stakeholders from academia, industry, and policymaking bodies have come together to collaborate and harness its immense possibilities.

Academia plays a crucial role in understanding, researching, and advancing the theoretical underpinnings of the metaverse. Through multidisciplinary studies and research projects, academics explore the technical, social, and ethical dimensions of this emerging landscape. By collaborating with industry and policymakers, academia can provide insights and guidance to shape the development and deployment of the metaverse in an inclusive, sustainable, and responsible manner.

Industry brings its expertise, resources, and innovation to the table, driving the development of metaverse technologies and applications. Collaborations between academia and industry facilitate the translation of theoretical concepts into practical solutions, enabling the creation of immersive experiences, virtual economies, and new business models. Industry also benefits from academia's research findings, as they inform the development of more robust and user-centered metaverse platforms and applications. Policymakers recognize the need for a regulatory framework that supports the growth of the metaverse while safeguarding privacy, security, and fair competition [41]. Collaboration with academia and industry allows policymakers to gain a comprehensive understanding of the metaverse's potential impact on society, economy, and governance. By working together, they can develop policies that foster innovation, protect user rights, address potential ethical challenges, and promote equitable access to the metaverse's benefits. The collaboration between academia, industry, and policymakers in the context of the metaverse is a dynamic and iterative process. It involves ongoing dialogues, knowledge sharing, and joint efforts to shape the development, adoption, and governance of this transformative digital realm. By pooling their expertise, resources, and perspectives, these stakeholders can create a metaverse that not only enhances human experiences but also aligns with societal values and aspirations.

7 Conclusion

The future development of the metaverse presents numerous exciting possibilities and potential challenges. As we embark on this journey, it is crucial to keep several key considerations in mind. First and foremost, inclusivity must be at the forefront. The metaverse should be accessible to people of all abilities, backgrounds, and socio-economic statuses. Efforts should be made to bridge the digital divide and ensure that no one is left behind in this immersive virtual world. Privacy and security are paramount. As the metaverse expands, users must have control over their personal data and be confident that their privacy is protected. Robust security measures should be in place to safeguard against cyber threats and unauthorized access. Transparency and clear user consent mechanisms are essential to build trust and maintain the integrity of the metaverse [42]. Interoperability and open standards should be encouraged. A fragmented metaverse with walled gardens would limit the potential of this interconnected virtual reality. Collaboration among different platforms and developers, along with the establishment of interoperable protocols and standards, would enable seamless experiences and foster innovation. Ethical considerations should guide the metaverse's development. Issues such as digital identity, virtual property rights, and content moderation need careful attention. Establishing ethical frameworks and guidelines can help address concerns related to online harassment, hate speech, and inappropriate content, while preserving freedom of expression and creative exploration. Sustainability is another critical aspect. The metaverse's environmental impact, including energy consumption and resource usage, should be minimized. Employing renewable energy sources and promoting eco-friendly practices can ensure the metaverse's long-term viability and reduce its ecological footprint. Education and digital literacy initiatives are vital to empower users within the metaverse. Promoting digital skills, critical thinking, and responsible behavior will help users navigate the virtual realm safely and responsibly. Collaboration with educators and organizations can contribute to the development of educational content and tools that enhance learning opportunities within the metaverse.

References

1. Ackerman MJ, Brodie J, McCallum MA (2010) Transforming the Metaverse: challenges and opportunities. In: Proceedings of the 10th international conference on virtual worlds and games for serious applications, pp 81–90
2. Ackerman MJ, Brodie J, McCallum MA (2012) Transforming the metaverse: overcoming challenges and shaping the future. IEEE Internet Comput 16(6):32–41
3. Ackerman MJ, Brodie J, McCallum MA (2012) Unifying the metaverse: a vision for the future. Int J Virtual Reality 11(2):1–14
4. Ackerman MJ, Brodie J, McCallum MA (2013) Building the metaverse: progress towards a unified virtual world. In: Proceedings of the 13th international conference on virtual worlds and games for serious applications, pp 3–9
5. Aichner T, Gruber B (2017) Managing customer touchpoints and customer satisfaction in B2B mass customization: a case study. Int J Indus Eng Manage 8(3):131
6. Alaimo C, Kallinikos J (2017) Computing the everyday: social media as data platforms. Inf Soc 33(4):175–191
7. Amorim T, Tapparo L, Marranghello N, Silva AC, Pereira AS (2014) A multiple intelligences theory-based 3D virtual lab environment for digital systems teaching. Proc Comput Sci 29:1413–1422
8. Aydoğan D (2021) Art exhibitions during the pandemic. In: Communication and Technology Congress–CTC, vol 2021, pp 49–55
9. Bakker RM, DeFillippi RJ, Schwab A, Sydow J (2016) Temporary organizing: promises, processes, problems. Organ Stud 37(12):1703–1719
10. Balcik B, Beamon BM, Krejci CC, Muramatsu KM, Ramirez M (2010) Coordination in humanitarian relief chains: practices, challenges, and opportunities. Int J Prod Econ 126(1):22–34
11. Bhattacherjee A (2001) Understanding information systems continuance: an expectation-confirmation model. MIS Q 351–370
12. Bonifacic I (2021) Project Cambria' is a high-end VR headset designed for Facebook's metaverse. Accessed on 13 Sept 2022

13. Brydges T (2021) Closing the loop on take, make, waste: Investigating circular economy practices in the Swedish fashion industry. J Clean Prod 293:126245
14. Casey P, Baggili I, Yarramreddy A (2019) Immersive virtual reality attacks and the human joystick. IEEE Trans Dependable Secure Comput 18(2):550–562
15. Fernandez CB, Hui P (2022, July) Life, the metaverse and everything: an overview of privacy, ethics, and governance in the metaverse. In: 2022 IEEE 42nd international conference on distributed computing systems workshops (ICDCSW). IEEE, pp 272–277
16. Gantz J, Reinsel D (2016) The digital universe in 2020: big data, bigger digital shadows, and biggest growth in the far reaches of the internet and beyond. IDC
17. Gartner, Inc. (2018, Dec 3). Gartner Says by 2021, Half of large enterprises will have their own digital twin. Press Release
18. Ivanov D, Dolgui A (2021) A digital supply chain twin for managing the disruption risks and resilience in the era of Industry 4.0. Prod Plann Control 32(9):775–788
19. Koos S (2022) Digital globalization and law. Lex Sci Law Rev 6(1):33–68
20. Mentzer JT, Foggin JH, Golicic SL (2000) Collaboration: the enablers, impediments, and benefits. Supply Chain Manag Rev 4(4):52–58
21. Moleta TJM (2017) DIGITAL EPHEMERA autonomous real-time events in virtual environments. In: Proceedings of the 22nd international conference of the association for computer-aided architectural design research in Asia
22. Nevelsteen KJ (2018) Virtual world, defined from a technological perspective and applied to video games, mixed reality, and the Metaverse. Comput Animation Virtual Worlds 29(1):e1752
23. Park SM, Kim YG (2023) Visual language navigation: a survey and open challenges. Artif Intell Rev 56(1):365–427
24. Robertson A (2022) Meta is adding a 'personal boundary' to VR avatars to stop harassment
25. Roe M, Spanaki K, Ioannou A, Zamani ED, Giannakis M (2022) Drivers and challenges of internet of things diffusion in smart stores: a field exploration. Technol Forecast Soc Chang 178:121593
26. Sharma K, Giannakos M (2020) Multimodal data capabilities for learning: what can multimodal data tell us about learning? Br J Edu Technol 51(5):1450–1484
27. Uricchio W (2011) The recurrent, the recombinatory and the ephemeral. Ephemeral media: transitory screen culture from television to YouTube, pp 23–36
28. Kshetri N (2005) Pattern of global cyber war and crime: a conceptual framework. J Int Manag 11(4):541–562
29. Kunthara S (2021) VCs will spend billions more to make the metaverse a reality. Crunchbase News
30. Labrecque LI (2014) Fostering consumer–brand relationships in social media environments: the role of parasocial interaction. J Interact Mark 28(2):134–148
31. Lau PL (2022) The metaverse: three legal issues we need to address. The conversation 1
32. Lee H, Hwang Y (2022) Technology-enhanced education through VR-making and metaverse-linking to foster teacher readiness and sustainable learning. Sustainability 14(8):4786
33. Li F (2020) Leading digital transformation: three emerging approaches for managing the transition. Int J Oper Prod Manag 40(6):809–817
34. Limayem M, Hirt SG, Cheung CM (2007) How habit limits the predictive power of intention: the case of information systems continuance. MIS Q 705–737
35. Lindenberg S (2001) Intrinsic motivation in a new light. Kyklos 54(2–3):317–342
36. Luangrath AW, Peck J, Hedgcock W, Xu Y (2022) Observing product touch: the vicarious haptic effect in digital marketing and virtual reality. J Mark Res 59(2):306–326
37. Makransky G, Mayer RE (2022) Benefits of taking a virtual field trip in immersive virtual reality: evidence for the immersion principle in multimedia learning. Educ Psychol Rev 34(3):1771–1798
38. Merre R (2022) Security will make or break the metaverse. Accessed online, 29, 2022
39. Quintana MGB, Fernández SM (2015) A pedagogical model to develop teaching skills. The collaborative learning experience in the Immersive Virtual World TYMMI. Comput Hum Behav 51:594–603

40. Quoring I, Chang SL (2020) Digital metaverse: a new perspective of cyber-physical systems. Comput Mater Continua 65(2):803–819
41. Sengupta D (2022) Nike acquired this company that makes virtual sneakers and NFTs for the Metaverse
42. Shi W, Zhang J (2020) Transforming the Metaverse: challenges and opportunities of virtual reality revolution. IEEE Access 8:1045–1056

Metaverse in Smart Cities

Akash Dogra

Abstract The integration of the Metaverse with smart city technology is an exciting prospect for urban planners, policymakers, and developers. The Metaverse can offer new and innovative ways to improve the quality of life for citizens in smart cities. Augmented reality can be used to enhance public services, such as providing directions or information on city landmarks. Smart transportation systems can also be improved, with the potential for real-time traffic updates and personalized navigation suggestions. Metaverse-based energy and resource management can help smart cities become more sustainable and cost-effective. By leveraging Metaverse technologies, cities can improve their energy and resource management strategies, reducing waste and increasing efficiency. The integration of the Metaverse into smart cities also presents several challenges. One such challenge is data privacy and security concerns. The Metaverse requires a significant amount of data, and ensuring that this data is secure and private is crucial. Access issues are another challenge, as not all citizens may have access to the necessary technology or infrastructure required to use Metaverse-enabled services. The integration of the Metaverse in smart city development offers numerous opportunities for improved urban planning, citizen engagement, and resource management. By identifying and addressing the challenges that arise with this integration, policymakers, planners, and developers can create more sustainable, innovative, and technologically advanced cities that meet the needs of all their citizens.

1 Introduction

In recent years, the idea of the Metaverse, a virtual environment made using digital technologies including virtual reality, augmented reality, and 3D graphics, has gained popularity. The Metaverse is envisioned as a setting where individuals can communicate with one another, share ideas, trade goods, and carry out business in a virtual setting [1]. Smart cities are starting to explore the potential of the Metaverse to

A. Dogra (✉)
Graphic Era Hill University, Uttrakhand, India
e-mail: akash.dogra1234@gmail.com

G. Chhabra and K. Kaushik (eds.), *Understanding the Metaverse*, Blockchain Technologies, https://doi.org/10.1007/978-981-97-2278-5_8

enhance their services and provide a better experience for their citizens [2]. The possible uses, advantages, difficulties, and future directions of the Metaverse in the creation of smart cities are all thoroughly explained in this article. The chapter investigates the idea of the Metaverse and how it might alter how we interact with technology. Additionally, it offers a thorough overview of smart cities and their distinguishing features. This chapter's major objective is to examine the potential uses of the Metaverse in the creation of smart cities, including the numerous ways in which it may be put to use to improve citizen involvement and participation. The article also analyzes the advantages and drawbacks of its integration, including potential exclusion of specific demographic groups and worries about data security and privacy. To give readers a more tangible grasp of the Metaverse's role in the creation of smart cities, the article offers case studies of successful Metaverse-based smart city projects from around the globe. The paper examines the prospects and future directions for the Metaverse in the creation of smart cities.

According to [1], the Metaverse is not a single platform, but rather a collection of interconnected digital spaces that can be accessed through various devices, including smartphones, tablets, computers, and virtual reality headsets. It encompasses a range of online experiences, such as social media, online gaming, and e-commerce platforms. Reference [3] suggests that one of the potential applications of the Metaverse in smart cities is in urban planning, where it can be utilized to create virtual models of cities to test various planning strategies. Another potential application of the Metaverse in smart cities is in e-commerce, as it can provide a virtual platform for businesses to sell their products and services to citizens. The Metaverse can also be used to enhance social interactions in smart cities and create a stronger sense of community. To enable the successful integration of the Metaverse into smart cities, Shin et al.[3], also notes potential difficulties that must be resolved. Given the Metaverse's reliance on technology, one of these is making sure it is accessible to all citizens. Given the dangers of cyberbullying, identity theft, and phishing schemes, making sure that users can utilize the Metaverse safely and securely is essential. While the Metaverse has the potential to improve smart cities in a number of ways, these issues must be resolved in order for it to be successfully included into the creation of smart cities.

It is important to note that the full potential of the Metaverse in smart city development is still largely unexplored and many questions remain unanswered. For instance, there are concerns about how the Metaverse will impact physical space, such as the need for large data centers and the potential for increased energy consumption. There are ethical and moral considerations to take into account, such as the potential for the Metaverse to further exacerbate issues of inequality and exclusion. Despite these challenges [4], the potential benefits of the Metaverse in smart city development cannot be ignored. The Metaverse has the ability to enhance citizen engagement, improve public services and infrastructure, and foster innovation and sustainability. Therefore, while creating smart city initiatives, city planners, and legislators need to take the Metaverse's potential into account. A new frontier in the creation of smart cities is the Metaverse. The Metaverse has the ability to completely alter how we think about urban planning, e-commerce, and social interactions by giving residents

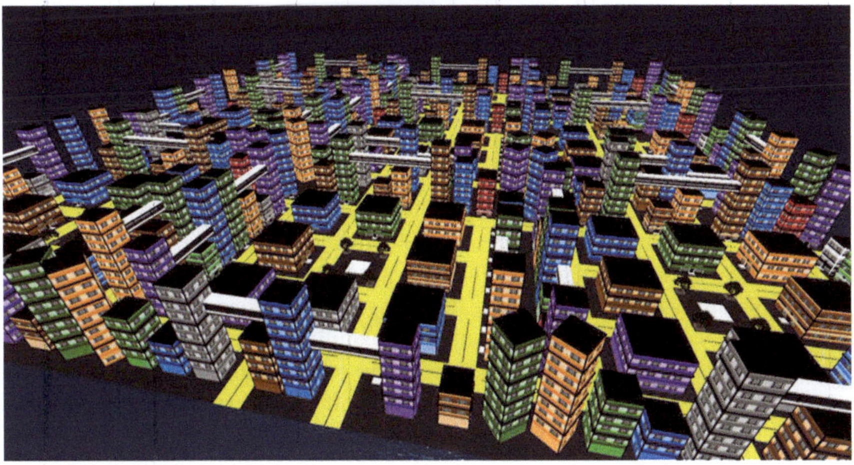

Fig. 1 "City Development" by Danko Whitfield, a digital artwork depicting a futuristic cityscape in a virtual world. Reproduced with permission from Journey To The Center Of The Metaverse—one avatar's battle with the reality of the virtual landscape [5]

a virtual platform to communicate with one another and their city's services. The integration of the Metaverse in the creation of smart cities presents a number of issues and factors that must be taken into account (Fig. 1).

1.1 Defining Smart Cities and Their Characteristics

Smart cities are metropolitan communities that use data and technology to raise the standard of living for residents. They are created with an emphasis on improving public services, infrastructure, and citizen engagement and are intended to be sustainable, innovative, and efficient. A variety of technologies, including sensors, data analytics, and automation, are used in "smart cities" to gather and analyze information in real-time, facilitating better resource allocation and decision-making [6].

Smart cities have several defining characteristics that set them apart from traditional urban areas. One of the primary characteristics of a smart city is its focus on sustainability. Smart cities aim to reduce their environmental footprint by implementing green initiatives, such as renewable energy sources, energy-efficient buildings, and sustainable transportation options. These initiatives not only benefit the environment but also improve the health and well-being of citizens by reducing pollution and promoting active lifestyles. Characteristic of smart cities is their focus on innovation. Smart cities leverage technology and data to drive innovation and creativity in all aspects of urban development [7]. This includes the use of digital technologies to enhance public services and citizen engagement, such as through the use of mobile apps, online platforms, and social media. Smart cities also prioritize efficiency in their operations and services. They use technology to automate processes, reduce waste, and improve resource allocation. For example, smart cities may use sensors to monitor energy and water usage in buildings, enabling them to identify areas for improvement and reduce waste [8].

With plans and projects under progress in cities all over the world, smart cities are a worldwide phenomena. Masdar City in Abu Dhabi is one of the most well-known smart city projects. A sustainable urban development, Masdar City strives to produce neither carbon dioxide nor garbage. To lessen its influence on the environment, it uses a variety of cutting-edge technologies, including water desalination, wind power, and solar electricity. In Barcelona, the "smart city" initiative has been underway since 2011 [9]. The initiative aims to improve public services, enhance citizen engagement, and promote sustainable urban development. Barcelona has implemented a range of innovative initiatives, such as smart lighting, intelligent waste management, and a digital platform that enables citizens to access and provide feedback on public services. In South Korea, the Songdo International Business District is a smart city development that leverages technology and data to enhance sustainability and efficiency. The city features a range of smart technologies, such as automated waste collection systems and intelligent transportation systems, to reduce energy usage and improve resource allocation [10].

In India, Dehradun, the capital city of Uttarakhand state in India, is one of the country's emerging smart cities. The city's smart city project focuses on improving its infrastructure, transportation, and technology to enhance the quality of life of its residents. Under the Smart City Mission, Dehradun has implemented several initiatives, including the installation of intelligent traffic management systems, the development of pedestrian-friendly areas, and the creation of smart parking solutions. The city has launched a mobile application that allows citizens to access various services, such as paying utility bills and lodging complaints. These efforts have helped transform Dehradun into a modern city that is efficient, sustainable, and technology-driven, making it a popular destination for both tourists and businesses.

While smart cities have many advantages, they also face a number of difficulties that must be overcome if they are to be successful. Data security and privacy are a major issue for smart cities [11]. Smart city technologies are producing a growing quantity of data, thus it is important to make sure that this data is gathered, kept, and shared in an ethical and secure manner. By offering a safe and decentralized platform for data sharing and storing, the Metaverse may be able to help with this problem. Smart cities must make sure that all residents, regardless of their socioeconomic level or other circumstances, have access to the advantages and services provided by these projects. The Metaverse could help address this challenge by enabling virtual participation in decision-making processes and public services, reducing the need for physical presence and providing more accessible channels for citizen engagement. Smart cities face a challenge in ensuring that their initiatives are sustainable in the long term [12]. The Metaverse could help address this challenge by enabling simulations and virtual environments for testing and optimizing smart city projects before they are implemented in the physical world. This could reduce the risk of costly mistakes and ensure that smart city initiatives are sustainable and effective.

2 Applications of the Metaverse in Smart Cities

The Metaverse, a virtual world that is created by the convergence of multiple virtual and physical worlds, has the potential to transform the way we live, work, and interact with our environment. In smart cities, the Metaverse can be used to enhance urban planning and development, improve public services, and promote sustainable resource management [13]. This article explores some of the key applications of the Metaverse in smart cities, including virtual city tours and simulations, augmented reality and mixed reality for public services and citizen engagement, smart transportation systems and traffic management, and energy and resource management. Simulations can also be used to test and optimize urban development projects before they are implemented in the physical world, reducing the risk of costly mistakes and improving the sustainability and effectiveness of smart city initiatives.

Urban planning and development is one of the most exciting uses of the Metaverse in smart cities. To give planners and developers a more accurate and engaging picture of the urban environment, the Metaverse can be utilized to construct virtual city tours and simulations. To evaluate different scenarios, such as the influence of new construction on traffic flow, the effectiveness of public transportation lines, and the placement of public amenities, virtual city tours and simulations can be employed.

Virtual city tours and simulations can also be used to engage citizens in the planning and development process. By providing a virtual platform for citizens to explore and provide feedback on proposed developments, planners, and developers can gain valuable insights into public preferences and concerns. This can lead to more inclusive and effective urban planning and development. Augmented reality (AR) and mixed reality (MR) [14] technologies can be used to enhance public services and engagement, providing citizens with more immersive and interactive experiences. For example, AR and MR technologies can be used to provide virtual guides for public transport, enabling citizens to navigate the city more easily. They can also be used to provide virtual tours of public amenities, such as parks and museums, providing citizens with a more engaging and informative experience. AR and MR technologies can also be used to enhance citizen engagement in the decision-making process. By providing virtual platforms for public consultations and feedback, citizens can provide input and feedback on proposed developments, policies, and public services. This can lead to more inclusive and effective decision-making processes. Smart transportation systems and traffic management are critical components of smart cities [15].

The Metaverse can be used to enhance these systems, providing more efficient and effective transportation solutions [16]. For example, Metaverse technologies can be used to create virtual traffic simulations, enabling planners and developers to test various traffic management scenarios. They can also be used to create virtual guides for public transport, providing citizens with real-time information on traffic flow and transport schedules. Metaverse technologies can also be used to enhance autonomous vehicles and drone delivery systems. By providing virtual platforms for testing and optimizing these technologies, developers can improve their efficiency and effectiveness. This can lead to more sustainable and efficient transportation solutions for smart cities. Energy and resource management is another critical component of smart cities. The Metaverse can be used to enhance energy and resource management systems, providing more efficient and effective solutions. For example, Metaverse technologies can be used to create virtual simulations of energy usage, enabling planners and developers to test various scenarios and optimize energy efficiency. They can also be used to create virtual simulations of waste management, enabling planners and developers to test various waste management scenarios and optimize resource allocation [17].

Metaverse technologies can also be used to create virtual platforms for energy and resource management, providing citizens with real-time information on energy usage and waste management. This can lead to more sustainable and efficient use of resources in smart cities.

3 Benefits of Incorporating the Metaverse in Smart City Development

The idea of the Metaverse is gaining traction quickly and sparking interest across a range of industries, including smart city construction. With its immersive and collaborative features, the Metaverse provides a number of advantages that can improve the effectiveness, efficiency, and sustainability of smart cities. In this chapter, we will examine the advantages of integrating the Metaverse into the development of smart cities, including improved citizen engagement and participation, improved infrastructure and public service efficiency and effectiveness, improved stakeholder communication and collaboration, and potential cost and revenue savings [18].

The improvement of citizen engagement and participation in civic life is one of the most important advantages of incorporating the Metaverse into the construction of smart cities. For residents to interact and engage with their city, its services, and infected structures, the Metaverse provides an immersive and engaging platform. It gives residents a more interesting and educational experience, allowing them to comprehend the advantages of the development of smart cities and actively participate in the decision-making process.

The Metaverse also offers a more inclusive platform for citizen engagement, enabling citizens to provide feedback and suggestions that can improve the efficiency, effectiveness, and sustainability of smart cities. This can lead to better-informed decision-making, more effective policies and programs, and greater citizen satisfaction with city services and infrastructure. Another significant benefit of incorporating the Metaverse in smart city development is the increased efficiency and effectiveness of public services and infrastructure [19]. The Metaverse offers a platform for testing and optimizing various scenarios, enabling planners and developers to identify and address potential issues before implementing them in the real world. This can lead to more efficient and effective public services and infrastructure, reducing costs and enhancing citizen satisfaction. For example, the Metaverse can be used to test traffic management scenarios, enabling planners and developers to optimize traffic flow and reduce congestion. It can also be used to test energy usage scenarios, enabling planners and developers to optimize energy efficiency and reduce waste [20].

The Metaverse also offers an enhanced platform for communication and collaboration between stakeholders in smart city projects. It provides a centralized platform for sharing data, information, and ideas, enabling stakeholders to work together more efficiently and effectively. This can lead to better-informed decision-making, more effective policies and programs, and greater citizen satisfaction with city services and infrastructure. For example, the Metaverse can be used to facilitate virtual meetings and workshops, enabling stakeholders to collaborate on smart city projects regardless of their physical location. It can also be used to create virtual platforms for public consultations and feedback, enabling citizens to provide input and feed- back on proposed developments, policies, and public services [21]. Incorporating the Metaverse in smart city development can also lead to potential cost savings and revenue generation through Metaverse applications. The Metaverse offers a range of applications that can enhance the efficiency, effectiveness, and sustainability of smart cities, leading to cost savings and revenue generation. For example, the Meta- verse can be used to create virtual platforms for advertising and marketing, enabling businesses to reach a wider audience and generate more revenue. It can also be used to create virtual platforms for education and training, enabling citizens to acquire new skills and knowledge that can enhance their employability and contribute to the economy.

The benefits of incorporating the Metaverse in smart city development are significant and far-reaching. From improving citizen engagement and participation to increasing the efficiency and effectiveness of public services and infrastructure, enhancing communication and collaboration between stakeholders, and potential cost savings and revenue generation, the Metaverse offers a range of opportunities for enhancing the sustainability and livability of smart cities. As smart city development continues to evolve and embrace new technologies, the Metaverse is poised to play an increasingly important role.

4 Challenges and Considerations for Implementing the Metaverse in Smart Cities

As discussed in the previous Sect. 3, there are several benefits to incorporating the Metaverse in smart city development. However, as with any new technology, some challenges and considerations need to be addressed. In this section, we will explore some of the major challenges and considerations related to implementing the Metaverse in smart cities.

Data privacy and security concerns related to Metaverse technologies: One of the biggest challenges of incorporating the Metaverse in smart city development is ensuring the privacy and security of data [22]. As we saw in the previous section, the Metaverse relies heavily on data collection and analysis to function effectively. With the rise of smart cities and the Metaverse, data privacy has become a pressing issue that needs to be addressed. As more and more devices become connected to the Internet and data is collected in real-time, the risk of sensitive information being compromised increases. In this article, we will examine the data privacy issues associated with the Metaverse in smart cities and what can be done to address them. Smart cities are designed to be efficient and convenient, with data being collected from different sources such as sensors, cameras, and other Internet-connected devices. The Metaverse provides a platform for this data to be processed, analyzed, and visualized in real-time, allowing city planners to make informed decisions about how to improve infrastructure and services. However, with this convenience comes the risk of data breaches and unauthorized access to sensitive information. One of the biggest concerns with the Metaverse in smart cities is the amount of data that is collected and stored. Personal information such as names, addresses, and even biometric data can be collected without the user's consent or knowledge [23]. This can lead to issues such as identity theft and unauthorized surveillance, which can have serious consequences for individuals and society as a whole.

Another issue is the lack of transparency in how data is collected, stored, and used [24]. Users may not be fully aware of the extent of data collection or how their information is being used. This can lead to a lack of trust between the public and the organizations responsible for managing the data, which can ultimately undermine the effectiveness of smart city initiatives. The Metaverse and smart cities offer many benefits, data privacy is a critical issue that needs to be addressed. By implementing strong security measures and being transparent about data collection and usage, we can create a more secure and trustworthy environment for users. It is up to all of us to work together to ensure that the benefits of the Metaverse and smart cities are not outweighed by the risks to user privacy.

To address these concerns, it is important to implement strong data privacy and security protocols [25]. This can include encrypting data to protect it from unauthorized access, using secure servers to store data, and limiting the amount of data collected to only what is necessary for the Metaverse application to function. Governments and regulatory bodies should establish clear guidelines and regulations around data privacy in smart cities. This can help ensure that user privacy is protected and that organizations are held accountable for any breaches or misuse of data. Potential ethical and social implications of Metaverse-based civic engagement and governance: Another challenge to consider is the potential ethical and social implications of using the Metaverse for civic engagement and governance [24]. As we saw in the previous section, the Metaverse can be used to facilitate citizen participation in smart city decision-making processes. There are concerns that this could lead to the exclusion of certain groups, such as those without access to the necessary technology or those who are less digitally literate.

It is crucial to make sure that everyone has access to the Metaverse, regardless of their financial situation or technological prowess, in order to allay these worries. This may entail offering guidance and assistance to residents as they navigate the Metaverse, as well as creating user-friendly interfaces that are simple to use and comprehend. The digital divide and access problems that might prevent some demographic groups from using Metaverse-enabled smart city services: Concerns have been raised over the possibility that the Metaverse would widen the digital divide, in which some sections of the population lack access to technology and related services. This could be particularly problematic in the context of innovative city development, where access to technology is often seen as a key factor in improving quality of life.

Legal and regulatory challenges in integrating the Metaverse with existing smart city infrastructure: Finally, there are legal and regulatory challenges to consider when integrating the Metaverse with existing smart city infrastructure [26]. This can involve concerns about the security and privacy of data as well as concerns over who owns and controls the infrastructure and services that are housed in the Metaverse. It is crucial to collaborate closely with governmental organizations and other stakeholders to create rules and regulations that promote the integration of the Metaverse with current smart city infrastructure in order to overcome these difficulties. This may entail creating clear rules for data security and privacy as well as legal frameworks for ownership and management of infrastructure and services based on the Metaverse.

As a result, there are a number of difficulties and things to think about when incorporating the Metaverse into the creation of smart cities. These include issues with data privacy and security, ethical and social ramifications of civic engagement and governance based on the Metaverse, access issues that might exclude specific population groups from the Metaverse-enabled smart city services, and legal and regulatory difficulties associated with integrating the Metaverse with currently in place smart city infrastructure [27].

5 Case Studies of Metaverse Integration in Real-World Smart City Projects

Innovative technologies are increasingly being included as the idea of smart cities develops to enhance resident quality of life and speed up urban growth. The Metaverse, a virtual world that enables immersive experiences and interaction with digital content, is one of the most promising technologies in this area. There have been a number of successful projects implemented globally, despite the fact that the use of the Metaverse in the construction of smart cities is still in its early phases. We will examine case studies of Metaverse integration in actual smart city projects in this chapter, evaluating their impact and efficacy and identifying lessons learned and best practices for doing so.

5.1 Case Study 1: Singapore's Virtual Singapore [28]

Singapore is recognized as one of the most technologically advanced smart cities in the world, and a major contributing element to its success is the usage of the Metaverse in urban planning and development. A 3D computer model of the city-state called Virtual Singapore was introduced in 2017 and enables cutting-edge urban planning and development. The National Research Foundation Singapore (NRF), the Singapore Land Authority (SLA), and the Government Technology Agency of Singapore (GovTech) worked together to construct Virtual Singapore, a digital duplicate of the city-state of Singapore. Users can explore and examine various facets of the city's infrastructure, buildings, and surroundings thanks to this 3D virtual model of the city-state that is updated in real-time.

The Virtual Singapore platform is designed to help planners, engineers, architects, and policymakers to make informed decisions by visualizing and simulating the impact of various changes and scenarios in a controlled and accurate environment. It can be used to simulate natural disasters, traffic patterns, and urban planning scenarios, allowing decision-makers to test different solutions and predict their impact on the city and its residents. By using Virtual Singapore, urban planners can simulate different scenarios and test the impact of proposed changes on the city's infrastructure, such as the addition of a new subway line or the construction of a new building. The project has also been used to create simulations of emergencies, such as flooding, to test the effectiveness of disaster management plans.

Virtual Singapore is a 3D digital model of the city-state of Singapore, designed to facilitate urban planning, management, and decision-making. The platform was developed by the National Research Foundation (NRF) and the Singapore Land Authority (SLA), in partnership with various government agencies and private sector companies. One notable case study involving Virtual Singapore is the development of a smart water management system for the city. The system was designed to improve the efficiency and effectiveness of Singapore's water management operations, which are critical to ensuring the city's long-term sustainability (Fig. 2).

The smart water management system involved the integration of Virtual Singapore with various data sources, such as sensors and weather forecasts, to provide real-time information on water demand, supply, and quality. The system used machine learning algorithms to analyze this data and generate insights that could be used to optimize water management operations. For example, the system could predict water demand based on factors such as weather patterns, population demographics, and economic activity, allowing water utilities to adjust supply accordingly. The system could also monitor water quality in real-time, detecting anomalies such as leaks or contamination and alerting authorities to take action.

Another key use case for Virtual Singapore is in urban planning and design. The platform allows planners and designers to create and test different scenarios for the city's development, using 3D models of buildings, infrastructure, and public spaces. For example, Virtual Singapore was used in the planning of the Jurong Lake District, a new mixed-use development in western Singapore. Planners used the platform

Fig. 2 City vision via Metaverse [28]

to simulate different scenarios for the district's layout, such as the placement of buildings and open spaces, and to test the impact of different design choices on factors such as pedestrian flow and energy consumption (Fig. 3).

The impact of Virtual Singapore has been significant, both in terms of improving the efficiency of urban planning and development and enhancing public engagement

Fig. 3 Singapore initiative to save water via use of VR [28]

with the city's development. By allowing citizens to explore and interact with the digital model of the city, Virtual Singapore has helped to increase awareness and understanding of urban planning and development issues among the general public. Virtual Singapore is also a powerful tool for researchers and academics, who can use the platform to study various aspects of the city, from the social and economic dynamics to environmental sustainability and climate change.

The Virtual Singapore platform is part of Singapore's Smart Nation initiative, which aims to harness the power of digital technologies to create a more efficient, sustainable, and livable city-state. It represents a significant step forward in urban planning and management, providing decision-makers with unprecedented levels of data and insights that can be used to improve the quality of life for Singaporeans and ensure the city's long-term success. Virtual Singapore is a digital twin of the city-state of Singapore that provides decision-makers with real-time data and insights to inform urban planning and management. Its ability to simulate and analyze different scenarios makes it an invaluable tool for researchers, policymakers, and planners, and it represents a significant step forward in Singapore's efforts to become a Smart Nation. The project has improved communication and collaboration between stakeholders in smart city projects, including government agencies, private developers, and citizens.

5.2 *Case Study 2: Helsinki XR Center [29]*

The Helsinki XR Center is a public–private partnership that aims to promote the use of extended reality (XR) technologies in a range of applications, including smart city development. Helsinki XR Center is a virtual and augmented reality (VR/AR) innovation hub based in Helsinki, Finland. The center was established in 2018 by Metropolia University of Applied Sciences, Helsinki Business Hub, and Finnish Virtual Reality Association (FIVR) to support and promote the development of the XR industry in Finland and beyond. The center provides resources and support for startups and other organizations working on XR projects and also serves as a hub for research and development in the field.

The center provides a collaborative environment for XR developers, businesses, researchers, and educators to connect, experiment, and develop XR technology and applications. It offers a wide range of services, including access to state-of-the-art XR equipment and software, research and development facilities, workshops, events, and networking opportunities. Helsinki XR Center's main focus is on developing XR applications in the fields of education, entertainment, healthcare, and industry. The center collaborates with leading Finnish and international companies, startups, and academic institutions to develop innovative XR solutions and applications that address real-world challenges and create new business opportunities (Fig. 4).

The XR City project, one of the most significant initiatives to emerge from the Helsinki XR Center, intends to build a virtual representation of Helsinki that may be utilized for urban planning and development. The project gives users an immersive

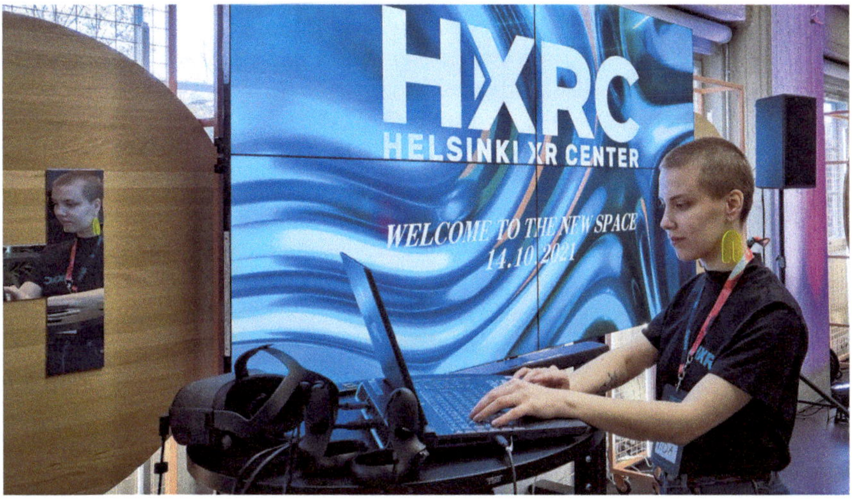

Fig. 4 Helsinki XR Center. Helsinki XR Center—Welcome to the Home of Extended Realities [29]

experience that enables them to explore and engage with the city in a way that is not feasible in the real world by combining virtual reality, augmented reality, and other XR technologies.

One of the key initiatives of Helsinki XR Center is the XR Hub Finland program, which supports XR startups and SMEs through coaching, funding, and networking opportunities. The program aims to create a thriving XR ecosystem in Finland, where companies can access the resources they need to develop and commercialize their XR products and services. One notable case study involving Helsinki XR Center is the development of the VR application called "MedVR". The application was developed by a team of researchers at the University of Helsinki, in collaboration with healthcare professionals and Helsinki XR Center. MedVR is a VR training tool for healthcare professionals, designed to improve their ability to perform medical procedures, such as inserting a chest tube or performing cardiopulmonary resuscitation (CPR).

The final product, MedVR, has been well-received by healthcare professionals and has the potential to significantly improve the quality of medical training and patient outcomes. The application has been tested extensively in simulated medical scenarios, and the results have shown that healthcare professionals who use MedVR perform better in real-life medical situations. Another case study involving Helsinki XR Center is the development of the "Helsinki XR Matchup" event. The event is a matchmaking event for XR professionals, startups, and investors, designed to promote collaboration and innovation in the XR industry. Helsinki XR Center played a key role in the development and organization of the event, providing the necessary resources and expertise to make it a success.

The Helsinki XR Matchup event brought together a diverse group of XR professionals and stakeholders, including investors, startups, and established companies.

The event featured presentations, workshops, and networking opportunities, allowing participants to connect and explore potential collaborations and business opportunities. The success of the Helsinki XR Matchup event has helped to raise the profile of Helsinki XR Center and the XR industry in Finland. It has also helped to promote collaboration and innovation in the industry, facilitating the development of new XR products and services.

Supporting XR development and innovation, Helsinki XR Center also aims to increase public awareness and understanding of XR technology and its potential. The center hosts various events and exhibitions, including the annual XR Expo, which showcases the latest XR products, services, and applications from around the world. Helsinki XR Center is a hub for XR development and innovation, offering a collaborative environment for XR professionals to connect, experiment, and develop new applications and solutions. Its focus on supporting XR startups and SMEs through the XR Hub Finland program makes it an important contributor to the growth of the XR industry in Finland and beyond. The impact of the XR City project has been significant, particularly in terms of improving citizen engagement and participation in smart city development. By creating an immersive and interactive experience, the project has helped to increase public awareness and understanding of urban planning and development issues and has encouraged more active participation from citizens in the decision-making process. The project has also been successful in promoting collaboration between stakeholders in smart city development, including government agencies, private developers, and academic institutions.

5.3 Case Study 3: Microsoft CityNext [30]

Microsoft CityNext is a global initiative aimed at helping cities around the world to develop and implement smart city solutions. The initiative provides a range of tools and resources for cities to use, including cloud-based software, data analytics tools, and other technologies (Fig. 5).

One of the key components of the Microsoft CityNext initiative is the use of the Metaverse in smart city development. By creating virtual replicas of cities, Microsoft CityNext allows for advanced urban planning and development, as well as enhanced citizen engagement and participation. The initiative has been used in a range of cities around the world, including Berlin, Barcelona, and Auckland. One notable case study involving Microsoft CityNext is the "Smart City Program" implemented in the city of Moscow, Russia. The program aimed to transform the city into a more sustainable and efficient metropolis by using technology to address a range of urban challenges.

Microsoft's Dynamics 365 customer relationship management (CRM) system, Azure cloud computing platform, and Power BI business analytics tool were all deployed as part of the Smart City Program. A variety of solutions, including an intelligent transportation system, a smart energy management system, and a digital healthcare system, were created using these technologies. The implementation of an intelligent transportation system (ITS) that employed real-time data and analytics to

Fig. 5 Microsoft CityNext—Winner of the Microsoft Partner of the Year Award. Ing. Punzenberger COPA-DATA, COPADATA [30]

optimize traffic flow and alleviate congestion was one of the primary elements of the Smart City Program. The system was created to enhance public transportation, shorten commutes, and enhance air quality.

The ITS system involved the deployment of a range of sensors and cameras throughout the city, which collected data on traffic patterns, road conditions, and air quality. This data was then analyzed using Microsoft's Azure cloud computing platform and the Power BI analytics tool, providing real-time insights that could be used to optimize traffic flow and reduce congestion. The Smart City Program also included the development of a smart energy management system, which used Microsoft's Dynamics 365 CRM system to manage energy consumption across the city. The system allowed city officials to monitor and analyze energy usage data in real-time, identifying areas where energy efficiency could be improved and implementing solutions to reduce energy consumption.

Another key component of the Smart City Program was the development of a digital healthcare system, which aimed to improve access to healthcare services for citizens. The system involved the deployment of a range of digital health tools, such as mobile health apps and telemedicine services, which allowed citizens to access healthcare services remotely. The Smart City Program in Moscow demonstrates the potential of Microsoft CityNext to help cities use technology to address urban challenges and improve quality of life for citizens. The deployment of Microsoft technologies, such as Azure, Dynamics 365, and Power BI, allowed the city to develop a range of smart and sustainable solutions that have the potential to transform urban living.

The case studies presented in this chapter demonstrate the potential for Metaverse technologies to enhance and improve smart city development. From urban planning and public services to transportation and energy management, the Metaverse offers a new avenue for innovation and collaboration in the field of urban development. As the technology continues to evolve and more projects are implemented, it will be important to carefully evaluate the impact and effectiveness of Metaverse integration in smart city projects and to identify best practices for successful implementation.

6 Future Directions and Opportunities for the Metaverse in Smart City Development

An attractive prospect for the growth of smart cities is the Metaverse. The Metaverse has the potential to promote creativity and sustainability in urban planning and development since it is a virtual world that replicates our actual reality. To fully realize the advantages of the Metaverse in smart cities, several risks and problems must be handled, just like with any new technology [31]. This chapter will explore the future directions and opportunities for the Metaverse in smart city development, including emerging trends and technologies, the potential for innovation and sustainability, challenges and risks, and recommendations for policymakers, planners, and developers.

The application of artificial intelligence (AI) and machine learning (ML) is one of the major themes in the creation of smart cities based on the Metaverse. Large volumes of data may be analyzed using these technologies, and city services and infrastructure can be improved. AI and ML can be used, for instance, to forecast traffic patterns and optimize travel routes or to assess energy use and improve energy systems. Utilizing blockchain technology, which may be used to build safe and open systems for managing city resources and services, is another developing trend [32].

Another emerging technology is the Internet of Things (IoT) [33], which refers to the interconnected network of devices and sensors that collect and share data. In a Metaverse-based smart city, IoT devices could be used to gather real-time data on traffic, energy usage, air quality, and other factors, allowing city planners and developers to make informed decisions about city services and infrastructure. One of the most promising aspects of the Metaverse in smart city development is its potential to drive innovation and sustainability. By using virtual simulations and models, city planners and developers can test new ideas and designs before implementing them in the physical world. This can help to minimize the environmental impact of new developments and ensure that they meet the needs of the community.

The Metaverse can facilitate collaboration between stakeholders in smart city projects, including city planners, developers, policymakers, and citizens. By providing a platform for virtual meetings and discussions, the Metaverse can help to ensure that all voices are heard and that decisions are made transparently and inclusively. Despite its potential, there are also challenges and risks associated with integrating the Metaverse into smart city development [34]. One of the main challenges is ensuring data privacy and security. As more data is collected and shared in the Metaverse, it is essential to have robust systems in place to protect personal information and prevent cyber attacks. Another challenge is ensuring that the benefits of Metaverse-based smart city projects are distributed equitably. As with any new technology, there is a risk that certain segments of the population may be excluded or left behind [35]. It is important to ensure that Metaverse-based services and infrastructure are accessible to all members of the community, regardless of socioeconomic status or other factors. There is a risk that the Metaverse could be used to perpetuate existing power structures or exclude marginalized voices from the decision-making process. It is essential to ensure that the Metaverse is used in a way that promotes transparency, inclusivity, and social justice [36].

Recommendations for Policymakers, Planners, and Developers:

To maximize the benefits of the Metaverse in smart city development, policymakers, planners, and developers need to take a proactive and collaborative approach. This may include:

1. Prioritizing equity and inclusion in Metaverse-based smart city projects, and ensuring that all members of the community have access to the benefits of these projects [37].
2. Establishing clear guidelines and regulations around data privacy and security in the Metaverse, and ensuring that these guidelines are followed by all stakeholders [37].
3. Investing in the development of Metaverse technologies and infrastructure, and encouraging public–private partnerships to accelerate innovation [38].

The Metaverse holds enormous potential for revolutionizing smart city development, enabling more efficient and effective public services, enhancing citizen engagement and participation, and driving innovation and sustainability. To fully realize these benefits, policymakers, planners, and developers must carefully consider the challenges and risks associated with Metaverse technologies and take steps to address them [39]. This includes ensuring data privacy and security, promoting digital equity and inclusion, and developing robust legal and regulatory frameworks to guide the integration of the Metaverse with existing smart city infrastructure.

As the Metaverse continues to evolve and mature, there will undoubtedly be new and exciting opportunities for its integration into smart city projects. Emerging technologies such as blockchain, AI, and the Internet of Things (IoT) are already being explored as potential complements to Metaverse-based smart city initiatives, offering new ways to collect, analyze, and utilize data to improve urban living [40]. The growing emphasis on sustainability and resilience in urban development is likely to drive the adoption of Metaverse technologies, as they offer new ways to monitor and manage resource use and environmental impacts.

In the coming years, we can expect to see more and more smart city projects incorporating Metaverse technologies, as policymakers, planners, and developers increasingly recognize the potential benefits of this approach. However, to ensure the success of these initiatives, it will be critical to approach Metaverse integration thoughtfully and strategically, taking into account the unique challenges and considerations associated with this emerging technology. By doing so, we can harness the power of the Metaverse to build more livable, sustainable, and equitable cities for all.

7 Conclusion

The potential for the Metaverse to transform the development and functioning of smart cities is immense. As we have explored in this chapter, Metaverse technologies have the potential to enhance citizen engagement and participation, improve the efficiency and effectiveness of public services and infrastructure, and foster collaboration and communication between stakeholders in smart city projects [41]. At the same time, there are also significant challenges and risks associated with the integration of the Metaverse in smart city development, including data privacy and security concerns, ethical and social implications, and legal and regulatory challenges.

As we look to the future of smart city development, it is clear that the Metaverse will play an increasingly important role. Emerging trends and technologies, such as virtual and augmented reality, artificial intelligence, and the Internet of Things, will continue to drive innovation and sustainability in smart city projects. These technologies have the potential to transform the way we interact with our urban environments and with each other, opening up new possibilities for creativity, collaboration, and social impact. Realizing the full potential of the Metaverse in smart city development will require a collaborative and inclusive approach. Policymakers, planners, developers, and citizens all have a role to play in shaping the direction and outcomes of Metaverse-based smart city projects [4, 42]. This means engaging with stakeholders from diverse backgrounds and perspectives, ensuring transparency and accountability in decision-making processes, and prioritizing equitable and sustainable outcomes for all citizens.

At the same time, it is important to be aware of the potential risks and downsides of Metaverse integration and to take steps to mitigate these risks. This includes implementing robust data privacy and security measures, ensuring ethical and responsible use of Metaverse technologies, and addressing the digital divide and access issues that could exclude certain segments of the population from Metaverse-enabled smart city services [43].

The Metaverse represents an exciting and transformative opportunity for smart city development. By leveraging the power of virtual and augmented reality, artificial intelligence, and the Internet of Things, we have the potential to create more livable, sustainable, and connected urban environments. However, realizing this potential will require a collaborative and inclusive approach, one that prioritizes equity, sustainability, and responsible use of these powerful technologies. By working together, we can build a future in which the Metaverse and smart cities work in harmony to create a better world for all [44].

In conclusion, the Metaverse presents a significant opportunity to enhance the development and operation of smart cities worldwide. Its potential to increase citizen engagement, improve public services, and facilitate sustainable urban growth is just beginning to be realized. While there are challenges and risks associated with integrating Metaverse technologies into smart city initiatives, policymakers, planners, and developers must work together to address these issues and maximize the benefits of the Metaverse in urban contexts. The success stories of Metaverse-based smart city projects from around the world demonstrate the potential for this technology to revolutionize urban planning and development [45]. From virtual city tours and simulations to augmented reality and mixed reality applications, the Metaverse offers a range of tools and possibilities for enhancing urban environments and citizen engagement [46]. The future of the Metaverse in smart cities is bright, and emerging trends and technologies are likely to enhance its potential even further.

As Metaverse technology continues to evolve, policymakers, planners, and developers must keep pace with these developments and remain aware of the risks and challenges associated with integrating the Metaverse into smart city initiatives. Data privacy and security concerns, ethical and social implications, and the digital divide are all important considerations that must be taken into account. With careful planning and collaboration, these issues can be addressed and overcome. Ultimately, the Metaverse has the potential to drive innovation and sustainability in smart city development, transforming the way cities are designed, built, and managed. As smart cities continue to grow and evolve, they must embrace emerging technologies like the Metaverse to create more livable, sustainable, and equitable urban environments [47]. With the right approach, the Metaverse can be a powerful tool for achieving these goals, and its potential for impact is only just beginning to be realized.

References

1. Castronova E (2020) The Metaverse: definition and relevance to the real world. In: Consalvo M, Dutton S (eds) The Oxford handbook of virtuality. Oxford University Press, pp 50–60. https://doi.org/10.1093/oxfordhb/9780190499624.013.4
2. Shin H (2022) Metaverse and smart cities. In: Chang V, Pasi G (eds) Handbook of smart cities: software services and cyber infrastructure. Springer, pp 1–15. https://doi.org/10.1007/978-3-030-93583-6_78-1.t
3. Whitfield D. "City development." Journey to the center of the Metaverse—one avatar's battle with the reality of the virtual landscape. Wordpress, 23 July 2019. https://journeymetaverse.wordpress.com. Accessed 31 Mar 2023
4. Allam Z, Jones DS (2020) Pandemic stricken cities on lockdown. Where are our planning and design professionals [now, then and into the future]? Land Use Policy 97:104805
5. Allam Z, Bibri SE, Jones DS, Chabaud D, Moreno C (2022) Unpacking the 15-minute city via 6G, IoT, and digital twins: towards a new narrative for increasing urban efficiency, resilience, and sustainability. Sensors 22:1369
6. Miles I (1993) Stranger than fiction: how important is science fiction for futures studies? Futures 25:315–321
7. Lawler DL (1980) Certain assistances: the utilities of speculative fictions in shaping the future. Mosaic J Interdiscip Study Lit 13:1–13
8. Taylor S, Soneji S (2022) Bioinformatics and the Metaverse: are we ready? Front Bioinform 2
9. Li VQT, Ma L, Wu X (2022) COVID-19, policy change, and post-pandemic data governance: a case analysis of contact tracing applications in East Asia. Policy Soc 41:1–14
10. Bibri SE, Allam Z (2022) The Metaverse as a virtual form of data-driven smart urbanism: on post-pandemic governance through the prism of the logic of surveillance capitalism. Smart Cities 5:715–727
11. Kitchin R (2020) Civil liberties or public health, or civil liberties and public health? Using surveillance technologies to tackle the spread of COVID-19. Space Polity 24:362–381

12. Sekalala S, Dagron S, Forman L, Meier BM (2020) Analyzing the human rights impact of increased digital public health surveillance during the COVID-19 crisis. Health Hum Rights 22:7–20
13. Allam Z, Jones DS (2021) Future (post-COVID) digital, smart and sustainable cities in the wake of 6G: digital twins, immersive realities and new urban economies. Land Use Policy 101:105201
14. Cassauwers T (2022) Is 5G bad for your health? It's complicated, say researchers. Available online. https://ec.europa.eu/research-and-innovation/en/horizon-magazine/5g-bad-your-health-its-complicated-say-researchers. Accessed on 30 May 2022
15. Facebook. Connect 2021: our vision for the Metaverse. Available online: https://tech.fb.com/connect-2021-our-vision-for-the-metaverse/. Accessed on 1 Mar 2023
16. Nordor Intelligence. Extended reality (XR) market—growth, trends, COVID-19 impact, and forecast (2021–2026). Available online: https://www.mordorintelligence.com/industry-reports/extended-reality-xr-market. Accessed on 1 Mar 2023
17. Markets and Markets. Extended reality market with COVID-19 impact analysis by technology (AR, VR, MR), Application (Consumer, commercial, enterprises, healthcare, aerospace and defense), offering, device type, and region (North America, Europe, APAC)—Global forecast to 2026. Available online: https://www.marketsandmarkets.com/Market-Reports/extended-reality-market-147143592.html. Accessed on 1 Mar 2023
18. Bibri SE, Allam Z, Krogstie J (2022) The Metaverse as a virtual form of data-driven smart urbanism: platformization and its underlying processes, institutional dimensions, and disruptive impacts. Comput Urban Sci (in press)
19. Bettencourt L (2014) The uses of big data in cities, vol 2. Santa Fe Institute, Santa Fe, New Mexico
20. Bibri SE (2021) Data-driven smart eco-cities of the future: an empirically informed integrated model for strategic sustainable urban development. World Futures 1–44
21. Caprotti F, Chang I, Catherine C, Joss S (2022) Beyond the smart city: a typology of platform urbanism. Urban Transform 4:1–21
22. Van Dijck J, Poell T, De Waal M (2018) The platform society: public values in a connective world. Oxford University Press, Oxford, UK
23. Gordon E, Manosevitch E (2011) Augmented deliberation: Merging physical and virtual interaction to engage communities in urban planning. New Media Soc 13:75–95
24. Wilkins G, Stiff A (2019) Hem realities: augmenting urbanism through tacit and immersive feedback. Archit Cult 7:505–521
25. Lee L-H, Braud T, Zhou P, Wang L, Xu D, Lin Z, Kumar A, Bermejo C, Hui P (2021) All one needs to know about Metaverse: a complete survey on technological singularity, virtual ecosystem, and research agenda. arXiv 2021. arXiv:2110.05352
26. Ameer S, Shah MA (2018) Exploiting big data analytics for smart urban planning. In: Proceedings of the 2018 IEEE 88th vehicular technology conference (VTC-Fall), Chicago, IL, USA, 27–30 Aug 2018, pp 1–5
27. "Meet Virtual Singapore, the city's 3D digital twin." GovInsider, 28 Jan 2018. https://govinsider.asia/intl-en/article/meet-virtual-singapore-citys-3d-digital-twin. Accessed 31 Mar 2023
28. "Helsinki XR Center." Helsinki XR center—welcome to the home of extended realities. https://helsinkixrcenter.com/. Accessed 31 Mar 2023
29. Microsoft CityNext. "Microsoft CityNext - Winner of the Microsoft Partner of the Year Award." Ing. Punzenberger COPA-DATA, COPADATA. https://www.copadata.com/en/industries/smart-city-solutions/smart-city-citynext/. Accessed 31 Mar 2023
30. Sharifi A, Khavarian-Garmsir AR, Kummitha RKR (2021) Contributions of smart city solutions and technologies to resilience against the COVID-19 Pandemic: a literature review. Sustainability 13:8018
31. Popper R (2010) Mapping foresight: revealing how Europe and other world regions navigate into the future. Publications Office of the European Union, Luxembourg
32. Salerno R (2014) City ideologies in techno-urban imaginaries. Urban 2014:185–192

33. Bibri SE (2022) The shaping of the Metaverse as an alternative to the imaginaries of data-driven smart cities: a study in science, technology, and society. Smart Cities (in press)
34. Jansen J (1994) Towards a sustainable future, en route with technology. In: The Dutch Committee for long-term environment policy—the environment: towards a sustainable future. Dordrecht, The Netherlands, Kluwer, pp 497–523
35. Martino JPJTF, Change S (2003) A review of selected recent advances in technological forecasting. Technol Forecast Soc Chang 70:719–733
36. Bibri SE (2018) Approaches to futures studies: a scholarly and planning approach to strategic smart sustainable city development. In: Bibri SE (ed) Smart sustainable cities of the future: the untapped potential of big data analytics and context-aware computing for advancing sustainability. Springer International Publishing, Cham, Switzerland, pp 601–660
37. Miola A (2008) Backcasting approach for sustainable mobility; European Commission, Joint Re- search Centre, Institute for Environment and Sustainability, Luxembourg
38. Dreborg KH (1996) Essence of backcasting. Futures 28:813–828
39. Pappenberger F, Cloke HL, Persson A, Demeritt D (2011) HESS Opinions "On forecast (in)consistency in a hydro-meteorological chain: curse or blessing?" Hydrol Earth Syst Sci 15:1225–1245
40. Porter AL, Cunningham SW, Banks J, Roper AT, Mason TW, Rossini FA (2011) Forecasting and management of technology. Wiley, NJ, USA
41. Stephenson N (2003) Snow crash: a novel. Random House Publishing Group, New York, NY, USA
42. Meta. Meta. Available online: https://about.facebook.com/meta/. Accessed on 1 Mar 2023
43. Shaw FX. Microsoft cloud at ignite 2021: Metaverse, AI and hyperconnectivity in a hybrid world. Available online: blogs.microsoft.com. Accessed on 1 Mar 2023
44. Adinarayan T. Metaverse only gets real when apple joins, Morgan Stanley Says. Available online: bloomberg. Accessed on 1 Mar 2023
45. Magic Leap. The Metaverse is already here. Available online: .magicleap.com. Accessed on 3 Mar 2023
46. Roblox. Nike is building its Metaverse inside of "Roblox". Available online: https://www.engadget.com/nike-roblox-nikeland-metaverse-192234036.html. Accessed on 3 Mar 2023
47. Radoff J. Clash of the Metaverse Titans: Microsoft, Meta and Apple. Available online: https://medium.com/building-the-metaverse/clash-of-the-metaverse-titans-microsoft-meta-and-apple-ce505b010376. Accessed on 1 Mar 2023

Metaverse in Military Defense Applications

Akash Dogra

Abstract The concept of the metaverse, a virtual and interconnected world where real and digital elements coexist, is rapidly emerging as a transformative technology in various sectors. One such sector where the metaverse is making substantial inroads is in military defense applications. This book chapter provides a comprehensive exploration of the metaverse's integration into military defense, drawing from real cases and highlighting the potential threats and challenges it presents. The metaverse continues to gain prominence in military defense, and it also brings forth a set of unique challenges and potential dangers. This chapter examines these emerging threats, including concerns related to cybersecurity, data privacy, and information warfare within the metaverse. It addresses the implications of virtual espionage, the manipulation of digital assets, and the potential for adversaries to exploit vulnerabilities within metaverse-based defense systems. By analyzing real cases and highlighting the associated risks, readers can better comprehend the transformative potential of this technology while remaining vigilant against its potential pitfalls. As military defense strategies evolve in the era of the metaverse, a thorough understanding of its applications and vulnerabilities becomes paramount for policymakers, military professionals, and technology enthusiasts alike.

1 Introduction

The metaverse, a concept that was once limited to science fiction, has now become a reality. The metaverse is a virtual world that is not just immersive but interactive, where people can meet and interact with each other, digital objects, and environments in a three-dimensional space. This is rapidly evolving and has the potential to revolutionize many aspects of our lives. One area where the metaverse can have a significant impact is in military defense applications [1].

A. Dogra (✉)
Graphic Era Hill University, Dehradun, Uttrakhand, India
e-mail: akash.dogra1234@gmail.com

G. Chhabra and K. Kaushik (eds.), *Understanding the Metaverse*, Blockchain Technologies, https://doi.org/10.1007/978-981-97-2278-5_9

The metaverse, a term coined by Neal Stephenson in his 1992 science fiction novel"Snow Crash," refers to a collective virtual shared space created by the convergence of virtually enhanced physical reality and physically persistent virtual reality. In simpler terms, it is a virtual world where users can interact with a computer-generated environment and other users. With the rapid advancements in technology, the concept of the metaverse has evolved from a mere science fiction idea to a potential reality [2]. The metaverse has the potential to revolutionize various industries, including entertainment, education, and business. However, one of the most significant and often overlooked applications of the metaverse is in the field of military defense.

According to a recent report titled "Military Augmented Reality Market to 2025," the AR/VR defense market will be worth $1.79 billion by 2025, up from $511 million last year. At that rate of growth, the military is one of the early sectors to fulfill the hype of commercial mixed reality.

The defense industry, it turns out, is harnessing advances first made in gaming. BISim uses game-based technology to develop low-cost training and simulation software products. The products boast many of the same technical aspects you'd find in popular video games, such as"whole-earth rendering" and"pre-programmed AI behaviors" [3].

BIS's products include a gunship crew trainer for the US Air Force and virtual parachute training, among other scenario-based modules [4]. Simulators have long played some role in military training, but the latest generation of VR headsets coupled with the advanced physics engines that power modern games have made the technology especially well suited to combat training.

"Where today's high-end simulators rely on large and expensive display environments using domes and collimated displays," says Burwell, "next generation training systems will benefit from emerging VR and AR technologies that enable solutions that are orders of magnitude less expensive, provide higher fidelity, and offer a smaller footprint supporting training at the point of need."

In a kind of development feedback loop, Burwell predicts that the development happening in the military market will in turn make its way back into the consumer market [5]. "Military AR/VR solutions are driven by specific requirements to meet training objectives. For example, high-resolution head mounted displays (HMD)s are needed to support flight training where pilots need to see fine detail at a distance. Today, these requirements are met with specialized hardware. But the road-maps of component suppliers show a path where within a few years the more expensive and esoteric technologies will become a commodity" [6].

With rapid adoption of these technologies in training scenarios in particular, there's concern that teaching war fighters using video games could have unintended ethical consequences. Burwell believes that it comes down to execution.

"To deploy AR/VR responsibly in simulation and training," he says, "trainers should provide trainees and personnel with clear objectives and context for how these tools are being used to prepare for real-world scenarios. The simulations behind these tools depict operational environments as accurately as possible and facilitate after

Fig. 1 Pilot being trained on simulator [1]

action review, so individuals and units can understand the impact of their actions" [7] (Fig. 1).

It's a consideration with growing importance. It's still too early to make sweeping assessments of the effectiveness of the technologies in defense applications, but it's clear that AR/VR will be playing a larger role in training soldiers here and elsewhere in the years ahead.

The use of technology has always played a vital role in military defense strategies. The military has always been on the forefront of adopting new technologies, and the metaverse is no exception. The potential for the metaverse to be used in military defense applications is vast, and the benefits that it could offer are significant [8]. One of the most significant advantages of the metaverse is its potential to provide military personnel with a realistic simulation of combat scenarios. This simulation could help train soldiers and prepare them for the realities of the battlefield, reducing the risk of casualties and increasing the chance of success. With the emergence of the metaverse, the military can now leverage this virtual world to enhance situational awareness, improve training, and streamline mission planning. The metaverse can also help the military to create more realistic simulations of combat scenarios, which can help to better prepare soldiers for the battlefield.

The metaverse could also provide a way to better understand complex battlefield environments. By creating a digital twin of a given area, military personnel can have a better understanding of the terrain, structures, and obstacles they will face in the field [9]. This could help them plan their tactics and strategies better, reducing the risk of mistakes and making operations more effective. Another advantage of the metaverse is the potential for collaboration and coordination between units. Military personnel can use the metaverse to communicate and share information, making it easier to work together to achieve their objectives. The metaverse could also provide a way for commanders to have a better understanding of their units' status and location, making it easier to coordinate their movements and allocate resources effectively.

Military defense strategies have always relied heavily on technology to provide an edge in combat. From the first use of the radio to coordinate troops during the World War I to the use of drones and other advanced technologies in modern warfare, technology has played a critical role in shaping military tactics and strategies [10]. With the emergence of the metaverse, the military can now leverage this virtual world to enhance situational awareness, improve training, and streamline mission planning. The metaverse can help military personnel create realistic simulations of combat scenarios, allowing them to test strategies and tactics in a safe and controlled environment [11]. This type of training can provide soldiers with the experience they need to better prepare for the realities of combat.

The metaverse can provide a new level of situational awareness, which is crucial for military personnel in the field. Soldiers can use the metaverse to gain a better understanding of their surroundings, anticipate potential threats, and plan their next moves more effectively. The metaverse can also facilitate better communication and coordination between units, making it easier to collaborate and work together to achieve their objectives. As the metaverse continues to evolve and become more advanced, it is likely that it will play an increasingly important role in military defense applications. By leveraging this technology [12], the military can improve its training, coordination, and situational awareness, ultimately leading to better outcomes on the battlefield.

In this chapter, we will explore the potential applications of the metaverse in military defense. We will provide an overview of how the metaverse can be used to improve military defense strategies, provide case studies of real-world examples, discuss the challenges and risks associated with using the metaverse in military defense, and provide a glimpse into the future of this technology. With the increasing adoption of the metaverse in many industries, including gaming and education, it is only a matter of time before its full potential is realized in military defense applications. Finally, we will look at the future of the metaverse in military defense applications, discussing potential areas for research and development and how the technology could evolve to enhance military strategies and tactics further. The metaverse has the potential to revolutionize military defense applications, from training soldiers to improving situational awareness and collaboration between units. While there are challenges and risks associated with using this technology, the potential benefits are significant, and the military is sure to be on the forefront of exploring its potential in the coming years.

2 Relevance of the Metaverse in Military Defense Applications

The relevance of the metaverse in military defense applications, focusing on its potential benefits, challenges, and ethical considerations. The use of the metaverse in training and simulation, intelligence gathering and analysis, command and control, and cyber warfare. And also the potential risks and ethical concerns associated with the use of the metaverse in military defense [13].

One of the most apparent applications of the metaverse in military defense is in the realm of training and simulation. The metaverse can provide a highly immersive and realistic environment for military personnel to train and practice various skills, ranging from basic combat tactics to complex mission planning and execution.

The use of virtual reality (VR) and augmented reality (AR) technologies can significantly enhance the effectiveness of military training by providing a more engaging and interactive experience for the trainees. For example, the metaverse can be used to create realistic virtual battlefields, allowing soldiers to practice their combat skills in a safe and controlled environment. This can help reduce the risks associated with live-fire exercises and other traditional training methods. Moreover, the metaverse can be used to simulate various scenarios and environments, enabling military personnel to adapt and respond to different situations effectively [14].

Another advantage of using the metaverse for military training is the ability to collect and analyze data on individual and team performance. This can help identify areas for improvement and provide personalized feedback to enhance the overall effectiveness of the training program. The metaverse can facilitate collaboration and communication among military personnel, fostering teamwork and coordination, which are crucial for successful mission execution [15].

The metaverse can also play a significant role in intelligence gathering and analysis. As the virtual world becomes more interconnected with the physical world, it can serve as a valuable source of information for military intelligence agencies. For example, the metaverse can be used to monitor and track the activities of potential adversaries, gather information on their capabilities and intentions, and even predict their future actions [16].

The metaverse can be used as a platform for conducting social engineering and psychological operations. By creating realistic virtual personas, military intelligence operatives can infiltrate online communities and gather valuable information from unsuspecting targets. Moreover, the metaverse can be used to spread disinformation and propaganda, influencing public opinion and shaping the narrative in favor of a particular military objective.

The use of artificial intelligence (AI) and machine learning algorithms can further enhance the capabilities of the metaverse in intelligence gathering and analysis. By processing vast amounts of data collected from the virtual world, AI-powered systems can identify patterns and trends, providing valuable insights for military decision-makers [17]. The metaverse can also revolutionize the way military command and control operations are conducted [18]. By providing a shared virtual environment, the metaverse can facilitate real-time communication and collaboration among military personnel, regardless of their physical location. This can help streamline decision-making processes and enhance situational awareness, enabling military commanders to make more informed decisions and respond to emerging threats more effectively.

Moreover, the metaverse can be used to create virtual command centers, where military leaders can monitor and manage various aspects of a military operation. By leveraging VR and AR technologies, these virtual command centers can provide a highly immersive and interactive experience, allowing military commanders to visualize complex data and information in a more intuitive and accessible manner [19].

As the metaverse becomes more interconnected with the physical world, it also becomes a potential battleground for cyber warfare. Military defense agencies can leverage the metaverse to conduct offensive and defensive cyber operations, targeting the critical infrastructure and information systems of potential adversaries. For example, the metaverse can be used as a platform for launching cyber attacks, exploiting vulnerabilities in the virtual world to gain unauthorized access to sensitive information and systems. Conversely, the metaverse can also be used to enhance the resilience of military networks and systems, by simulating various cyber threats and testing the effectiveness of defensive measures. While the metaverse offers numerous benefits for military defense applications, it also presents several risks and ethical concerns that need to be addressed. One of the primary concerns is the potential for misuse of the metaverse for malicious activities [20].

3 Background

Military defense technologies have a long and rich history, dating back to ancient times when early civilizations used weapons like swords, spears, and bows and arrows to protect their territories. As human societies became more complex and wars became more frequent, military technologies evolved to become more sophisticated and effective. Military defense technologies have undergone significant evolution over the centuries, driven by a variety of factors including advances in science and technology, changes in warfare tactics, and geopolitical developments [21]. The earliest recorded military technologies were simple weapons like the sling and the bow and arrow, used by ancient civilizations such as the Egyptians and Greeks. Over the centuries, military defense technologies have undergone numerous advancements and innovations. In the medieval period, gunpowder was developed in China and later spread to Europe, leading to the invention of the first muskets and the development of artillery. The Industrial Revolution saw the introduction of new weapons technologies such as the machine gun and the submarine, and the use of new materials like steel and iron transformed the construction of fortifications and warships [22].

The First and Second World Wars were significant turning points in the evolution of military defense technologies. The development of tanks, aircraft, and submarines transformed the way wars were fought, and new technologies such as radar, sonar, and cryptography played a critical role in intelligence gathering and communication. The use of atomic bombs at the end of the World War II marked the beginning of the nuclear era, and the development of intercontinental ballistic missiles (ICBMs) and other delivery systems created new threats and challenges [23]. In recent years, technology has become an even more essential component of military defense strategies [24]. With the advent of digital and computer technologies, military forces have gained unprecedented access to powerful tools for intelligence gathering, communication, and combat. Today, the use of drones, artificial intelligence, and cyber weapons has become increasingly common in military defense. These technologies have the potential to enhance situational awareness, improve decision-making, and increase the effectiveness of military operations.

Since the end of the Cold War, the focus of military defense technologies has shifted toward precision-guided munitions, unmanned aerial vehicles (UAVs), and other digital and computer technologies. These technologies have transformed the way wars are fought, allowing for greater precision and control in targeting, and providing new capabilities for intelligence gathering and communication [25]. The increasing use of technology in military defense strategies has also led to new risks and challenges. Cyberattacks, for example, can compromise sensitive military data and disrupt communication networks, while the deployment of autonomous weapons and artificial intelligence raises ethical concerns [26]. The use of the metaverse in military defense is still in its early stages, with most applications focused on training and simulation. However, there have been some early examples of the use of virtual reality and augmented reality technologies in military training [27].

In 2015, the US Army began using a virtual reality simulation called the Dismounted Soldier Training System (DSTS) to train soldiers in urban combat scenarios. The system uses a VR headset and a specially designed rifle with motion sensors to simulate realistic combat scenarios, allowing soldiers to practice their skills in a safe and controlled environment [28].

In 2019, the US Marine Corps began using an augmented reality system called the Augmented Immersive Team Trainer (AITT) to train small unit leaders in tactical decision-making. The system uses AR goggles and a specially designed vest to create a virtual training environment, allowing marines to practice and refine their decision-making skills in a realistic and challenging scenario [29].

Other military organizations around the world have also begun exploring the potential applications of the metaverse in military defense. For example, the Chinese military has reportedly been developing a virtual reality simulation for military training, while the Israeli Defense Forces have used VR simulations to train soldiers in battlefield medicine [30].

While the use of the metaverse in military defense is still in its early stages, it is clear that virtual and augmented reality technologies have the potential to transform the way military training is conducted. As these technologies continue to advance, it is likely that their applications in military defense will expand, providing new capabilities for intelligence gathering, communication, and combat. However, it will also be important to address the potential risks and challenges associated with the use of the metaverse in military defense, including cybersecurity and data privacy concerns [31]. The use of technology in military defense also poses significant challenges and risks. Cyberattacks can compromise sensitive military data and disrupt communication networks, while the deployment of autonomous weapons and artificial intelligence raises ethical concerns. Despite these challenges, the importance of technology in modern military defense strategies cannot be overstated. As the nature of warfare continues to evolve, military forces will continue to rely on innovative technologies to maintain their edge on the battlefield.

Military defense technologies have a long and rich history of evolution and innovation, from the earliest weapons to the most advanced digital and computer technologies of today [32]. Technology has become a critical component of modern military defense strategies, providing new capabilities for intelligence gathering, communication, and combat. As technology continues to evolve, military forces will need to adapt and innovate to maintain their edge on the battlefield [33].

4 Metaverse and Military Defense

The metaverse, a term coined by Neal Stephenson in his 1992 science fiction novel Snow Crash, refers to a collective virtual shared space created by the convergence of virtually enhanced physical reality and physically persistent virtual reality. In simpler terms, it is a virtual world where users can interact with each other and their environment in real-time. With the rapid advancements in technology, the concept of the metaverse is becoming increasingly relevant and applicable to various industries, including military defense.

The military has always been at the forefront of adopting and integrating new technologies to enhance its capabilities. The metaverse offers a new frontier for military defense applications, with the potential to revolutionize training, mission planning, and situational awareness. This essay will explore the potential benefits of using the metaverse in military defense and provide examples of how it has been used in military defense applications [18].

Situational awareness is a critical aspect of military operations, as it allows commanders and soldiers to understand the battlefield environment and make informed decisions. The metaverse can help improve situational awareness by providing a virtual environment that accurately represents the real world, allowing military personnel to visualize and interact with the environment and other users in real-time. For example, the US Army has been developing a system called the Synthetic Training Environment (STE), which aims to provide a realistic virtual environment for soldiers to train in. The STE can replicate real-world environments, such as urban areas, forests, and deserts, allowing soldiers to familiarize themselves with the terrain and practice tactics before deploying to the actual location. This can help improve situational awareness by giving soldiers a better understanding of the environment they will be operating in [34].

The metaverse can be used to create virtual command centers, where commanders can monitor and manage operations in real-time. By integrating data from various sources, such as satellite imagery, drone feeds, and intelligence reports, the metaverse can provide a comprehensive and up-to-date picture of the battlefield [35]. This can help commanders make more informed decisions and improve overall situational awareness. One of the most significant potential benefits of the metaverse in military defense is its ability to enhance training. Traditional military training methods, such as live exercises and simulations, can be expensive, time-consuming, and limited in scope [36].

The metaverse offers a cost-effective and scalable alternative, allowing military personnel to train in a realistic virtual environment that can be easily customized and adapted to various scenarios. For example, the US Air Force has been using a virtual reality platform called the Distributed Mission Training (DMT) system to train pilots in realistic combat scenarios. The DMT system allows pilots to fly virtual aircraft in a simulated environment, where they can practice tactics, communication, and decision-making skills. This type of training can help pilots become more proficient and confident in their abilities, ultimately improving their performance in real-world situations [36].

The metaverse can also be used to facilitate joint training exercises between different branches of the military or even between different countries. By connecting users in a shared virtual environment, the metaverse can enable large-scale training exercises that would be difficult or impossible to conduct in the real world [37]. This can help improve interoperability and cooperation between different military units, ultimately enhancing overall military effectiveness. The metaverse can also play a crucial role in improving mission planning by providing a virtual environment where military personnel can visualize and interact with the mission area. By integrating data from various sources, such as satellite imagery, intelligence reports, and terrain data, the metaverse can create a realistic and accurate representation of the mission area. For example, the US Navy has been using a system called the Virtual Environment for Submarine Tactics and Analysis (VESTA) to help plan submarine missions. VESTA allows submarine commanders to visualize the underwater environment, including the location of enemy vessels and potential obstacles, in a 3D virtual environment. This can help commanders develop more effective tactics and strategies, ultimately improving the success rate of submarine missions [37].

The metaverse can be used to conduct virtual rehearsals of missions, allowing military personnel to practice their roles and responsibilities in a realistic environment. This can help identify potential issues or challenges before the actual mission, allowing for adjustments and improvements to be made. Ultimately, this can lead to more effective and successful missions.

5 Case Studies

5.1 Project MAVEN [2]

Project MAVEN is an innovative platform that leverages the latest virtual and augmented reality technologies to provide military personnel with a more immersive and intuitive way to analyze complex data. Project MAVEN represents a significant advancement in the use of virtual and augmented reality technologies to support military operations.

Project Maven is a program initiated by the US Department of Defense in April 2017 to develop and deploy artificial intelligence (AI) and machine learning (ML) technologies for military purposes. The project's official name is the Algorithmic Warfare Cross-Functional Team (AWCFT), and it aims to use machine learning algorithms to analyze and interpret video footage from drones and other aerial surveillance platforms.

The project was launched in response to the growing demand for intelligence, surveillance, and reconnaissance (ISR) capabilities by the US military. The objective of Project Maven is to develop AI and ML technologies that can assist military analysts in identifying and tracking potential threats in real-time, thereby enhancing the effectiveness of military operations.

Project Maven is a collaboration between the Department of Defense and various technology companies, including Google, which was contracted to provide AI and ML expertise to the project. However, Google pulled out of the project in June 2018 following protests from its employees, who were concerned about the company's involvement in military applications of AI (Fig. 2).

Despite the controversy surrounding its development, Project Maven continues to be a priority for the Department of Defense, and it has expanded its scope to include other AI and ML applications, such as natural language processing and predictive analytics. The project is considered to be a key component of the Department of Defense's efforts to modernize its technological infrastructure and maintain its strategic advantage in an increasingly complex and dynamic global security environment.

By providing military personnel with an immersive and interactive way to analyze complex data sets, MAVEN can greatly improve situational awareness and enhance decision-making. By creating a 3D virtual environment, MAVEN allows users to visualize and interact with data in a way that traditional 2D displays cannot match. This can help military personnel gain a better understanding of the situation on the ground, improve situational awareness, and make more informed decisions.

Another important feature of MAVEN is its ability to support collaboration and communication among military personnel. By allowing multiple users to access the same 3D virtual environment, MAVEN can facilitate real-time communication and coordination, even in remote or decentralized environments. This can be particularly useful in complex military operations where multiple teams or units need to work together to achieve a common objective.

Fig. 2 Project Maven drone sight [1, 2]

MAVEN has a wide range of potential applications in military operations. For example, it can be used to analyze satellite imagery and sensor data to identify potential threats or track the movement of enemy forces. It can also be used to plan and coordinate complex operations, such as troop movements and logistics. In addition, MAVEN can be used to train military personnel, allowing them to practice and simulate various scenarios in a virtual environment.

The development of MAVEN is part of a broader trend toward the use of advanced technologies in military operations. As technology continues to advance, it is likely that we will see more platforms like MAVEN that use virtual and augmented realities to support military operations. However, it is also important to consider the ethical implications of these technologies and ensure that they are used in a responsible and transparent manner. In terms of contribution, Project MAVEN has the potential to revolutionize the way that military operations are conducted. By providing military personnel with a more intuitive and interactive way to analyze data, MAVEN can greatly enhance their ability to make informed decisions and respond to rapidly changing situations. This can help to reduce the risk of casualties and improve the overall effectiveness of military operations.

At the same time, it is important to consider the ethical implications of using advanced technologies like MAVEN in military operations. For example, there are concerns about the potential for these technologies to be misused or to cause unintended harm. It is therefore important to ensure that MAVEN and similar platforms are used in a responsible and transparent manner, with appropriate safeguards in place to protect the safety and well-being of all individuals involved. Project MAVEN represents an exciting development in the use of virtual and augmented reality technologies in military operations. With continued research and development, it is likely that we will see even more advanced platforms emerge in the coming years, further enhancing the capabilities of military personnel and improving the overall effectiveness of military operations.

5.2 Virtual Battlespace 3 [3]

Virtual Battlespace 3 (VBS3) is an advanced military simulation platform developed by Bohemia Interactive Simulations (BIS) that provides military personnel with a realistic training environment to enhance their skills and preparedness for various scenarios. The platform is widely used by military organizations around the world, including the US Army, Australian Defense Force, and the British Army.

The history of Virtual Battlespace 3 (VBS3) can be traced back to the early 2000s when Bohemia Interactive Simulations (BIS) first began developing military simulation platforms. The company's first product, Virtual Battlespace 1 (VBS1), was released in 2002 and quickly gained popularity among military organizations around the world.

BIS continued to develop and refine its military simulation technology, releasing Virtual Battlespace 2 (VBS2) in 2007. This platform was designed to provide a more realistic and immersive training experience, with advanced features such as dynamic weather and terrain modeling, as well as improved physics and ballistics simulations (Fig. 3).

In 2013, BIS released Virtual Battlespace 3, the latest iteration of its military simulation technology. VBS3 builds on the success of its predecessors, incorporating the latest advancements in graphics, physics, and artificial intelligence to provide an even more realistic and immersive training experience. VBS3 is designed to simulate a wide range of military operations, from small-unit tactics to large-scale battlefield scenarios. The platform includes a comprehensive set of tools that allow users to create custom scenarios and missions, as well as modify existing ones. The platform provides realistic physics and ballistics models, which allow users to experience the same challenges and conditions they would encounter in real-world situations.

One of the primary advantages of VBS3 is that it allows military organizations to conduct training exercises in a safe and controlled environment, without the risks and costs associated with live-fire exercises. The platform enables users to practice and fine-tune their skills in a virtual environment before applying them to real-world operations. This reduces the risk of injury and equipment damage, while also saving time and resources. VBS3 also provides a high degree of flexibility and customization, allowing users to tailor the training experience to their specific needs. The platform supports a variety of training scenarios, including urban combat, convoy operations, and air assault missions. It also allows users to simulate different weather conditions and terrain types, further enhancing the realism of the training experience.

Virtual Battlespace 3 is a powerful military simulation platform that provides military personnel with a safe, realistic, and customizable training environment. Its

Fig. 3 Training on virtual battlespace 3 [3]

advanced features and flexibility make it an essential tool for military organizations around the world to enhance their readiness and preparedness for a range of scenarios. Since its release, VBS3 has become a popular training tool for military organizations around the world. The platform has been used by a wide range of military branches, including the United States Army, Australian Defense Force, and the British Army, to train personnel for a variety of scenarios and mission types. VBS3 has also been used in a variety of research and development projects, including studies on the effectiveness of different training methods and the development of new technologies and tactics. The platform's flexibility and customization options make it an ideal tool for exploring new ideas and testing new strategies in a safe and controlled environment.

Virtual Battlespace 3 is one of ongoing development and refinement, with each iteration of the platform building on the successes and lessons learned from previous versions. The platform continues to be a key tool for military organizations around the world to enhance their readiness and preparedness for a range of scenarios.

5.3 HoloLens 2 [4]

The HoloLens 2 is an advanced augmented reality (AR) headset developed by Microsoft that has potential military applications. It is the successor to the original HoloLens, which was released in 2016. The HoloLens 2 was officially launched in February 2019 and has since been used in a variety of industries, including healthcare, education, and manufacturing. The HoloLens 2 is a cutting-edge technology that has its roots in Microsoft's research and development efforts in the field of augmented reality (AR). The company began exploring the potential of AR technology in the early 2010s, with the launch of the Kinect sensor for the Xbox gaming console. The Kinect used advanced sensors and cameras to track users' movements and provide an immersive gaming experience (Fig. 4).

Building on the success of the Kinect, Microsoft began developing the HoloLens, an AR headset that could overlay digital information onto the physical environment. The first version of the HoloLens was released in 2016 and was primarily targeted at developers and early adopters. The HoloLens quickly gained a reputation as one of the most advanced AR headsets on the market, with its ability to provide a high degree of immersion and interactivity.

The HoloLens 2 was released in 2019 and represented a significant improvement over the original HoloLens. The headset featured a more comfortable design, improved sensors and cameras, and a higher field of view. It also included advanced features such as eye tracking and hand tracking, which allowed users to interact with digital objects in a more natural and intuitive way.

One of the key advantages of the HoloLens 2 is its ability to provide soldiers with real-time situational awareness. The headset uses advanced sensors and cameras to

Fig. 4 Strategy planning via HoloLens [4]

map the physical environment and overlay digital information onto it. This means that soldiers can see important information such as maps, target locations, and other critical data overlaid onto the physical environment, providing them with a more complete picture of the battlefield. The HoloLens 2 can also be used for training purposes, allowing soldiers to practice their skills in a realistic virtual environment. The headset can simulate a range of scenarios and mission types, allowing soldiers to practice their tactics and techniques in a safe and controlled environment. This can help to improve their skills and preparedness for real-world operations.

Another advantage of the HoloLens 2 is its versatility and customization options. The headset can be customized to meet the specific needs of different military units and mission types. For example, it can be used to display information in different languages or to provide specialized training for specific roles such as medics or engineers. The HoloLens 2 has the potential to make a significant contribution to military operations. By providing soldiers with real-time situational awareness and advanced training capabilities, the headset can help to enhance their effectiveness and preparedness for a range of scenarios. It can also be used for a variety of other military applications, such as logistics and maintenance.

One of the most interesting facts about the HoloLens 2 is its use of advanced AI and machine learning algorithms. The headset uses these technologies to analyze the user's environment and provide a more immersive and interactive experience. For example, it can detect the user's hand gestures and respond appropriately, such as allowing them to manipulate digital objects or navigate menus. Another interesting fact about the HoloLens 2 is its potential as a tool for collaboration and communication. The headset can be used to connect users in different locations and allow them to work together in a shared virtual environment. This can be particularly useful for military operations that involve multiple units and teams working together. The HoloLens 2 is a powerful technology that has the potential to transform military operations and enhance the capabilities of soldiers and other personnel.

Its advanced features, versatility, and potential for collaboration make it an essential tool for military organizations looking to stay ahead of the curve in an increasingly complex and dynamic global security environment. The HoloLens 2 has already been used in a variety of military applications. For example, the US Army has been working with Microsoft to develop a custom version of the headset that can be used by soldiers in the field. The headset has also been used in a variety of military research projects, including studies on the use of AR technology in combat training.

The HoloLens 2 is a powerful tool that has the potential to enhance situational awareness and improve training for military personnel. Its advanced features, versatility, and customization options make it an ideal tool for military organizations looking to enhance their capabilities and preparedness for a range of scenarios.

5.4 Virtual Paratrooper Training Simulator

The Virtual Paratrooper Training Simulator [5] is an advanced virtual reality (VR) simulation developed by the US Army to provide soldiers with a safe and realistic training environment for paratrooper training. The simulator uses advanced physics models and simulated wind conditions to create a highly realistic paratrooper training experience. The Virtual Paratrooper Training Simulator has its roots in the US Army's ongoing efforts to develop advanced training technologies to enhance the capabilities and preparedness of soldiers. The simulator was developed by the US Army's Research, Development, and Engineering Command (RDECOM) in collaboration with industry partners (Fig. 5).

The development of the Virtual Paratrooper Training Simulator began in the early 2000s, with the goal of creating a safe and realistic training environment for paratrooper training. The simulator was designed to provide soldiers with a highly realistic training experience, using advanced physics models and simulated wind conditions to create a Virtual Paratrooper Training environment. The Virtual Paratrooper Training Simulator was officially launched in 2011 and has since been used by the US Army and other military organizations around the world to train soldiers for a range of scenarios and mission types. The simulator has been particularly useful for training new recruits, who may not have prior experience with paratrooper training.

One of the key advantages of the Virtual Paratrooper Training Simulator is its ability to provide soldiers with a safe training environment. Paratrooper training is inherently dangerous, with the risk of injury or death from parachute malfunctions or

Fig. 5 Virtual paratrooper training simulator [4]

other mishaps. By using the simulator, soldiers can practice their skills in a safe and controlled environment, without the risk of injury or death. The simulator has also been praised for its realism and flexibility. The physics models and simulated wind conditions are designed to closely mimic real-world conditions, allowing soldiers to experience the same challenges and conditions they would encounter in a real paratrooper training exercise. The simulator can also be customized to meet the specific needs of different military units and mission types, allowing soldiers to practice their skills in a variety of scenarios. This provides soldiers with a more immersive and effective training experience, allowing them to practice and refine their skills in a more realistic and practical way.

Another advantage of the Virtual Paratrooper Training Simulator is its flexibility and customization options. The simulator can be customized to meet the specific needs of different military units and mission types. For example, it can be used to simulate different weather conditions, terrain types, and other variables, allowing soldiers to practice their skills in a variety of scenarios. The Virtual Paratrooper Training Simulator has been used by the US Army and other military organizations around the world to train soldiers for a range of scenarios and mission types. The simulator has been particularly useful for training new recruits, who may not have prior experience with paratrooper training.

The Virtual Paratrooper Training Simulator is a powerful tool that provides soldiers with a safe and effective training environment for paratrooper training. Its advanced features, realism, and flexibility make it an essential tool for military organizations looking to enhance their capabilities and preparedness for a range of scenarios. The Virtual Paratrooper Training Simulator has made a significant contribution to military training and preparedness. Its advanced features, realism, and safety have made it an essential tool for military organizations looking to enhance their capabilities and preparedness for a range of scenarios. The simulator is a testament to the ongoing efforts of the US Army and other military organizations to develop advanced training technologies to enhance the effectiveness and safety of soldiers.

5.5 Project Athena [5]

Project Athena is an advanced virtual reality (VR) training program developed by the US Navy for submarine crews. The program is designed to provide a realistic training environment for submarine crews, allowing them to practice a range of scenarios and mission types in a safe and controlled environment.

The development of Project Athena can be traced back to the mid-2010s, when the US Navy began exploring the potential of virtual reality (VR) technology for submarine training. The goal was to create a more effective and realistic training program for submarine crews, allowing them to practice a range of scenarios and mission types in a safe and controlled environment. The development process for

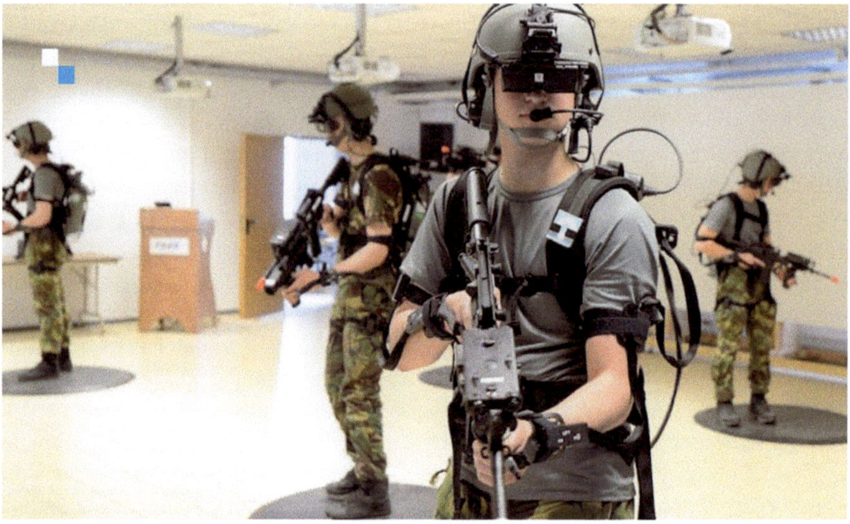

Fig. 6 Soldier experiencing real life situations in Project Athena [5]

Project Athena was a collaborative effort between the US Navy and industry partners, including VR technology companies and defense contractors. The program was designed to provide a high degree of immersion and interactivity, using advanced VR technology to create a virtual environment that closely mimics the real-world conditions that submarine crews would encounter during operations (Fig. 6).

The development of Project Athena was a complex and challenging process, requiring the integration of a range of advanced technologies and the creation of highly realistic and detailed virtual environments. The program also required extensive testing and evaluation to ensure that it met the needs of submarine crews and provided an effective training experience. Project Athena was officially launched in the early 2020s and has since been used by the US Navy and other military organizations around the world to train submarine crews for a range of scenarios and mission types.

The program has been praised for its realism and effectiveness, providing submarine crews with a safe and controlled training environment that closely mimics the real-world conditions that they would encounter during operations. Project Athena includes a range of training scenarios, from basic operation of the submarine to more complex scenarios such as emergency situations and combat operations. The program also includes a range of customization options, allowing submarine crews to tailor the training experience to their specific needs and requirements.

One of the key advantages of Project Athena is its ability to provide a safe and controlled training environment for submarine crews. Submarine operations are inherently dangerous, with the risk of equipment malfunctions, collisions, and other mishaps. By using the program, submarine crews can practice their skills in a safe and controlled environment, without the risk of injury or damage to equipment.

The program has also been praised for its realism and effectiveness. The VR technology used in Project Athena provides a high degree of immersion and interactivity, allowing submarine crews to practice their skills in a more realistic and practical way. This can help to improve their readiness and preparedness for real-world operations, as well as enhance their ability to work as a team.

Project Athena is a powerful tool that provides submarine crews with a safe and effective training environment. Its advanced features, realism, and customization options make it an essential tool for the US Navy and other military organizations looking to enhance their capabilities and preparedness for a range of scenarios.

6 Challenges

The use of the metaverse in military defense presents a range of potential challenges and risks. These include cybersecurity concerns, ethical considerations, and the need for proper training and support for those using the technology [38]. The use of the metaverse in military defense presents a range of potential challenges and risks that need to be considered and addressed. One of the main challenges is cybersecurity, as any advanced technology is vulnerable to cyber attacks. The metaverse is no exception, and military organizations using the technology need to take steps to secure their systems and protect them from cyber threats.

One of the main challenges associated with the use of the metaverse in military defense is cybersecurity. As with any advanced technology, the metaverse is vulnerable to cyber attacks, which could compromise sensitive military information or disrupt military operations. It is essential that military organizations take steps to secure their metaverse systems and protect them from cyber threats [39]. The use of the metaverse in military defense also raises ethical considerations. The metaverse has the potential to blur the line between reality and simulation, and this raises questions about the ethical implications of using advanced simulation technologies in military operations.

Military organizations need to carefully consider the ethical implications of using the metaverse and ensure that its use is consistent with ethical and legal guidelines. Another potential challenge associated with the use of the metaverse in military defense is the potential for psychological effects on soldiers. The highly immersive and realistic nature of the metaverse could lead to soldiers experiencing psychological stress or trauma from virtual combat scenarios. Military organizations need to consider the potential psychological effects of using the metaverse and provide appropriate support and resources to soldiers who may be affected [40]. The use of the metaverse in military defense also raises concerns about data privacy. The metaverse could be used to collect data on soldiers' performance, behavior, and decision-making processes. Military organizations need to ensure that the collection and use of such data are transparent and comply with relevant laws and regulations. Moreover, the use of the metaverse in military defense could also have geopolitical implications. If certain countries or militaries have access to more advanced and effective metaverse technologies, this could create a power imbalance and potentially lead to increased tensions or conflicts [41]. Another challenge associated with the use of the metaverse in military defense is the potential for technical glitches and system malfunctions. The highly complex nature of metaverse technologies means that they are prone to technical issues, which could compromise the effectiveness and safety of military operations. Military organizations using the metaverse need to have contingency plans in place to address technical issues and ensure that they do not compromise military readiness or operations. In addition to technical challenges, the use of the metaverse in military defense also raises legal and regulatory concerns [42]. The use of advanced simulation technologies in military operations could raise questions about the legality and ethical implications of such practices. Military organizations need to ensure that their use of the metaverse complies with relevant laws and regulations and is consistent with ethical and moral standards.

Another potential challenge associated with the use of the metaverse in military defense is the potential for interoperability issues. The metaverse technologies used by different military organizations may not be compatible with each other, which could lead to communication and coordination problems in joint operations. Military organizations need to ensure that their metaverse technologies are compatible with those of other organizations to ensure effective collaboration and coordination [43]. The use of the metaverse in military defense could also raise questions about the role of human decision-making in military operations. The highly automated and advanced nature of metaverse technologies could potentially lead to a reduction in the role of human decision-making in military operations. Military organizations need to ensure that the use of the metaverse is consistent with ethical and legal standards and that human decision-making remains a key component of military operations. Another potential challenge associated with the use of the metaverse in military defense is the issue of access [44]. The development and implementation of metaverse technologies require significant investment and resources, and this could lead to access issues for smaller military organizations or those with limited resources.

This could potentially create a power imbalance and limit the ability of these organizations to effectively train their soldiers and prepare for real-world operations. The use of the metaverse in military defense could also raise questions about the role of accountability in military operations. The highly advanced and automated nature of metaverse technologies could potentially lead to a lack of accountability for the actions of soldiers and military organizations. Military organizations need to ensure that the use of the metaverse is consistent with legal and ethical standards and that there is appropriate accountability for the actions of soldiers and organizations [45].

The use of the metaverse in military defense also presents challenges in terms of personnel selection and training. The advanced nature of metaverse technologies requires specialized skills and expertise, and military organizations need to ensure that they have the right personnel in place to effectively use the technology. Moreover, military organizations need to provide comprehensive training and support to ensure that those using the metaverse are able to use it effectively and safely.

Another potential challenge associated with the use of the metaverse in military defense is the issue of interoperability with other military technologies [46]. The metaverse technologies used by military organizations may need to be integrated with other technologies, such as communications and surveillance systems. Military organizations need to ensure that their metaverse technologies are compatible with other military technologies to ensure effective collaboration and coordination. Another challenge is the ethical considerations associated with the use of the metaverse in military defense. The metaverse has the potential to blur the line between reality and simulation, raising questions about the ethical implications of using advanced simulation technologies in military operations. It is important for military organizations to carefully consider the ethical implications of using the metaverse and ensure that its use is consistent with ethical and legal guidelines. Proper training and support are also essential for those using the metaverse in military defense. The metaverse is a complex and advanced technology that requires specialized training and expertise. It is important for military organizations to provide comprehensive training and support to ensure that those using the technology are able to use it effectively and safely [47]. The use of the metaverse in military defense has the potential to create new forms of battlefield challenges. For example, the metaverse could be used to create virtual battlefields in which soldiers engage in combat. This raises questions about the safety and effectiveness of using such virtual environments in real-world military operations. There are also potential risks associated with the use of the metaverse in military defense. For example, the metaverse could be used to create highly realistic simulations of military operations, which could be used to train soldiers for real-world combat. However, there is a risk that soldiers could become too reliant on these simulations and that this could lead to a lack of readiness for actual combat situations. The use of the metaverse in military defense presents a range of challenges and risks. It is important for military organizations to carefully consider these issues and take steps to address them, including securing the metaverse systems, considering the ethical implications of using the technology, providing proper training and support, and balancing the use of simulations with real-world training and readiness. The use of the metaverse in military defense could have economic implications. The development and implementation of metaverse technologies require significant investment and resources, and this could lead to financial strain on military organizations [48].

There are several differences between the metaverse and reality that cannot be overcome in the context of military operations. The metaverse is a virtual world that operates within the confines of a computer system, while reality has physical limitations such as terrain, weather, and natural obstacles that can impact military operations [49]. The metaverse cannot replicate these physical limitations in a truly realistic way. The metaverse may be able to simulate human behavior to some extent, but it cannot fully replicate the nuances of human decision-making and emotional responses in high-pressure situations. In reality, soldiers must deal with the stress and uncertainty of combat, which can impact their decision-making and performance. In the metaverse, actions and decisions have no real-world consequences, while in reality, military operations can have significant real-world consequences, including loss of life and damage to infrastructure and property. This can impact the decision-making and behavior of soldiers in ways that cannot be replicated in the metaverse [50]. The metaverse may be able to simulate a wide range of military equipment and weapons, but it cannot fully replicate the limitations and capabilities of real-world equipment. This can impact the effectiveness and safety of military operations. The metaverse may be able to simulate communication systems to some extent, but it cannot fully replicate the complexities and limitations of real-world communication systems in high-pressure situations. This can impact the coordination and effectiveness of military operations [51].

While the metaverse has many potential benefits for military operations, there are several differences between the metaverse and reality that cannot be overcome. Military organizations need to ensure that the use of the metaverse is balanced with real-world training and readiness to ensure that soldiers are fully prepared for real-world combat situations [52].

7 Future Outlook

The use of the metaverse in military applications is a rapidly evolving area that is expected to have significant implications for the future of military operations. While the technology is still in its early stages of development, it has the potential to revolutionize the way that military organizations train their soldiers, plan their operations, and conduct their missions. In this essay, we will explore the future outlook for metaverse technology in military applications, including its potential benefits, challenges, and ethical considerations.

The metaverse in military applications is its ability to provide highly realistic and immersive training simulations for soldiers. The metaverse can simulate a wide range of combat scenarios, allowing soldiers to gain valuable experience and skills in a safe and controlled environment. This can help to reduce the risk of injury and death during real-world military operations, while also improving the effectiveness and efficiency of military training.

In addition to training, the metaverse can also be used to assist in the planning and execution of military operations. The technology can be used to create detailed simulations of operational environments, allowing military planners to test different strategies and tactics before deploying troops into the field. This can help to reduce the risk of errors and improve the success rate of military operations.

Another potential benefit of the metaverse in military applications is its ability to enhance communication and coordination between soldiers and military organizations. The technology can be used to create virtual command centers, allowing military leaders to monitor and direct operations in real-time. This can help to improve the speed and accuracy of decision-making, while also reducing the risk of miscommunication and mistakes. However, the use of the metaverse in military applications also presents a range of challenges and ethical considerations that must be carefully considered and addressed. One of the most significant challenges is the issue of cybersecurity. The metaverse is a highly interconnected and complex system, making it vulnerable to cyber attacks and other forms of digital threats. Military organizations need to ensure that their metaverse technologies are secure and protected from external threats.

One of the challenges associated with the use of the metaverse in military applications is the issue of bias and discrimination. The advanced nature of metaverse technologies could potentially lead to biases and discrimination against certain groups or individuals. Military organizations need to ensure that their use of the metaverse is fair and unbiased and does not discriminate against any group or individual based on factors such as race, gender, or ethnicity [53]. The use of the metaverse in military applications also raises questions about the role of human decision-making in military operations. The highly automated and advanced nature of metaverse technologies could potentially lead to a reduction in the role of human decision-making in military operations. Military organizations need to ensure that the use of the metaverse is consistent with ethical and legal standards and that human decision-making remains a key component of military operations.

Another potential challenge associated with the use of the metaverse in military applications is the issue of interoperability. The metaverse technologies used by different military organizations may have different standards and protocols, which could lead to communication and coordination problems in joint operations. Military organizations need to ensure that their metaverse technologies are standardized and interoperable to ensure effective collaboration and coordination. The use of the metaverse in military applications presents both significant benefits and challenges. While the technology has the potential to revolutionize the way that military organizations train their soldiers, plan their operations, and conduct their missions, it also raises a range of ethical considerations and challenges that must be carefully considered and addressed. Military organizations need to ensure that their use of the metaverse is consistent with ethical and legal standards and that the technology is used in a responsible and effective way. Only by addressing these challenges can the full potential of the metaverse be realized in military applications.

8 Conclusion

The power of the metaverse in military applications lies not just in its technological capabilities, but in our ability to use it in a way that is consistent with our highest values and ethical standards. The use of the metaverse in military applications is a rapidly evolving area that presents both significant benefits and challenges. The technology has the potential to revolutionize the way that military organizations train their soldiers, plan their operations, and conduct their missions [54]. The metaverse can provide highly realistic and immersive training simulations, allowing soldiers to gain valuable experience and skills in a safe and controlled environment. It can also assist in the planning and execution of military operations, creating detailed simulations of operational environments and allowing military leaders to monitor and direct operations in real-time.

The metaverse has the potential to enhance communication and coordination between soldiers and military organizations, creating virtual command centers and improving the speed and accuracy of decision-making. However, the use of the metaverse in military applications also presents a range of challenges and ethical considerations that must be carefully considered and addressed. These challenges include issues of cybersecurity, bias and discrimination, the role of human decision-making, and interoperability [55]. The use of the metaverse in military applications is a complex and multifaceted issue that requires careful consideration of a wide range of factors. One of the most significant factors that must be considered is the social and cultural implications of the technology. The metaverse has the potential to fundamentally change the way that military organizations operate, and this could have important consequences for the social and cultural dynamics of military organizations. For example, the use of the metaverse in military applications could potentially lead to a shift in the balance of power within military organizations. The advanced nature of metaverse technologies could give certain individuals or groups within military organizations a significant advantage over others, potentially leading to power imbalances and conflicts within the organization.

The use of the metaverse in military applications could also raise questions about the relationship between soldiers and their commanders. The highly immersive and realistic nature of metaverse simulations could potentially lead to soldiers becoming too reliant on simulations and not adequately prepared for real-world combat situations. This could impact the relationship between soldiers and their commanders, potentially leading to a breakdown in trust and confidence. Another potential social and cultural implication of the use of the metaverse in military applications is the impact on the broader military culture. The use of the metaverse could potentially change the way that military organizations approach training, planning, and execution of operations. This could impact the broader military culture, potentially leading to changes in the way that military organizations view and approach their missions. The use of the metaverse in military applications also raises important ethical considerations. Military organizations need to ensure that their use of the metaverse is consistent with ethical and legal standards and that the technology is used in a way that is fair, unbiased, and non-discriminatory.

The need to ensure that the use of the metaverse does not lead to a reduction in the role of human decision-making in military operations. The highly automated and advanced nature of metaverse technologies could potentially lead to a reduction in the role of human decision-making, raising important ethical questions about the use of such technologies in military operations.

Military organizations need to ensure that their use of the metaverse is consistent with ethical and legal standards and that the technology is used in a responsible and effective way. They need to address the challenges that the use of the metaverse presents, including data privacy and security, access, accountability, personnel selection and training, interoperability, fairness and non-discrimination, cultural sensitivity, international cooperation, technical reliability, legal and ethical compliance, and social and cultural implications [56]. It is clear that the metaverse has the potential to significantly enhance the capabilities and effectiveness of military organizations.

However, the technology must be used in a responsible and ethical manner, with careful consideration given to its potential risks and challenges. Military organizations need to ensure that they are prepared to address these challenges and to use the metaverse in a way that is consistent with the highest standards of ethical and legal conduct. Only by doing so can the full potential of the metaverse be realized in military applications.

References

1. Lasserre S (2022) "4 use cases for virtual reality in the military and defense industry. TechViz. Blog AR/VR for engineering, 2 November. https://blog.techviz.net/4-use-cases-for-virtual-reality-in-the-military-and-defense-industry. Accessed 12 May 2023
2. Jeffrey C (2018) Founder of project Maven 'alarmed' at Google's decision to walk away. TechSpot, 26 June. https://www.techspot.com/news/75257-founder-project-maven-alarmed-google-decision-walk-away.html. Accessed 12 May 2023
3. VBS3—PAXsims (2017) PAXsims. https://paxsims.wordpress.com/tag/vbs3/. Accessed 12 May 2023
4. Holger D (2017) The Australian air force is now testing the Microsoft holoLens. VRScout, 23 January. https://vrscout.com/news/the-australian-air-force-is-now-testing-the-microsoft-hololens/. Accessed 12 May 2023
5. Steiner JG, Geer DE (1988) Network services in the Athena environment. In: Proceedings of the Winter 1988 Usenix Conference, 21 July
6. Metaverse in military—military metaverse development. Antier Solutions, 19 September 2022. https://www.antiersolutions.com/why-the-military-needs-a-metaverse/. Accessed 12 May 2023
7. Wang Y, Liu C (2022) Metaverse and the military: opportunities and challenges. In: International Conference on Intelligent Computing. Springer, pp 131–144
8. Beasley M, Yan Z (2022) A survey on the applications of metaverse in military defense. In: 2022 IEEE International Conference on Artificial Intelligence and Computer Applications (ICAICA), pp 118–122. IEEE. https://doi.org/10.48550/arXiv.2210.07990
9. Porter M, Heppelmann J (2018) Whiteboard session: "why every organization needs an AR strategy?, July 16
10. Scheutz M (2018) Enterprise augmented reality success with ThingWorx Studio. ARiA (AR in ACTION), February
11. Research and Markets (2018) Report: "Military augmented reality market to2025—global analysis and forecasts by components, product type & functions, February
12. Dodevska Z, Putnik G (2018) A young researcher's view of augmented reality based on quantitative analysis of articles at Google Scholar in the last 30 years. In: 7th International Symposium on Industrial Engineering—SIE 2018 (Accepted for publication)
13. Kipper G, Rampolla J (2013) Augmented reality: an emerging technologies guide to AR. Elsevier, Waltham, MA
14. Carmigniani J, Furht B (2011) Augmented reality: an overview. In: Handbook of augmented reality. Springer, New York, pp 3–46
15. Craig AB (2013) Understanding augmented reality: concepts and applications. Elsevier, Waltham, MA. https://doi.org/10.1016/C2011-0-07249-6
16. You X, Zhang W, Ma M, Deng C, Yang J (2018) Survey on urban warfare augmented reality. ISPRS Int J Geo Inf 7(2):46. https://doi.org/10.3390/ijgi7020046
17. Defense Advanced Research Projects Agency of USA (U.S. Department of Defense, State of Washington, USA) (2008) Urban leader tactical response, awareness visualization (ULTRA-Vis). Defense Advanced Research Projects Agency of USA, Arlington, VA, USA

18. Argenta C, Murphy A, Hinton J, Cook J, Sherrill T, Snarski S (2010) Graphical user interface concepts for tactical augmented reality. Proceedings of the SPIE 7688:76880I
19. Julier S, Baillot Y, Lanzagorta M, Brown D, Rosenblum L (2000) BARS: battlefield augmented reality system. In: Paper presented at the RTO IST Symposium on "New Information Processing Techniques for Military Systems" held in Istanbul, Turkey, 9–11 October 2000, and published in RTO MP-049
20. Azuma R, Baillot Y, Behringer R, Feiner S, Julier S, MacIntyre B (2001) Recent advances in augmented reality. IEEE Comput Graphics Appl 21(6):34–37. https://doi.org/10.1109/38.963459
21. General Dynamics (2018) Whitepaper: "augmented reality on the battlefield", February
22. Livingston MA, Rosenblum LJ, Julier SJ, Brown D, Baillot Y, Swan JE, Gabbard JL, Hix D (2002) An augmented reality system for military operations in urban terrain. In: Proceedings of the Interservice / Industry Training, Simulation, & Education Conference (I/ITSEC '02), Orlando, FL, December 2–5
23. Livingston MA, Rosenblum LJ, Brown DG, Schmidt GS, Julier SJ, Baillot Y, Swan JE, Ai Z, Maassel P (2011) Military applications of augmented reality. In: Handbook of augmented reality. Springer, New York, pp 671–706
24. Roberts D, Menozzi A, Clipp B, Menozzi A, Clipp B, Russler P (2013) Soldier-worn augmented reality system for tactical icon visualization. In: Head- and Helmet-Mounted Displays XVII; and Display Technologies and Applications for Defense. SPIE 2012, 8383
25. Livingston MA, Swan JE, Simon JJ (2004) Evaluating system capabilities and user performance in the battlefield augmented reality system. In: Proceedings of the NIST/DARPA Workshop on Performance Metrics for Intelligent Systems, Gaithersburg, MD, USA, 24–26 August
26. Colbert HJ, Tack DW, Bossi LCL (2005) Augmented reality for battlefield awareness; DRDC Toronto CR-2005-053; HumanSystems® Incorporated, Guelph, ON, Canada
27. Le Roux W (2010) The use of augmented reality in command and control situation awareness. Sci Mil S Afr J Mil Stud 38:115–133
28. Zysk T, Luce J, Cunningham J (2012) Augmented reality for precision navigation-enhancing performance in high-stress operations. GPS World 23:47
29. Kenny RJ (2015) Augmented reality at the tactical and operational levels of war; naval war college Newport United States: Newport. RI, USA
30. Julier S, Baillot Y, Lanzagorta M (2000) BARS: battlefield augmented reality system. In: Nato Symposium on Information Processing Techniques for Military Systems. In Proceedings of the NATO Symposium on Information Processing Techniques for Military Systems, Istanbul, Turkey, 9–11 October, pp 9–11
31. Gans E, Roberts D, Bennett M (2015) Augmented reality technology for day/night situational awareness for the dismounted Soldier. In: Proc SPIE 2015, 9470, 947004:1–947004:11
32. Cameron A (2009) The application of holographic optical waveguide technology to the Q-SightTM family of helmet-mounted displays. Proceedings of the SPIE 7326:73260H
33. Konstantopoulos D, Johnston J (2006) Data schemas for net-centric situational awareness. CCRTS
34. Youzhao G, Xueting J, Yingru Y, Qiang W, Linyuan L (2023) Construction and prospect of metacosmic technology system. J Univ Electronic Sci Tech 3
35. Wenxi W, Fang Z, Yue W, Huansheng N (2022) Summary of metacosmic technology 10
36. Xuemei L (2022) Application of blockchain technology in the development of Metaverse. Communication Info Tech
37. Kunliang C (2022) Interactive technology that can be used in the metaverse. Industrial Control Computer
38. Qiang W, Xueting J, Linyuan L (2022) Artificial intelligence technology and application in metaverse. J Intell Sci Tech
39. Danwen Z (2022) On the metaverse and its technical basis. China Media Technology
40. Ye Y, Erwei Y, Haoyang Z, Wei Q (2022) Exploration of the military application potential of the metaverse. J Command Control

41. Weidong S, Ning M, Qingliang M, Yue P, Yunxin L (2022) Exploration on the application of military simulation in meta-space. Guidance and Fuze
42. Kaiyue Z, Jiechun L (2022) Military training and the metaverse: opportunities and challenges coexist. Military Digest
43. Shengjie H, Mingke X, Jinghao L, Kaicheng L, Jingcong Y, Qi X (2022) Military equipment digital asset management architecture in the metaverse. J Comm Cont
44. Jiabao W, Linlin R (2022) Wargame deduction: simulating the "battlefield Metaverse" of future war. In: China's conversion from military to civilian
45. Boya J, Wenhao S (2022) Opportunities and challenges of military network propaganda in the era of the yuan universe. In: Network information military-civilian integration
46. Jiahui Y, Yuxiang S, Qi X, Xinlei Z, Xianzhong Z (2022) The Metaverse empowers command and control: the operational deduction of the fusion of virtual and real in the future. J Command Cont
47. Lele A (2013) Virtual reality and its military utility. J Ambient Intell Humaniz Comput 4(1):17–26
48. Luan XD, Xie YX, Ling-Da WU et al (2003) Application of virtual reality in military affairs. Acta Simulata Systematica Sinica
49. Luo SY(2016) The research about military application of virtual reality technology. Computer Knowledge Technology
50. Xing AN, Gang LI, Linwei XU et al (2011) A survey on application of virtual reality technology in U.S. military simulation training. Electronics Optics Control
51. Zhang M, Xu L, Yu W et al (2017) Research on high-tech ammunition training system based on virtual reality technology, 01012
52. Weiss PL, Jessel AS (1998) Virtual reality applications to work. Work 11(3):277–293
53. Anderson F, Annett M, Bischof WF (2010) Lean on Wii: physical rehabilitation with virtual reality Wii peripherals. Stud Heal Tech Inform 154(154):229
54. Satava RM, Jones SB (1998) Current and future applications of virtual reality for medicine. Proc IEEE 86(3):484–489
55. Yin Y (2010) Application of virtual reality in marine search and rescue simulator. Int J Virtual Reality
56. Deng HY, Fang WU, Yin C (2002) Virtual reality geographic information system (VRGIS)—a new field of the research of GIS. Appl Res Comp 19(9):33–35

Futuristic Blockchain Applications of the Metaverse

Nitesh Sureja and Heli Sureja

Abstract The following Facebook's official rebranding to Meta in October 2021, the metaverse has taken center stage as the de facto norm for virtual 3D environments and online social platforms. The metaverse aspires to deliver fascinating 3D and customized experiences to consumers by utilizing a number of practical technologies. Users' digital data and material security in the metaverse is an understandable concern, despite the many benefits and the general interest in the subject. Blockchain technology offers a potential answer due to its decentralization, transparency, and immutability. We hope that by providing a comprehensive overview of blockchain applications for the metaverse, we will help you better understand blockchain's role in this space. We start by outlining the basics of blockchain technology, the metaverse, and why it might be useful for the metaverse. Subsequently, we delve deeply into the technical elements of metaverse solutions based on blockchain, covering topics such as privacy protection, storage, data collection, interoperability, and sharing. We first describe the metaverse's technical challenges from each perspective, and then we show how blockchain may support. Furthermore, we examine the potential effects of blockchain on key metaverse enablers such as big data, multisensory and immersive applications, artificial intelligence, Internet of Things, and digital twins. Along with introducing some noteworthy initiatives, we aim to highlight the use of blockchain in the services and applications of the metaverse. Finally, we conclude by outlining several promising avenues for further research, development, and innovation regarding blockchain's application in the metaverse.

Keywords Metaverse · Blockchain · Privacy · Internet of things · Big data

N. Sureja (✉)
Department of Computer Science and Engineering, KSET, KPGU, Vadodara, Gujarat, India
e-mail: dir.kset@kpgu.ac.in

H. Sureja
Department of Computer Science and Engineering, BIT, GTU, Vadodara, Gujarat, India

G. Chhabra and K. Kaushik (eds.), *Understanding the Metaverse*, Blockchain Technologies, https://doi.org/10.1007/978-981-97-2278-5_10

1 Introduction

By expanding the range of services available beyond conventional systems with Internet connectivity, the metaverse has the potential to considerably accelerate the rate of digital adoption, making it the natural progression of digital technology. The digitization of services is a relatively new phenomenon that has emerged in the last several decades. Its goal is to boost efficiency in every sector that can be linked to the Internet, including commerce, leisure, education, and any system that can support it. These services and technologies were able to reach their full potential, thanks to digital systems, online processing and storage capabilities at remote data centers, and cloud platforms. Now that service availability, performance, and efficiency are at an all-time high, attention is shifting to the customer experience. Consequently, there is a growing desire for better and more engaging customer support, and service providers are eager to improve upon current standards. The realism and haptic aspects offered by quickly developing technologies, like virtual reality (VR), augmented reality (AR), mixed reality (MR), and extended reality (XR) are meeting consumer demand for computer interfaces [1]. To bring these vital technologies together, the metaverse is the perfect global solution. This idea allows prosumers to fully immerse themselves in an artificially manufactured setting. By utilizing their digital personalities, users can engage with this virtual ecosystem in accordance with the duality principle [2]. Specifically, in the metaverse, users' digital representations, known as avatars, have the same complete legal status as real-life persons. All transactions that occur in the virtual world are the avatar's responsibility, and they cannot be undone. With the minimum capacity, the content is accessible to everyone with a virtual reality or augmented reality immersive device, such as a headset or glasses [3]. But full-body haptic bodysuits like the Holosuit or Teslasuit may overcome biometrics, track motion and extract haptic information to immerse the user in the experience to the fullest.

Although the metaverse was first designed to enhance the capabilities of social media, it has tremendous promise for various commercial, industrial, social, educational, medical, military, and political sectors. Online remote access and control systems are notorious for their lack of immersion, which is a major drawback, especially when controlling remote automation systems based on SCADA or PLC [4], when trying on clothes, when seeing commercial real estate or architectural plans, when learning about medicine, engineering, or architecture, when controlling unmanned aerial, naval, or ground vehicles from a distance, and when enjoying three-dimensional digital entertainment. Virtual reality (VR) and augmented reality (AR) each offered solutions to these issues on their own, but no combined platform or environment existed. The metaverse makes this digital ecosystem accessible to everyone, increasing the variety of possible results. Industrial and military organizations can benefit by taking help of digital twins (DTs) since it improves visualization and coordination for remote operation and controlling of machinery or vehicles [5]. Due to its ability to enhance context comprehension and accuracy, three-dimensional visualization finds utility in educational and entertainment applications. Here are the survey's findings. In addition, there will always be compatibility, interoperability, legal, and ethical challenges when trying to put ideas like digital biometrics [6], explainable artificial intelligence (XAI) [7], and cryptocurrencies [8] into practice. Integrating these strategies throughout the planning phase will enhance the service experience of consumers by providing improved privacy and security assurance.

The metaverse has been touted as the answer to the problem of digital growth in the future, yet there are real worries and problems with it. Without a fully operational digital infrastructure, the promised services and applications will not be able to take advantage of the necessary processing and networking resources. Despite the existence of such infrastructure, the specified access methods are currently only feasible with the still-emerging 5G mobile technology, which is not yet extensively used and is still in the experimental stages. The standards for compatibility and interoperability between the real and virtual worlds need to be established before the metaverse can be implemented. The enormous processing power of the metaverse engines is obviously not going to be enough to satisfy demand, especially considering the potential and scalability of that need in light of the social media infrastructure. It is essential to employ optimal processing and operating techniques to lessen processing, storage, networking, and monetary expenditures. Such strategies necessitate more research and incorporation into the area since they are exclusively applicable to automated AI-based procedures. The increased personal investment is a benefit in place of a universally accessible system as, at the very least, one needs a headset or augmented reality glasses to enter the metaverse. Furthermore, it is critical to take user security and privacy into account. Notable biometrics from the physical world can be translated into the digital realm, even though certain privacy regulations that are relevant in the physical world may not be applicable in the virtual one. Therefore, before the metaverse can be employed in a practical setting, a lot more research and suitable standardization are required.

If we want to encourage the widespread use of virtual entities and their ultra-realistic integration, we need to improve current augmented and mixed reality technologies to the point where they enable XR. Artificial intelligence is essential for automating the metaverse ecosystem and fully delegating power to digital governance, as mentioned earlier. Users will have even greater confidence in the security of their digital assets and content associated with their avatars when AI is used. Improvements in image/video/3D rendering technology, together with the incorporation of artificial intelligence into existing computer vision processing, will allow for faster query processing of visual and telemetry data, and better 3D image processing overall. The use of XAI techniques during the design phase guarantees that the product is compatible with all devices. Since the current processing and storage infrastructure based on cloud computing lacks the networking capability, edge computing—a new paradigm—is crucial for hosting metaverse applications. Additionally, edge computing enables features that are context and location aware, and greater access capacity due to their close proximity. Network slicing, in conjunction with the eight enablers listed in [2], can also be employed to arrange and construct the flow of applications in the metaverse.

After finding its first use with the bitcoin cryptocurrency, blockchain quickly gained notoriety for its exceptional capacity to create a shared economy and lay the groundwork for the current digital currency market. When it comes to protecting personal information, some see blockchain technology as a huge step forward [9]. "Blockchain" is short for "distributed ledger technology," which might dramatically improve the security and management of a company's digital assets. By linking the recorded transactions or records into blocks, cryptographic methods—more especially, hashing algorithms—secure the ledgers. This makes it possible to share securely even in untrusted environments. The ability to handle content on decentralized ledgers removing central authorization, is a defining feature of blockchain [10]. Blockchain maintains that it is more secure and appropriate for e-commerce platforms because to its reliance on proof of work as its consensus mechanism. Blockchain technology is a crucial facilitator for the metaverse's digital environment since it guarantees accountability and transparency.

The imminent requirement to secure the digital content that each user of the metaverse holds is the principal application for blockchain technology. Users' privacy, security, and authenticity are guaranteed in the metaverse ecosystem through the use of blockchain technology, which records all transactions and information. Here, we take a look at how the data, network, consensus, incentive, contract, and application layers of blockchain could interact with the metaverse. Although the authors do mention four uses made possible by blockchain, their primary emphasis is on the technology's commercial utility. The use and reuse of very rich good quality data, maintaining the decentralized network's stability, assured data privacy, and handling of data related to the economy are only a few of the numerous benefits of blockchain and AI that are briefly discussed in [11] as they pertain to the metaverse. Blockchain technology is an essential part of the metaverse, although it hasn't been well-investigated in other studies [12–14].

We are unaware of any other studies that have investigated blockchain's potential uses in the metaverse. As a result, our chapter presents the effects of blockchain on enabling technologies and a number of possible metaverse applications that could benefit from blockchain integration.

Here are the key points from this chapter:

1. First, we provide a brief overview of blockchain technology and its role in the metaverse. We then go on to explain why it was included.
2. Furthermore, we cover how blockchain can be used to solve problems with privacy preservation, interoperability, data acquisition, sharing, and storage, in the metaverse.
3. As a final note on this chapter, we provide some possible directions for further research.

The remainder of this chapter will follow this format. Blockchain technology, the metaverse, and its inner workings are introduced in Sect. 2. Section 3 of the chapter delves into the technical uses of blockchain in the metaverse. Several possible directions for further research are outlined in Sect. 4, which concludes the chapter.

2 Metaverse and Blockchain: Foundations

The function of blockchain in the metaverse is presented in this section after a brief introduction to metaverse and the blockchain.

2.1 The Metaverse Foundations

What exactly is the metaverse?

First appearing in Greek as the prefix "meta," meaning "more comprehensive" or "transcending," is where the English term "meta" originated. An abbreviation for "universe," which means a container of space and time, the word "Verse" is used. The combination of these two words gives rise to the term "Metaverse," which describes a new digital living environment where traditional social systems undergo a transformation. Using today's technology, we can build the metaverse, a parallel realm that mirrors our real world [13]. Blockchain, digital twins, virtual reality (VR), and other similar technologies are all part of this category.

People in the metaverse can do things like live, work, and play together regardless of their physical location. In 1992, with his popular science fiction novel Snow Crash, Neal Stephenson initially proposed the concept of the metaverse [15]. In this world, individuals utilize computer avatars to compete and manipulate one another for social advancement. Having said that, the metaverse is in its early stages and lacks both practical applications and widely accepted standards.

What are the metaverse's supporting technologies?

The metaverse is a convergence of numerous cutting-edge technologies, including 6G, AI, VR, and digital siblings. The following key tools are necessary in the metaverse:

- Developing the metaverse relies heavily on extended reality technologies, such as augmented and virtual reality.
- Virtual reality allows users to immerse themselves in a digital world, while augmented reality makes it possible to superimpose digital information on top of the actual world [16]. Both of these strategies are crucial to the developing of the metaverse, an online environment where people can simulate interactions in the actual world.
- The other significant technology is the digital twin (DT), which uses real-world data to predict how an object will behave in the actual world and then uses that information to create a virtual replica of the thing [5, 17, 18]. We can mirror the real world in a virtual one easily through the digital twin in the metaverse. In a similar vein, the metaverse may provide some early solutions to the real-world mysteries.

- In the metaverse, blockchain technology—the third—performs two vital purposes. Users are able to store information anywhere in the metaverse using blockchain technology. However, blockchain technology can provide a comprehensive economic system that connects the virtual and physical worlds of the metaverse. Of particular note is the fact that the aforementioned NFTs make it possible to materialize digital goods. Just like with real-world items, users can trade virtual ones. The blockchain establishes a link between the both actual and virtual worlds [13].

The metaverse relies on faster transmission rates, which could be achieved by enhancing the computation, sensing, location, and communication capabilities of upcoming 6G wireless systems. Many fields, including gaming, education and entertainment, and engineering can benefit from participants obtaining immersive experiences in the metaverse without significant latency [19].

What are the metaverse's applications?

The following list includes some of the metaverse's most used uses.

- **Video Conferences and Telecommuting**: Many small firms are surviving the strange COVID-19 epidemic climate through Internet video conferences and telecommuting. However, 70% of an individual's expression is nonverbal, therefore face-to-face contact is crucial. Problems with telecommuting include delayed interaction, poorly understood feedback, and inefficient cooperation, in contrast to traditional face-to-face teamwork [20]. On the other hand, users can navigate and transact with a sociable avatar in the metaverse. Using nonverbal cues like eye contact and body language to convey different points of view to remote workers would greatly enhance the telecommuting experience.
- **Digital Real Estate**: Typically, real estate encompasses land and buildings that can be utilized for building, living, investing, renting, selling, or acquiring. It is also feasible to do the things listed above in the metaverse. And just like in the real world, factors like location, amenities, and accessibility to public transportation all play a role in determining house prices online. In the metaverse, users can do things like organize art exhibitions, concerts, and video game tournaments, as well as buy and sell houses to the public [21].
- **Digital Arts**: The creation of 3D images in digital arts has typically involved the use of modeling software such as ZBrush and Maya. But the metaverse is all about the display layer, so we can create figures directly and explore new forms of artistic expression. One positive development is the growing interest in digital art forms, particularly AI painting, which is bringing it into the public eye. Conversely, freshly developed blockchain technology has also digitalized previously offline artworks. Users are able to view the virtual gallery from any angle because it is hosted in the metaverse [22].

2.2 Blockchain Foundations

In 2008, the concept of blockchain was initially put forth by Nakamoto Satoshi in a white paper [23]. The distributed ledger, or blockchain, is characterized by its ability to store transactions in consecutive blocks. The blocks are linked to each other through the hash values assigned to each blocks.

Along with the inevitable cryptographic hash, a block also includes transaction data, a nonce, and a date [24]. The timestamp of the block decides the genuinity of the block so that potential adversaries cannot manipulate the blockchain. Remember that network adjusted time is the mean of all the timestamps from all the nodes that are linked. For the blockchain to function properly, every node in the network must follow a standard consensus protocol while creating and verifying blocks. This proves that the blame does not lie with a single or small number of nodes. At its heart, blockchain technology rests on the consensus system, which regulates both legal conduct and overarching ideals [25]. One of the most famous algorithms used by Bitcoin, known as Proof of Work (PoW), is one that asks miners to pool their processing resources to solve a seemingly arbitrary mathematical problem [26]. To avoid a concentration of processing power, the difficulty—also called the nonce of the next block generation—is changed dynamically every 10 min. While PoW does reduce the processing power of most attackers, it also causes substantial energy consumption and a less transaction rate.

Proof of Stake (PoS) solves the problems that Proof of Work (PoW) introduced; so, the winning miner is decided by their bitcoin holdings rather than their processing power [27]. Participants are required to provide storage space in order to prove a challenge presented by the service provider, as part of the Proof of Space (PoSpace) consensus that is being spread via the recently established interplanetary file system (IPFS) [28]. The transaction data in each block is organized in a Merkle tree, which makes the authentication procedure more effective. Be aware that users can still access any branch of the Merkle tree for inspection purposes, even in the absence of full transaction data. Both the aforementioned fundamental blockchain components and the blockchain's transaction processing are illustrated in Figs. 1 and 2, respectively.

Since bitcoin just decentralizes transaction records, it exemplifies a first-generation blockchain. Later on, professionals found out that blockchain can manage assets and family trusts in addition to being a ledger. The outcome was the second version of the blockchain that was based on Ethereum.

Ethereum mainly introduced the concept of smart contracts [29]. The rules for smart contracts are stored in code on the blockchain. A smart contract's corresponding function can be triggered by a transaction to automatically transfer funds or send a notification to a specified account. The proliferation of apps is directly proportional to the accessibility of the smart contract. To illustrate the point, smart contracts enhance the security of the voting process, making it harder for hackers to manipulate and understand. Furthermore, insurance firms are coordinating with hospitals to record patient data on the blockchain, which will allow businesses to quickly access their

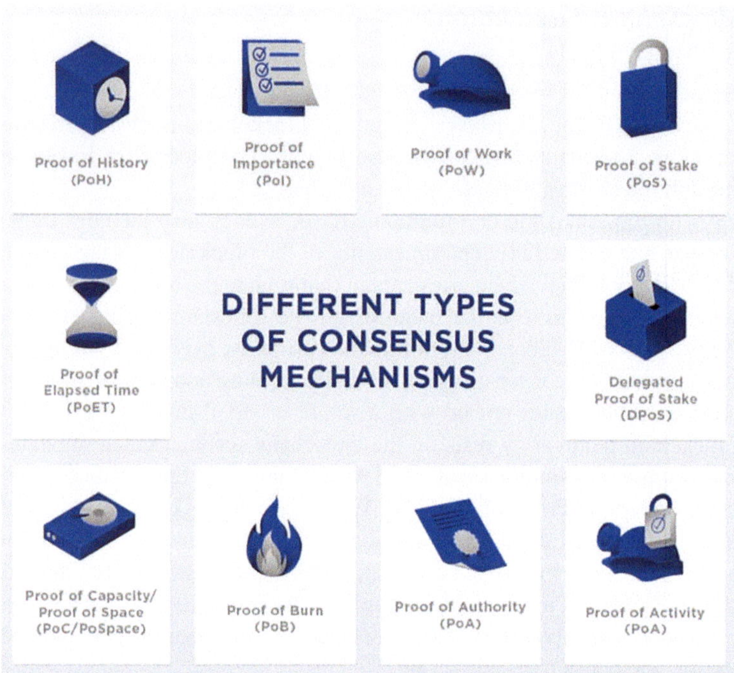

Fig. 1 Consensus mechanisms [31]

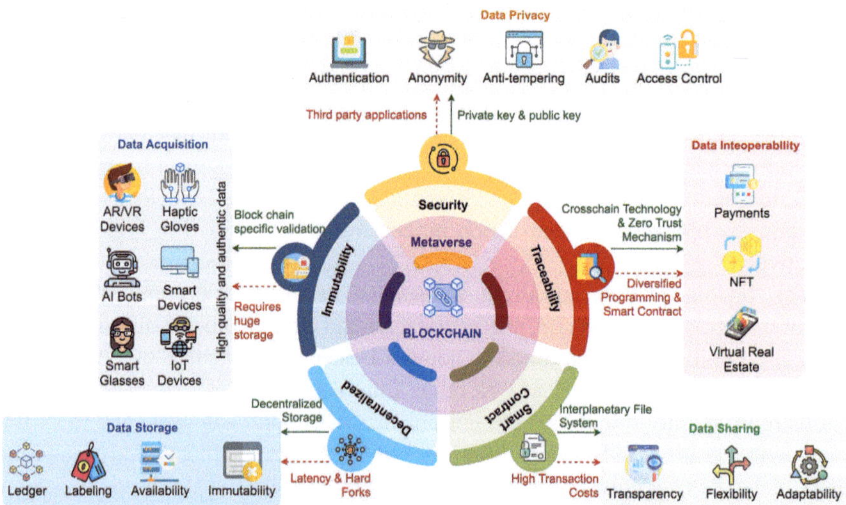

Fig. 2 Use of blockchain for metaverse technical elements [38]

funds using smart contracts. NFTs, or non-fungible tokens, are a variant of smart contracts that have lately been popular around the world. To define the features of NFT-based assets, Ethereum has published two specifications, ERC-721 and ERC-1155. In contrast to Ethereum and bitcoin, each NFT is non-transferable and cannot be divided. Numerous sectors, including copyright, insurance, sports, games, and the arts contribute to the NFT market's current valuation of over $7 billion [30].

Popular consensus methods

A consensus mechanism in blockchain is a fault-tolerant method used to reach the necessary consensus across decentralized nodes on a single data value or a single network state. In addition to being described in this subsection, Fig. 1 also shows the common consensus processes.

- **Proof of Work**: The majority of cryptocurrency networks prefer the Proof of Work consensus method [32]. This method is used to add a new block to the database and confirming the transaction. In the 2008, it is utilized in bitcoin essay. The term "proof of work" was initially used in a 1999 publication by Ari Juels and Markus Jakobsson. After adding a new block to the blockchain, the transaction is validated via the Proof-of-Work consensus mechanism. This approach uses a group of people called "minors" to compete for the completion of the network transaction. Mining is essentially a game of competition. Successfully creating a legitimate block is what pays miners. Among PoW implementations, bitcoin is the most famous. It's possible that PoW manufacturing is a low-probability stochastic process. A trustworthy evidence of labor is the result of extensive trial and error. The foundation for the evidence of work is a mathematical puzzle with a clearly demonstrable solution. Implementing PoW on a blockchain is possible with the Hashcash PoW technique [32].
- **Proof-of-Stake (PoS)**: The concept of Proof-of-Stake (PoS) was initially discussed at "Bitcoin-Talk" forum in 2011. Before anyone else, in 2012, Nadal and King introduced PoS to address the issue of extreme use of energy at PoW mining's. In 2012, Peercoin is introduced as the first working PoS cryptocurrency. Then came other digital currencies such as Algorand, Cardano, Nxt, and Blackcoin.
- **Proof of Work (PoW)**: The consensus algorithm known as Proof of Work (PoW) is utilized by the bitcoin and Ethereum networks. In this case, miners (also called block adders) need to change the block's nonce and then execute complicated mathematical operations to find the proper hash. The miner whose hash is found to be less challenging than the required level gets the chance to add his block to the network. Consequently, takes the award. It uses robust computation to reach a consensus in a way that is conducive to puzzles. After that, the miner's existing network users legitimately transacted in the block.
- **Proof of Stake (PoS)**: PoS consensus eliminates the high-energy demand of PoW, which is known as Proof of Stake. In a proof-of-stake (PoS) network, validators, also known as miners, are selected to add blocks by staking a portion of their obtained currency. It wasn't the first algorithm for reaching a consensus in a

network. Implementation is contingent upon the network having an adequate number of nodes.

- **Delegated Proof of Stake (DPoS)**: Voting for representatives is an aspect of delegated Proof of Stake (DPoS) that improves the PoS method. Users of the network choose the reliable delegates (or miners) by voting with their coins. Then, at random, one delegate will have the chance to add their block.
- **Proof of Importance (PoI)**: One block harvester is selected from all participants according to their important evaluations in the Proof of Importance (PoI) process. Its goal is to eliminate privileged access to PoS consensus for the wealthy. An individual's significance level is influenced by their reputation within the network and the caliber of their transactions.
- **Proof of Capability (PoC)**: The Proof of Capability (PoC) function in a decentralized network involves mining a block utilizing the storage capacity or disk. The computation factor is exchanged for disk room. As a prerequisite to actual mining, the proof-of-concept instructs miners to catalogue all possible nonces and block hashes. Currently, all a miner does is upload files with potential hashes that have been calculated to the network. Proof-of-concept transactions are added and validated more quickly.
- **Proof of Time Spent**: The PoET method is based on the concepts of time-lottery. The amount of time miners must wait is distributed at random. While the miner node is waiting, the first node to wake up, also known as the node with a low waiting time, gets to contribute its block to the network. Once the block has been validated by the network validators, it is inserted.
- **Evidence of Activity (PoA)**: PoA combines PoW and PoS approaches. Prior to adding a header, re-ward address, and an empty block to the network, miners must complete the difficult computation. After then, each player's currency balance is utilized to choose an empty block. After then, the miner of that partial block might choose to include their transactions in the partial block. Network validators also verify the transactions.
- **Proof of Authority (PoA)**: The Proof of Authority (PoA) agreement mechanism is utilized by private or authorized blockchain networks as evidence of authorship. Important to Proof of Authority (PoA) is the picture of the network user (miner) who wants to add a new block of transactions. Here, miners bet their reputations rather than their wallets.
- **Proof of Burn (PoB)**: To enable miners to contribute their block, Proof of Burn (PoB) allows them to move some of their funds to an unavailable account. In cryptocurrency, "burning the coins" is moving your obtained funds to an escrow account. The burned coins could no longer be used in regular transactions after PoB burned them. Consequently, not even their owner can spend them. The more coins a miner burns, the more likely it is that the network will accept his fresh block of transactions. By destroying coins, the miner has the ability to mine digitally.
- **Byzantine Fault Tolerance (BFT)**: The goal of Byzantine Fault Tolerance (BFT) is to resolve the dilemma that plagued the Byzantine generals while making decisions. The foundation of this is the communication problem that generals of different armies may face while deciding to advance or withdraw at the same

time. Metadata, digital signatures, and hashes are the tools used by the BFT mechanism to regulate communication between nodes. Decentralized network components can be synchronized with its help.

2.3 Blockchain's Role in the Metaverse

When we think about the metaverse, our minds might go to all sorts of exciting adventures and fun activities.

In a parallel universe where ecological ecology is inevitable, the only scenario that closely resembles our own is this one. Furthermore, the main services blockchain provides are digital assets, including nonhomogenized and homogenized tokens based on ERC-1155, ERC-721, and ERC-20. Because it can maintain the seamless operation of the metaverse's economy, blockchain technology is the metaverse's lifeblood. By facilitating the exchange of virtual products for actual world currency, blockchain empowers metaverse users to engage with the cryptoeconomy.

The blockchain has made it possible for digital currencies like Bitcoin to transact in the metaverse. All virtual worlds have the ability to transfer and use assets, including money, to the metaverse. Users in the crypto-enabled metaverse can purchase virtual goods, take virtual trips, and even attend virtual performances. Blockchain has the potential to turn metaverse into a decentralized, open-source public platform. The metaverse, an open-source blockchain architecture, will give decentralized wallets and exchanges an intuitive user experience.

Below are the reasons why blockchain technology is being used in the metaverse.

- **Safeguarding Personal Information**: In the metaverse, a large amount of personal data is stored to ensure that individuals have the greatest possible experiences. In order to develop effective targeting systems, these firms or programs need this information. If the wrong people get their hands on the data, they might potentially go after users in the real world as well. Blockchain technology safeguards users' private information by granting them complete control over their data through its access control, authentication, and consensus methods. Protected metaverse data is a feature of the blockchain thanks to hash algorithms and asymmetric key encryption.
- **Enabling Data Integrity**: The integrity of the data is guaranteed by the metaverse, which receives data from many applications like healthcare, financial operations, entertainment, and many more. The production of items in the metaverse is greatly affected by the quality of the real-world data contributed by users. With blockchain's detailed audit trails of all transactions, individuals and businesses can verify every single one [33, 34]. The outcome will be an improvement in the data fidelity of the metaverse.
- **Enabling Safe and Effortless Information Exchange**: The metaverse makes use of augmented and virtual reality technologies to build a more linked and lifelike setting. When combined with augmented reality, the metaverse's genuine value becomes apparent for both digital and real-world items. For the metaverse to be

a success, augmented and virtual reality data must be easily shared. Only then can innovative, state-of-the-art apps be developed to aid with real-world problem-solving. Safe and easy data transfer across the metaverse is made possible by the blockchain's cutting-edge encoding information technology.

- **Encouraging Data Interoperability**: Metaverse users need to be able to access and maintain assets in different virtual worlds, as well as use different apps. Data interoperability is limited across these virtual worlds because of the diverse circumstances in which they are developed. A cross-chain protocol is used to transfer the data across the blockchains positioned separately. It makes easier for consumers to move across diverse virtual environments.
- **Data Security**: Data in the metaverse needs to be regularly and accurately updated to keep it secure. The metaverse data is securely stored in every block of the blockchain, thanks to its immutability. No changes or deletions can occur without the consent of the majority of participants [35]. This blockchain system ensures the metaverse's data integrity.

Here are a few ways that blockchain technology can be used in the metaverse.

- **Economic System**: Blockchain's four essential features in an economic system are decentralization, immutability, transparency, and disclosure. It is crucial to guarantee the efficacy and safety of the millions of commodity exchange transactions that take place in the metaverse every second [36]. Cryptocurrencies built on the blockchain have all the makings of an ideal building block for a massively scalable virtual economy, according to the qualities already mentioned. Cryptocurrencies built on the blockchain allow for decentralized, peer-to-peer transactions in the metaverse. Fast, simple, and cheap online shopping is now possible in the metaverse with cryptocurrency. The fact that not every metaverse uses the same coin makes the services of cryptocurrency platforms such as Binance and Coinbase indispensable.
- **Smart Contract Deployment**: The blockchain allows smart contracts to be programmable, verifiable open, automated, and visible. Because of this, a third-party verification platform is no longer necessary for trustworthy interactions to take place on the blockchain. The metaverse's financial system can take advantage of smart contracts to decentralize contract operations in a trustworthy, verifiable, traceable, and non-custodial way. With widespread use in the gaming, social, and financial sectors, this has the potential to greatly diminish any undesirable behaviors, such as shady deals, corruption, or rent seeking, that can occur within the financial organization.
- **NFTs**: NFTs are perfect for representing identities, such as exclusive and indivisible assets that may be freely traded and transferred, because their most important features are indivision and uniqueness. Notional tokens, or NFTs, serve as proof of ownership for these virtual properties in the metaverse [30].
- **Healthcare**: The use of blockchain technology can enhance the security and efficiency of healthcare facilities. It provides a safe distributed ledger for storing sensitive information, such as patient records. Insurance companies and providers may soon be able to make automated payouts using smart contracts built on the

blockchain, eliminating the need for middlemen and human processing. To ensure that only authorized individuals have access to medical information, decentralized identification protocols on blockchains like Ethereum could allow patients to have control over their health information [37].

3 Using Blockchain in the Metaverse from a Technical Standpoint

This part examines the most advanced blockchain methods for the metaverse from a technological standpoint, such as gathering data, storing, sharing, and compatibility, and protection of personal privacy. Figure 2 shows an example of abovementioned elements in the metaverse.

3.1 Data Collection

Collecting data is an essential part of the metaverse ecosystem. Information such as users' biometric whereabouts and movements, financial and credit card details, and bank and credit card details are gathered [39]. Look at Fig. 2 for reference. The metaverse is an online marketplace where people may use a variety of devices to play games, interact, and work. Consequently, massive volumes of diverse data will be generated [40]. Data collection in the metaverse makes use of a wide variety of tools, including web forms, bots, high-definition cameras, and many more.

Problems with data storage in the metaverse

The decentralized systems, such as Etheria, 4G Capital, Ampliative Art, and WeiFund produce vast amount of real-time unstructured data. The massive amounts of data being produced provide a formidable obstacle to acquisition. When developing metaverse applications such as recommender systems, data integrity, reliability, and assurance from unknown sources are of the utmost importance [41]. Data collection systems are being tested by the increasing use of virtual reality headsets and entertainment streaming [42]. Acquiring inaccurate or duplicate data can potentially affect data quality [43].

Applications of blockchain technology

With blockchain technology in place, applications like social networks will have an easier time obtaining valid metaverse data. The real data and records are attained as transactions of a distributed ledger by the blockchain. With a timestamp, cryptographic hash of the block before it, and metadata, a blockchain records every action as a transaction [44]. This means that the security of other blocks could be jeopardized if the data within a block were to be changed. The data retrieved from any block cannot be accessed [45]. Data collected here is secure because all the nodes in the ledger has to agree to any alterations to the data [46]. Through consensus processes, every data collected in the metaverse is verified using a method specific to blockchains [47, 48]. This keeps metaverse data from being duplicated. With blockchain technology, the metaverse is less likely to have data duplication during acquisition. The data collected from the metaverse using acquisition techniques provided by blockchain will be accurate because every block in the blockchain is authorized [49].

Securing accurate and high-quality data becomes a challenge while acquiring metaverse data. The distributed and complicated structure of blockchain makes it potentially slow, but it will allow data collection methods to overcome these restrictions [50]. It can take a lot longer to execute a transaction that is powered by a blockchain. This entire procedure can take many days. Consequently, transaction costs are higher than normal, and the network has a maximum capacity [51]. To store all of the data recorded on a blockchain, it is necessary to duplicate it along the chain. As data collection increases, so does the need for storage space [52]. To get around these issues with data collection methods, there is still work to be done on building a strong blockchain for the metaverse.

3.2 Storing Data

People govern the metaverse, a virtual area that exists alongside the physical world. The metaverse will consist of all the things, places, and experiences that can be accessed through the Internet. Metaverse data storage is essential. The data file that every user joins the metaverse creates grows as a consequence of interactions with other users.

The real world's processing capacity will be severely taxed once the metaverse is operational due to the massive amounts of data that will be generated. Data storage must be given top importance if the metaverse is to be utilized [34].

Problems with data storage in the metaverse

In the metaverse, both the physical and virtual worlds coexist. More and more people will join the virtual worlds, which means the metaverse will produce a massive data which will overwhelm world's data storage ability once the metaverse is fully functional. Consequently, there would be major storage issues when implementing metaverse applications like as healthcare, real estate, entertainment, gaming, etc. [53]. The risk of data loss, corruption, or leaking increases if the metaverse uses a centralized storage mechanism. The metaverse's provision of sensitive data-dependent biometrics, vocal inflections, and vital signs is at risk due to the possible data loss [54] and corruption in centralized applications [55]. Yet another enormous challenge will be data labeling and management in light of the enormous datasets produced by metaverse applications [56].

Using blockchain technology

Because a fresh block is created for every transaction, the storage in the metaverse is untouchable [57]. Data becomes more trustworthy and transparent in the metaverse when stored across the chain as an exact replica of the original blocks [13]. All sorts of metaverse applications, from digital goods to real estate, are at threat if the data storing process is hacked [58]. The utilization of blockchain increases the quantity of available for use in metaverse applications such as life support alerts and vital monitoring since numerous blocks contribute to data distribution. The decentralized blockchain allows data scientists to work collectively to purify the data, which in turn reduces the expenses and time needed to label data and prepare datasets for analysis [59].

Data scientists will be able to work together more efficiently thanks to blockchain's decentralized structure, which will help speed up data identification and labeling. On top of that, in Fig. 2, the blockchain offers reliability, data availability, and transparency. Data backups are included in every block of the blockchain. With the use of a distributed ledger based on consensus, metaverse data will be harder to replicate and manipulate [60]. Nevertheless, additional research is necessary to address the delay problem, as all new data must be replicated along the entire chain. While a hard split is certainly something to think about, blockchain technology makes data tampering impossible.

3.3 *Information Sharing*

Many different groups with a stake in the metaverse can benefit from exchanging data. Figure 2 shows that, utilizing the same platform may facilitate better interactions between people and applications. Data sharing helps everybody like engineers, scientists, doctors, etc., who are part of the metaverse [61]. The data gathered by the metaverse's different devices (AR/VR and IoT) was utilized to construct some tailored systems that react to human actions.

Because of this, many apps will be able to improve the user experience [62]. Businesses will have the ability to analyze the data through the metaverse by exchanging data through different systems. Metaverse product development, advertising, content personalization, content strategy, and consumer understanding will all benefit from shared data [57].

Problems with exchanging metaverse data

The metaverse is a dangerous place for data owners to share information because of centralized platforms for data sharing [63, 64]. Data availability and latency are both negatively impacted by the high degree of data mutability in the traditional sharing environment. Scaling data that can be changed is more challenging than data that cannot be changed [65]. Many metaverse applications, like healthcare, media, entertainment, traffic optimization, and other industries work real-time and produce massive volumes of data. Data exchanging becomes an issue in a traditional data-sharing context as the demand for real-time data increases.

Using blockchain technology

For uses like online education and cryptocurrency trading, blockchain technology might make metaverse transactions more accurate and transparent [64]. A distributed ledger containing all transactions generated by programs such as finance and governance would be accessible to all stakeholders. Consequently, the parties involved in the metaverse will benefit from more open data [66]. By eliminating gray market transactions, blockchain technology will boost user confidence and allow programmers and users to understand how external programs manage data. Examples of such applications are Thunderbird, the Bat, and Pegasus [67]. Additionally, the owner of the data will have full authority over the data. By eliminating the requirement for data validation, blockchain technology can save both time and money [68]. Thanks to smart contracts, data sharing will become more flexible. Typically, they are employed to automate the contract execution process, allowing all parties involved to immediately know the outcome without wasting time or energy on an intermediary. Blockchain enables the various scripting of smart contracts. Nmusik, Ascribe, Tracr, UBS, and Applicature are among the programs that will gain from this [69].

Because of blockchain technology, metaverse data will be more versatile and adaptable. It takes more time to transfer information using a blockchain because data must be duplicated before it can be sent down the chain [70]. A massive amount of computing power will be needed to ensure that the number of blocks in the metaverse grows in tandem with the number of users [71]. Users might expect to pay a premium for the convenience of verifying shared transactions. For data sharing in the metaverse to work, next-generation blockchains will need to address this issue.

3.4 Compatibility of Data

Metaverse interoperability is going to be the driving factor. Financial and healthcare applications are just two of many that will be able to interconnect and share data. The metaverse is going to be a place where people can interact socially and culturally in virtual spaces. As virtual bridges are gradually built, users will be able to move freely between virtual worlds, preserving their avatars and possessions. The purpose of an identity standard is to provide users with unique credentials that may be used both inside and outside of a virtual environment [55, 72, 73]. This might be the same as the numbers we see on a regular basis, such as our driver's license, social security card, passport, etc.

Metaverse Data Compatibility Concerns

The concept of the metaverse will emerge from the convergence of many online realms. Disorganized and fractured are the current centralized traditional digital platforms. Accounts, avatars, hardware, and payment infrastructure must be set up before individuals may engage in different domains [53]. Users have limited options for transferring their digital assets, including NFT and avatars, to different virtual worlds. Due to the lack of transparency in virtual worlds, transitioning may be challenging; for example, you cannot use the same Roblox account in Decentraland and Roblox. The interoperability of the virtual worlds is what decides if an application may be used there. Regardless of their physical location or the technology they use, applications in the digital world should have no trouble exchanging data with one another. For the metaverse to work together, we need a way to regulate how different virtual worlds interact with one other, which is a major limitation of the current method.

Using blockchain technology

To ensure interoperability among metaverse virtual worlds, a cross-chain protocol is the way to go [74, 75]. This paves the way for the exchange of virtual products such as avatars, NFTs, and money among different virtual realms. The groundwork for widespread use of the metaverse will be laid by this protocol. The need for intermediaries in the metaverse will be eliminated when virtual worlds are able to communicate with each other through cross-blockchain technology [76]. Blockchain technology will simplify the process of linking users and applications in the metaverse.

Additional study is required, but blockchain technology shows promise for enhancing virtual world interoperability in a wide variety of metaverses. The biggest problem with cross-blockchain-enabled metaverse interoperability is that there are several public blockchains in different virtual worlds, and they don't work together [77].

3.5 Safeguarding Personal Information

The metaverse will make use of cutting-edge HCI technology to let users communicate with one another and their virtual environments [78]. The centralized nature of Web 2.0 raises concerns about data privacy. As Web 3.0—the Internet of the Metaverse—grows in size and complexity, the lines between the real and virtual worlds will blur [79]. The impact of Web 2.0 on the protection of individual rights is far from over. Consequently, data privacy will pose an even greater challenge in the soon-to-be-released Web 3.0. The risk of a data breach will be exacerbated by apps' inadequate security measures due to the exponential development in data generation in the metaverse [80]. Concern about maintaining the confidentiality of individual records is a major issue [81].

Problems with data privacy in the virtual world

At first, the metaverse ecosystem would be difficult to adapt to since attackers can trick consumers and steal crucial data. Because users won't know they're interacting with an AI bot like promobot, they could be deceived into thinking they're talking to a real person. Personally identifiable information (PII) is at the heart of privacy concerns [82].

Adding validity information to the metaverse will make managing large amounts of data much more challenging.

Using blockchain technology

By allowing users to govern their data using private and public keys, blockchain technology effectively gives metaverse users ownership of their data. The blockchain-enabled metaverse has a strict policy on third-party intermediaries collecting or abusing user data. When it comes to the blockchain-enabled metaverse, the owners of personal data will have complete control over who can access their data and when [83]. As seen in the figure, blockchain ledgers typically include an audit trail, which guarantees that all transactions in the metaverse are thorough and consistent. 4. The adoption of zero-knowledge proof on the blockchain has made it easy to identify vital data in the metaverse while protecting users' privacy and commodities ownership. Users are able to persuade applications of anything using blockchain technology's zero-knowledge proofs [84].

The safety of blockchain technology and metaverse data privacy could be at risk from a single human mistake, such as losing a private key. Blockchain technology, on the other hand, can aid users in protecting the privacy of their data. Due to the fact that third-party programs in the metaverse often use inadequate security measures, which permit the compromise of private data, attackers can easily target these programs [85]. Research exploring the potential applications of blockchain technology in the metaverse to safeguard user data privacy is still in its early stages.

This section covered many problems with data gathering, storage, sharing, compatibility, and privacy protection in the metaverse. Look at the Fig. 2 for reference. By utilizing features like as encryption, cross-chain technology, the zero trust mechanism, interplanetary file system, blockchain-specific validation, and decentralized storage, the blockchain assists the metaverse in overcoming these challenges.

4 Conclusion

A thorough examination and analysis of blockchain technology's functions and impacts on the development and expansion of metaverse apps and services has been carried out in this study. The study began by outlining the fundamental ideas of blockchain technology and the metaverse, and then moved on to discuss the role of blockchain in creating and developing the metaverse. Afterward, this study thoroughly examined numerous important technical aspects and real-world uses of blockchain technology within the metaverse. In addition, helpful insights were offered by a comprehensive examination of difficulties and a conversation on blockchain's appropriateness in this setting. Technical upgrades to the blockchain have been implemented in the metaverse, making it more suitable for use in virtual apps and services and improving its overall performance. Following the formulation of the conclusion, we provide numerous possible directions for future research. It is clear from a thorough analysis of blockchain technology from both a technical and use case standpoint that it has the ability to revolutionize the metaverse by allowing for the creation of various virtual applications and services, thus enhancing the immersive experience. Many studies have been concentrating on different practical and technological aspects of modern blockchain versions. Consensus mechanisms, network management, and interoperability among blockchain systems are all part of these dimensions. To keep authorized nodes in a distributed network's data states in sync with one another, consensus techniques are important. Numerous consensus algorithms have been refined throughout the years in pursuit of the sweet spot between throughput and delay. Striking a balance between decentralization, scalability, and security all at once is no easy feat.

References

1. Lee Y, Moon C, Ko H, Lee S-H, Yoo B (2020) Unified representation for XR content and its rendering method. In: The 25th International Conference on 3D Web Technology, pp 1–10
2. Lee Y, Moon C, Ko H, Lee S-H, Yoo B () Unified Representation for XR Content and its Rendering Method. In: The 25th International Conference on 3D Web Technology. https://doi.org/10.1145/3424616.3424695
3. Yousefpour A, Fung C, Nguyen T, Kadiyala K, Jalali F, Niakanlahiji A, Kong J, Jue JP (2019) All one needs to know about fog computing and related edge computing paradigms: a complete survey. J Syst Architect 98:289–330. https://doi.org/10.1016/j.sysarc.2019.02.009
4. Bolger RK (2021) Finding wholes in the metaverse: posthuman mystics as agents of evolutionary contextualization. Religions 12:768. https://doi.org/10.3390/rel12090768
5. Thepmanee T, Pongswatd S, Asadi F, Ukakimaparn P (2022) Implementation of control and SCADA system: case study of Allen Bradley PLC by using WirelessHART to temperature control and device diagnostic. Energy Rep 8:934–941. https://doi.org/10.1016/j.egyr.2021.11.163
6. Bouri E, Saeed T, Vo XV, Roubaud D (2021) Quantile connectedness in the cryptocurrency market. J Int Fin Mark, Institut Money 71:101302. https://doi.org/10.1016/j.intfin.2021.101302
7. Bisogni C, Iovane G, Landi RE, Nappi M (2021) ECB2: a novel encryption scheme using face biometrics for signing blockchain transactions. J Inform Security Appl 59:102814. https://doi.org/10.1016/j.jisa.2021.102814
8. Ramu SP, Boopalan P, Pham Q-V, Maddikunta PKR, Huynh-The T, Alazab M, Nguyen TT, Gadekallu TR (2022) Federated learning enabled digital twins for smart cities: concepts, recent advances, and future directions. Sustain Cities Soc 79:103663. https://doi.org/10.1016/j.scs.2021.103663
9. Wang S, Qureshi M, Miralles P, Huynh-The T, Gadekallu T, Liyanage M (2021) Explainable AI for B5G/6G: technical aspects, use cases, and research challenges
10. Gadekallu TR, Pham Q-V, Nguyen DC, Maddikunta PKR, Deepa N, Prabadevi B, Pathirana PN, Zhao J, Hwang W-J (2022) Blockchain for edge of things: applications, opportunities, and challenges. IEEE Internet Things J 9:964–988. https://doi.org/10.1109/jiot.2021.3119639
11. Yang Q, Zhao Y, Huang H, Xiong Z, Kang J, Zheng Z (2022) Fusing blockchain and AI with metaverse: a survey. IEEE Open J Comp Soc 3:122–136. https://doi.org/10.1109/ojcs.2022.3188249
12. Jeon H, Youn H, Ko S, Kim T (2022) Blockchain and AI meet in the metaverse. Blockchain Potent AI. https://doi.org/10.5772/intechopen.99114

13. Mystakidis S (2022) Metaverse. Encyclopedia 2:486–497. https://doi.org/10.3390/encyclopedia2010031
14. Wang F-Y, Qin R, Wang X, Hu B (2022) MetaSocieties in metaverse: metaEconomics and metaManagement for metaEnterprises and metaCities. IEEE Trans Computat Soc Syst 9:2–7. https://doi.org/10.1109/tcss.2022.3145165
15. Park S-M, Kim Y-G (2022) A metaverse: taxonomy, components, applications, and open challenges. IEEE Access 10:4209–4251. https://doi.org/10.1109/access.2021.3140175
16. Stephenson N (2003) Snow crash. Spect
17. Koutitas G, Smith S, Lawrence G (2020) Performance evaluation of AR/VR training technologies for EMS first responders. Virtual Real 25:83–94. https://doi.org/10.1007/s10055-020-00436-8
18. Tao F, Zhang H, Liu A, Nee AYC (2019) Digital twin in industry: state-of-the-art. IEEE Trans Industr Inf 15:2405–2415. https://doi.org/10.1109/tii.2018.2873186
19. Alazab M, Khan LU, Koppu S, Ramu SP, Boobalan P, Baker T, Maddikunta PKR, Gadekallu TR, Aljuhani A (2023) Digital twins for healthcare 4.0—recent advances, architecture, and open challenges. IEEE Cons Elect Magazine 12:29–37. https://doi.org/10.1109/mce.2022.3208986.
20. Tang F, Chen X, Zhao M, Kato N (2023) The Roadmap of communication and networking in 6G for the metaverse. IEEE Wirel Commun 30:72–81. https://doi.org/10.1109/mwc.019.2100721
21. Chang Y, Chien C, Shen L-F (2021) Telecommuting during the coronavirus pandemic: Future time orientation as a mediator between proactive coping and perceived work productivity in two cultural samples. Personality Individ Differ 171:110508. https://doi.org/10.1016/j.paid.2020.110508
22. Duan H, Li J, Fan S, Lin Z, Wu X, Cai W (2021) Metaverse for social good. In: Proceedings of the 29th ACM International Conference on Multimedia. https://doi.org/10.1145/3474085.3479238
23. Ko SY, Chung H, Kim KJ, Shin Y (2021) A study on the typology and advancement of cultural leisure-based metaverse. KIPS Trans Software Data Eng 10:331–338. https://doi.org/10.3745/KTSDE.2021.10.8.331
24. Nakamoto S (2008) Bitcoin, a peer-to-peer electronic cash system
25. Huo R, Zeng S, Wang Z, Shang J, Chen W, Huang T, Wang S, Yu FR, Liu Y (2022) A comprehensive survey on blockchain in industrial internet of things: motivations, research progresses, and future challenges. IEEE Comm Surv Tut 24:88–122. https://doi.org/10.1109/comst.2022.3141490
26. Dotan M, Pignolet Y-A, Schmid S, Tochner S, Zohar A (2021) Survey on blockchain networking. ACM Comput Surv 54:1–34. https://doi.org/10.1145/3453161
27. Alangot B, Reijsbergen D, Venugopalan S, Szalachowski P, Yeo KS (2021) Decentralized and lightweight approach to detect eclipse attacks on proof of work blockchains. IEEE Trans Netw Serv Manage 18:1659–1672. https://doi.org/10.1109/tnsm.2021.3069502
28. Thomsen SE, Spitters B (2021) Formalizing Nakamoto-style proof of stake. In: 2021 IEEE 34th Computer Security Foundations Symposium (CSF). https://doi.org/10.1109/csf51468.2021.00042
29. Jian X, Leng P, Wang Y, Alrashoud M, Hossain MS (2021) Blockchain-empowered trusted networking for unmanned aerial vehicles in the B5G era. IEEE Netw 35:72–77. https://doi.org/10.1109/mnet.011.2000177
30. Zarir AA, Oliva GA, Jiang ZM (Jack) Hassan AE (2021) Developing cost-effective blockchain-powered applications. ACM Trans Soft Eng Method 30:1–38. https://doi.org/10.1145/3431726
31. What is consensus? a beginner's guide. Available online: https://crypto.com/university/consensus-mechanisms-explained. Accessed on 31 October 2023
32. Nadini M, Alessandretti L, Di Giacinto F, Martino M, Aiello LM, Baronchelli A (2021) Mapping the NFT revolution: market trends, trade networks, and visual features. Scient Rep 11. https://doi.org/10.1038/s41598-021-00053-8

33. Cai L, Li Q, Liang X (2022) Introduction to blockchain basics. Adv Blockchain Tech, 3–43. https://doi.org/10.1007/978-981-19-3596-1_1
34. Xiong H, Jin C, Alazab M, Yeh K-H, Wang H, Gadekallu TR, Wang W, Su C (2022) On the design of blockchain-based ECDSA with fault-tolerant batch verification protocol for blockchain-enabled IoMT. IEEE J Biomed Health Inform 26:1977–1986. https://doi.org/10.1109/jbhi.2021.3112693
35. Lian Z, Zeng Q, Wang W, Gadekallu TR, Su C (2023) Blockchain-based two-stage federated learning with non-IID data in IoMT system. IEEE Trans Comput Soc Syst 10:1701–1710. https://doi.org/10.1109/tcss.2022.3216802
36. Arafeh M, El Barachi M, Mourad A, Belqasmi F (2019) A blockchain based architecture for the detection of fake sensing in mobile crowdsensing. In: 2019 4th International Conference on Smart and Sustainable Technologies (SpliTech). https://doi.org/10.23919/splitech.2019.8783092
37. Kim T, Kim S (2021) Digital transformation, business model and metaverse. J Digital Conv 19:215–224. https://doi.org/10.14400/JDC.2021.19.11.215
38. Huynh-The T, Gadekallu TR, Wang W, Yenduri G, Ranaweera P, Pham Q-V, da Costa DB, Liyanage M (2023) Blockchain for the metaverse: a review. Futur Gener Comput Syst 143:401–419. https://doi.org/10.1016/j.future.2023.02.008
39. Chengoden R, Victor N, Huynh-The T, Yenduri G, Jhaveri RH, Alazab M, Bhattacharya S, Hegde P, Maddikunta PKR, Gadekallu TR (2023) Metaverse for healthcare: a survey on potential applications, challenges future directions. IEEE Access 11:12765–12795. https://doi.org/10.1109/access.2023.3241628
40. Rauschnabel PA, Babin BJ, Dieck T, Krey MC, Jung NT (2022) What is augmented reality marketing? Its definition, complexity, and future. J Bus Res 142:1140–1150. https://doi.org/10.1016/j.jbusres.2021.12.084
41. Tao H, Bhuiyan MZA, Abdalla AN, Hassan MM, Zain JM, Hayajneh T (2019) Secured data collection with hardware-based ciphers for IoT-based healthcare. IEEE Internet Things J 6:410–420. https://doi.org/10.1109/jiot.2018.2854714
42. Brunschwig L, Campos-López R, Guerra E, de Lara J (2021) Towards domain-specific modelling environments based on augmented reality. In: 2021 IEEE/ACM 43rd International Conference on Software Engineering: New Ideas and Emerging Results (ICSE-NIER). https://doi.org/10.1109/icse-nier52604.2021.00020
43. Jeong J-B, Lee S, Ryu E-S (2022) Rethinking fatigue-aware 6DoF video streaming: focusing on MPEG immersive video. In: 2022 International Conference on Information Networking (ICOIN). https://doi.org/10.1109/icoin53446.2022.9687247
44. Shiau W-L, Huang L-C (2022) Scale development for analyzing the fit of real and virtual world integration: an example of Pokémon Go. Inf Technol People 36:500–531. https://doi.org/10.1108/itp-11-2020-0793
45. Luo Y, Su Z, Zheng W, Chen Z, Wang F, Zhang Z, Chen J (2021) A novel memory-hard password hashing scheme for blockchain-based cyber-physical systems. ACM Trans Internet Technol 21:1–21. https://doi.org/10.1145/3408310
46. Zhang L, Zhang Z, Wang W, Jin Z, Su Y, Chen H (2022) Research on a covert communication model realized by using smart contracts in blockchain environment. IEEE Syst J 16:2822–2833. https://doi.org/10.1109/jsyst.2021.3057333
47. Xu C, Qu Y, Luan TH, Eklund PW, Xiang Y, Gao L (2022) A lightweight and attack-proof bidirectional blockchain paradigm for internet of things. IEEE Internet Things J 9:4371–4384. https://doi.org/10.1109/jiot.2021.3103275
48. Bouraga S (2021) A taxonomy of blockchain consensus protocols: a survey and classification framework. Expert Syst Appl 168:114384. https://doi.org/10.1016/j.eswa.2020.114384
49. Lashkari B, Musilek P (2021) A comprehensive review of blockchain consensus mechanisms. IEEE Access 9:43620–43652. https://doi.org/10.1109/access.2021.3065880
50. Guo J, Ding X, Wu W (2022) Reliable traffic monitoring mechanisms based on blockchain in vehicular networks. IEEE Trans Reliab 71:1219–1229. https://doi.org/10.1109/tr.2020.3046556

51. Xu X, Sun G, Luo L, Cao H, Yu H, Vasilakos AV (2021) Latency performance modeling and analysis for hyperledger fabric blockchain network. Inf Process Manage 58:102436. https://doi.org/10.1016/j.ipm.2020.102436
52. Alrubei SM, Ball EA, Rigelsford JM, Callum AW (2020) Latency and performance analyses of real-world wireless IoT-blockchain application. IEEE Sensors J 20:7372–7383. https://doi.org/10.1109/jsen.2020.2979031
53. Chen L, Fu Q, Mu Y, Zeng L, Rezaeibagha F, Hwang M-S (2022) Blockchain-based random auditor committee for integrity verification. Futur Gener Comput Syst 131:183–193. https://doi.org/10.1016/j.future.2022.01.019
54. Bian Y, Leng J, Zhao JL (2022) Demystifying metaverse as a new paradigm of enterprise digitization. Big Data, 109–119. https://doi.org/10.1007/978-3-030-96282-1_8
55. Wang X, Liu X, Cheng C-T, Deng L, Chen X, Xiao F (2021) A joint user scheduling and trajectory planning data collection strategy for the UAV-assisted WSN. IEEE Commun Lett 25:2333–2337. https://doi.org/10.1109/lcomm.2021.3067898
56. Sonnen J (2022) Metaverse for beginners 2023. Justin Sonnen
57. Scargill T, Chen Y, Eom S, Dunn J, Gorlatova M (2022) Environmental, user, and social context-aware augmented reality for supporting personal development and change. In: 2022 IEEE Conference on Virtual Reality and 3D User Interfaces Abstracts and Workshops (VRW). https://doi.org/10.1109/vrw55335.2022.00042
58. Liang W, Fan Y, Li K-C, Zhang D, Gaudiot J-L (2020) Secure data storage and recovery in industrial blockchain network environments. IEEE Trans Industr Inf 16:6543–6552. https://doi.org/10.1109/tii.2020.2966069
59. Yang D, Zhou J, Chen R, Song Y, Song Z, Zhang X, Wang Q, Wang K, Zhou C, Sun J, Zhang L, Bai L, Wang Y, Wang X, Lu Y, Xin H, Powell CA, Thüemmler C, Chavannes NH, Chen W, Wu L, Bai C (2022) Expert consensus on the metaverse in medicine. Clinical eHealth 5:1–9. https://doi.org/10.1016/j.ceh.2022.02.001
60. Xie J, Yu FR, Huang T, Xie R, Liu J, Liu Y (2019) A survey on the scalability of blockchain systems. IEEE Netw 33:166–173. https://doi.org/10.1109/mnet.001.1800290
61. Kraus S, Kanbach DK, Krysta PM, Steinhoff MM, Tomini N (2022) Facebook and the creation of the metaverse: radical business model innovation or incremental transformation? Int J Entrep Behav Res 28:52–77. https://doi.org/10.1108/ijebr-12-2021-0984
62. Jeon JE (2021) The effects of user experience-based design innovativeness on user-metaverse platform channel relationships in South Korea. J Distribution Sci 19:81–90. https://doi.org/10.15722/JDS.19.11.202111.81
63. Liu L, Feng J, Pei Q, Chen C, Ming Y, Shang B, Dong M (2021) Blockchain-enabled secure data sharing scheme in mobile-edge computing: an asynchronous advantage actor-critic learning approach. IEEE Internet Things J 8:2342–2353. https://doi.org/10.1109/jiot.2020.3048345
64. Egliston B, Carter M (2021) Critical questions for Facebook's virtual reality: data, power and the metaverse. Internet Pol Rev 10. https://doi.org/10.14763/2021.4.1610
65. Yu K, Tan L, Aloqaily M, Yang H, Jararweh Y (2021) Blockchain-enhanced data sharing with traceable and direct revocation in IIoT. IEEE Trans Industr Inf 17:7669–7678. https://doi.org/10.1109/tii.2021.3049141
66. Rashid A, Masood A, Abbas H, Zhang Y (2021) Blockchain-based public key infrastructure: a transparent digital certification mechanism for secure communication. IEEE Network 35:220–225. https://doi.org/10.1109/mnet.101.2000532
67. Vashistha N, Hossain MM, Shahriar MR, Farahmandi F, Rahman F, Tehranipoor MM (2022) EChain: a blockchain-enabled ecosystem for electronic device authenticity verification. IEEE Trans Consum Electron 68:23–37. https://doi.org/10.1109/tce.2021.3139090
68. Min T, Cai W (2022) Portrait of decentralized application users: an overview based on large-scale Ethereum data. CCF Trans Pervasive Comp Interact 4:124–141. https://doi.org/10.1007/s42486-022-00094-6
69. Ali O, Jaradat A, Kulakli A, Abuhalimeh A (2021) A comparative study: blockchain technology utilization benefits, challenges and functionalities. IEEE Access 9:12730–12749. https://doi.org/10.1109/access.2021.3050241

70. Luo Y, Jin H, Li P (2019) A blockchain future for secure clinical data sharing. Proceedings of the ACM International Workshop on Security in Software Defined Networks & Network Function Virtualization. https://doi.org/10.1145/3309194.3309198
71. Gao Y, Wu W, Si P, Yang Z, Yu FR (2021) B-ReST: blockchain-enabled resource sharing and transactions in fog computing. IEEE Wirel Commun 28:172–180. https://doi.org/10.1109/mwc.001.2000102
72. Sparkes M (2021) What is a metaverse? New Scientist 251:18. https://doi.org/10.1016/s0262-4079(21)01450-0
73. Stokel-Walker C (2022) Welcome to the metaverse. New Scientist 253:39–43. https://doi.org/10.1016/s0262-4079(22)00018-5
74. Belchior R, Vasconcelos A, Guerreiro S, Correia M (2021) A survey on blockchain interoperability: past, present, and future trends. ACM Comput Surv 54:1–41. https://doi.org/10.1145/3471140
75. Madine M, Salah K, Jayaraman R, Al-Hammadi Y, Arshad J, Yaqoob I (2021) AppXchain: application-level interoperability for blockchain networks. IEEE Access. 9:87777–87791. https://doi.org/10.1109/access.2021.3089603
76. Jabbar R., Fetais N, Krichen M, Barkaoui K (2020) Blockchain technology for healthcare: enhancing shared electronic health record interoperability and integrity. In: 2020 IEEE International Conference on Informatics, IoT, and Enabling Technologies (ICIoT). https://doi.org/10.1109/iciot48696.2020.9089570.
77. Wibowo S, Sandikapura T (2019) Improving data security, interoperability, and veracity using blockchain for one data governance, case study of local tax big data. In: 2019 International Conference on ICT for Smart Society (ICISS). https://doi.org/10.1109/iciss48059.2019.8969805
78. Siyaev A, Jo G-S (2021) Towards aircraft maintenance metaverse using speech interactions with virtual objects in mixed reality. Sensors 21:2066. https://doi.org/10.3390/s21062066
79. ARVAS, İ.S (2020) Gutenberg Galaksisinden Meta Evrenine: Üçüncü Kuşak İnternet. Web 3.0. AJIT-e online academic J Inform Techn 13:53–71. https://doi.org/10.5824/ajite.2022.01.003.x
80. Kostenko OV (2022) Electronic jurisdiction, metaverse, artificial intelligence, digital personality, digital avatar, neural networks: theory, practice, perspective. World Sci. https://doi.org/10.31435/rsglobal_ws/30012022/7751
81. Xi N, Chen J, Gama F, Riar M, Hamari J (2022) The challenges of entering the metaverse: an experiment on the effect of extended reality on workload. Inf Syst Front. https://doi.org/10.1007/s10796-022-10244-x
82. Hughes I (2022) The metaverse: is it the future? ITNOW 64:22–23. https://doi.org/10.1093/itnow/bwac011
83. Kumar P, Kumar R, Srivastava G, Gupta GP, Tripathi R, Gadekallu TR, Xiong NN (2021) PPSF: a privacy-preserving and secure framework using blockchain-based machine-learning for IoT-driven smart cities. IEEE Trans Netw Sci Eng 8:2326–2341. https://doi.org/10.1109/tnse.2021.3089435
84. Sedlmeir J, Völter F, Strüker J (2021) The next stage of green electricity labeling. ACM SIGEnergy Energy Inform Rev 1:20–31. https://doi.org/10.1145/3508467.3508470
85. Hassan MU, Rehmani MH, Chen J (2019) Privacy preservation in blockchain based IoT systems: integration issues, prospects, challenges, and future research directions. Futur Gener Comput Syst 97:512–529. https://doi.org/10.1016/j.future.2019.02.060

Comprehensive Metaverse Design Concept Using Augmented Reality, Virtual Reality, and Mixed Reality

Ashwani Kumar Yadav and Shri Prakash Dwivedi

Abstract The metaverse is an evolving paradigm for the next-generation Internet, intended to provide 3D immersive experiences and self-sustaining virtual shared space by utilizing a wide range of relevant technologies such as AR, MR and VR in real-time applications. Augmented reality (AR) is a technology that enhances the real-world environment by superimposing digital content, such as graphics, sounds, or videos onto the user's view of the real world. It is described as a system that combines three fundamental elements: real-time interaction, realistic 3D registration of virtual and real objects, and a combination of the real and virtual worlds. By superimposing digital content onto the real world, augmented reality can create an immersive experience that enhances a user's perception of their surroundings. An immersive sense of a virtual world is provided to users by pose tracking and 3D near-eye displays used in virtual reality (VR). VR headsets or multi-projected environments can be used for entertainment, education, medical or military training, and businesses. It commonly includes auditory and video feedback, however, might also permit different forms of sensory and pressure inputs via haptic technology. Mixed reality (MR) is typically visible as "bridging the physical and digital world." Mixed reality has many applications in the field of entertainment, from television shows to game consoles. This chapter describes the application of augmented reality, virtual reality, and mixed reality as a precursor for the metaverse development. This work will help developers create useful tools for the metaverse environment. It also brings to light various challenges and opportunities for the metaverse design using AR, VR, and mixed reality.

Keywords Metaverse · Blockchain · Edge computing · Augmented reality · Virtual reality · Mixed reality

A. K. Yadav (✉) · S. P. Dwivedi
Department of Information Technology, G.B. Pant University of Agriculture and Technology, Pantnagar, Udham Singh Nagar 263145, Uttarakhand, India
e-mail: ashwaniittalks@gmail.com

G. Chhabra and K. Kaushik (eds.), *Understanding the Metaverse*, Blockchain Technologies, https://doi.org/10.1007/978-981-97-2278-5_11

1 Introduction

The word "verse" with the prefix "meta" is used to form the phrase "metaverse" which describes an imaginary synthetic environment that resembles the real world. The metaverse is a technologically advanced, highly realistic virtual world where individuals will be able to recognize, play, work, and feel as though they were in the real world. It is an artificial universe consisting of computer-generated items, digital objects, virtual environments, and user-controlled avatars. Neal Stephenson used the term "metaverse" for the first time in his science fiction novel "Snow Crash," published in 1992 [1]. The metaverse is described by Stephenson as the enormous virtual universe coexisting with the actual world and where people converse using digital avatars. Since its inception, the concept of the metaverse as a computer-generated universe has been recognized by a wide variety of ideas, including lifelogging [2], collective space in virtuality, a mirror world, and an embodied spatial Internet. We view the metaverse as a virtual world that combines the physical and digital worlds, rendered by the amalgamation of extended reality (XR), Internet, and web technologies. The Milgram and Kishino's Reality Virtuality Continuum model [3] tells about XR, which includes AR, VR, and MR, which combines digital and physical components to varying degrees. Similar to this, Snow Crash's metaverse scene depicts the contrast between the real world and its exact replica in digital settings. Individual users in the metaverse own distinct avatars, which are analogous to the user's physical self and allow them to experience an alternate existence in a virtual reality that is a metaphor for their real realities (Fig. 1).

To achieve such duality, the metaverse must move through three stages: digital twins, digital natives, and eventually the coexistence of physical-virtual reality, or surreality. Large-scale, highly accurate digital replicas of real-world objects and entities are referred to as "digital twins." The digital twins mimic their physical counterparts [4] by using temperature, object motion, and even function accordingly. This is implemented by the parallel transmission of data between the virtual and physical twins [5]. A number of applications are already in use, including computer-aided design (CAD) for designing buildings and product design, smart urban planning, industrial systems assisted by artificial intelligence (AI), and robot-supported hazardous activities. The second stage after the creation of digital copy is native content creation. The content creator represented by an avatar could be connected to physical counterparts or may only exist in virtual worlds. Meanwhile, related ecosystems, such as laws and regulations, businesses and cultures (e.g. data ownership), and social norms may contribute to facilitating these digital creations. Such ecosystems are related to existing standards and laws in real-world society, enabling the development of physical objects and intangible contents [6]. However, research on such

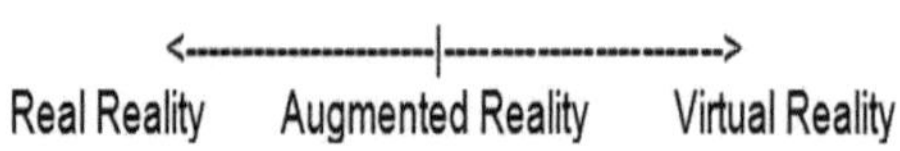

Fig. 1 Milgram and Kishino's reality virtuality continuum

applications is still in its nascent phase, with a focus on the first point of interaction with users, such as input techniques and authoring systems for creating content [7–10]. The third or the last stage of metaverse development could see its transformation into a persistent, self-sustaining virtual world that simultaneously engages with the physical world.

As a result, the avatars, which represent human users in the actual world, may engage in a variety of real-time activities, with a theoretically vast number of concurrent users across multiple virtual worlds [5]. Apart from this, the metaverse can enable interoperability between platforms representing varying virtual worlds, allowing users to produce content and circulate it extensively across virtual worlds. For example, a user may produce content in one game, such as Fortnite, and then transfer that material to another platform or game, such as Minecraft or Roblox, still preserving their own identity and experience.

To a greater extent, the platform can interact with the physical world through a variety of channels, including content, avatars, and computer agents in the metaverse interacting with smart gears and robots. Users can access information through headmounted wearable displays or mobile headsets (like the Microsoft Hololens). One could contend that we already exist in the metaverse in light of the diversified concepts of computer-mediated universes. However, this is only partially true, and we look at a number of instances to support this claim while taking into account the three-stage metaverse development roadmap. While social networks allow users to generate content, it is only possible to do so with texts, photographs, and videos, and there are few opportunities for user participation (e.g. post sharing and liking posts). The Earth 3D map offers frames for images of the real world but has GPS data as the only physical feature. Video games have become highly vibrant and realistic. Users may enjoy wonderful visuals and in-game physics, for instance, in Call of Duty scenario of Black Ops Cold War provide a great sense of realism that closely reflects the real world in many aspects. Second Life completed its eighteen years in virtual world, is widely recognized as the largest 3D universe created by users themselves. Users may create and design their own 3D environments and live extravagantly in such a virtual world. Moreover, video games are still lacking the skill of interaction. Emerging platforms that make use of Microsoft Mesh and VRChat environments provide enhanced surroundings that mimic virtual areas for social gatherings and online meetings. These virtual environments, however, are transient and disappear after the gatherings and meetings. AR games like Pokémon Go have virtual objects that have also been linked to actual reality without reflecting any digital twin concept.

In designing the metaverse, other than the Internet, social networks, gaming, and virtual environments, other technologies should also be considered. The metaverse is being built on the foundation of the emergence of AR and VR, less latency networks, artificial intelligence, edge computing, and blockchain-based distributed ledgers (hyperledgers). As a critical lens for creating the metaverse design space, this chapter examines the current technologies and technological infrastructures. The perceived metaverse virtual universe is formed by perpetual, shared, concurrent, and 3D virtual spaces. This chapter can be seen as an attempt to provide a thorough understanding of the metaverse, taking into account both the technological and ecosystem segments. In 2021, metaverse, revived people's dreams of creating a perfect virtual society in which everyone are closely connected. Then, major players began to commit themselves to creating metaverse design resulting in the involvement of large corporations in the development of metaverses like Microsoft Mesh [11] and NVIDIA's Omniverse [12]. From prehistoric times to the modern digital era, humans have a long history of forming close ties and reducing distances between one another. However, creating a metaverse is an arduous endeavour, despite being fascinating and exciting.

2 Metaverse Standards

To make the notions of the metaverse a reality, industry and standards communities are working together in developing standards. As a result, Table 1, the perfect integration of the physical and virtual worlds is the main focus of the current metaverse standards.

- **ISO/IEC 23005**: In order to achieve interoperability, simultaneous reaction, and smooth information flow, it standardized the interfaces between the virtual and physical worlds [13]. A comprehensive range of metaverse business services that allow for the integration of audiovisual data, rendered sensory effects, and virtual item properties into interactions between the real and virtual worlds are covered under ISO/IEC 23005.
- **IEEE 2888**: IEEE 2888 defines standardized interfaces for synchronizing the physical world and cyberspace [14]. In order to control actuators and gather sensory data and enable interaction between the virtual and physical worlds, IEEE 2888 standards establish information formats and application programme interfaces (APIs).

Table 1 Table of current metaverse standard

Standard	Description
ISO/IEC 23005	Provides guidelines for the design, development, and deployment of Metaverses
IEEE 2888	Standardizes interfaces for the physical and virtual words, and defines data formats and APIs
IEEE P1589	It is for AR learning experiences that facilitate the creation of experience repositories and online marketplaces
IEEE P2048	Provides terminology, definitions, and taxonomies in the Metaverse, and ensures that the Metaverse can develop sustainably
IEEE 7016	Presents an overview of the techno-social aspects of Metaverses, as well as a methodology for evaluating their ethical viability

Source [14–17]

- **IEEE P1589**: is suggested for the augmented reality learning experience, which outlines how activities, the learning context, particular surroundings, and possibly other components of AR-enhanced learning activities would be represented in a standardized format, along with data standards [15] by utilizing sensors and computer vision. It is made simpler for users to develop learning experiences that combine real-world interactions with web application running on a browser.
- **IEEE P2048**: Since the metaverse-related business is relatively young, there has been a lot of hype, muddle, and misinformation about it. Early adopters may be misled and unnecessarily hindered by a lack of consensus on key terminology, definitions, and taxonomies. IEEE P2048 is designed to be used as a standard [16] that will set forth the definitions, classes, and levels of a metaverse, as well as make it easier for activities linked to metaverse sustainable development to grow and encourage the healthy expansion of the metaverse market.
- **IEEE P7016**: This standard's objectives are to give a high-level overview of the technosocial aspects of metaverse systems and to specify a methodology for conducting ethical assessments during the design and operation of these systems [17]. It offers recommendations to metaverse developers on how to give an ethically sound system a top design priority.

3 Metaverse Architecture

Since the metaverse is still in its nascent stages, neither academia nor industry can agree on a definition of its structure. Considering [18, 19], we suggest that the architecture of the metaverse connects the physical and digital environments. As such, the suggested architecture can be utilized to include, following metaverse elements:

- **Infrastructure**: The infrastructure consists of the core technologies that enable the metaverse's functionality, that is, MEMS voice interfacing, GPUs managing data, Wi-Fi, 5G, and 6G networks, storage, microservices, cloud-edge computing, and communication technologies.
- **Interactivity**: It acts as the link between the virtual and real worlds. It serves as an interface between the virtual and real worlds, allowing people in the real world to communicate with the metaverse and the realization of the metaverse. These include AR/MR/VR/XR technologies, digital twins, immersive user experiences, sensing, computer vision, 3D modelling, and geospatial mapping.
- **Ecosystem**: It simplifies the operation. It makes the metaverse collaborate more smoothly, including avatars, creating content, data exchange among twins, virtual economy, social acceptability, e-sports, gaming, commerce, tourism, and other activities (Fig. 2).

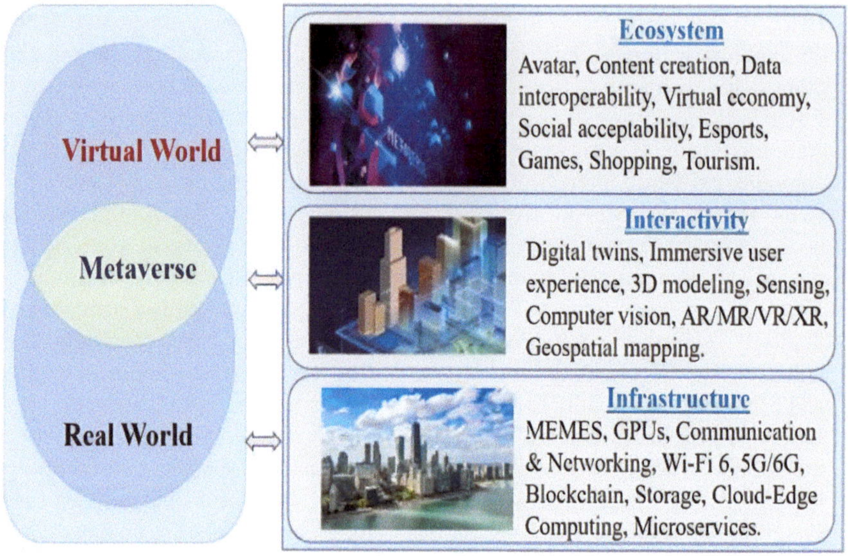

Fig. 2 Typical architecture of the metaverse. *Source* [20]

4 Metaverse Design Space and Design Tensions

4.1 Computing and Experience

The reality-shaping nature of digital technology has been observed in recent information systems research. The way that digital technologies link to human experience is an important part of how they shape reality [21]. In the context of experiential computing, the phenomenon refers to how computing technology is becoming more permeated by and intertwined with human experience [21], how important aspects of human experience have to be examined.

Humans use digital technologies for simple tasks such as conversing and shopping, which are not computational in the traditional sense. Digital technologies also integrate various aspects of the human experience. (Fig. 3).

The metaverse provides experiences in an alternate reality, while experiential computing is concerned with the experiences in a physical environment pervaded by digital technologies. But next to experience, a crucial component of the metaverse is experience's interconnectedness—the synthesis of multiple experiences. Next, we pay attention to the crucial design choices regarding the metaverse as a variety of interconnected immersive experiences with unparalleled depth using these four aspects of human experience. We can add another important dimensional aspect to the idea of metaverse that is the transition, which conceptualizes the fusion of many design spaces and the movement between those design spaces.

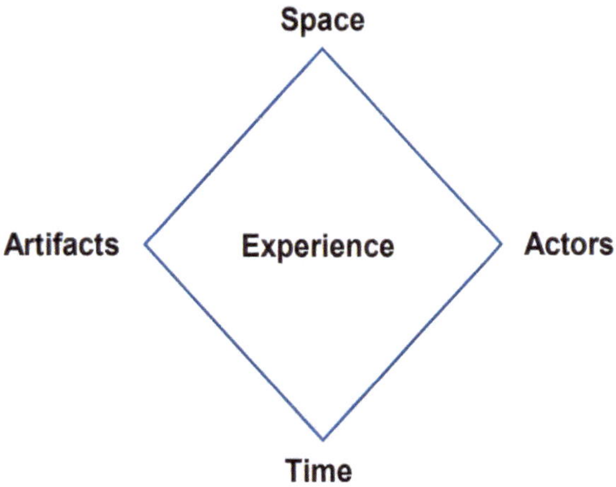

Fig. 3 Key dimensions of human experiences [21]

4.2 The Metaverse: Design Spaces and Design Tensions

Building the metaverse is a challenge that requires designers to (1) determine and workout feasible techniques for creating distinctive experiences and (2) consider these as components of a meta design space that enables a convergence of several different experiences. The first component is represented from the fundamental dimensions of human experience, and the second as an added transition dimension.

In other words, any unique human experience has its own design challenge problem, even so, the overall design of the metaverse. There are tensions because the different design dimensions vary over their dimensional ranges—space, time, artefacts, actors, and transitional tensions. Each of these dimensions describes a different experience area, and together they make up the entire problem space in which feasible solutions can be developed. Figure 4 visualizes the metaverse as a meta design space of interconnected design spaces.

While the differences along the artefactual, spatial, temporal, and actor dimensions give birth to distinctive experiences in each design space, it is the meta design space that establishes crucial elements like the overall governance of the metaverse, including issues of control and ownership. Thus, tensions that may exist across the various facets of the human experience occur in a variety of configurations, each of which reflects a particular design decision made by the designers.

4.2.1 Spatial Tensions (Where)

Human experience exists in space, both in the actual world and in virtual realities produced by digital technologies. Modern video games usually consist of massive three-dimensional spaces, as opposed to early computer games, which mostly offered two-dimensional experiences. Physically constraints do not exist theoretically in the

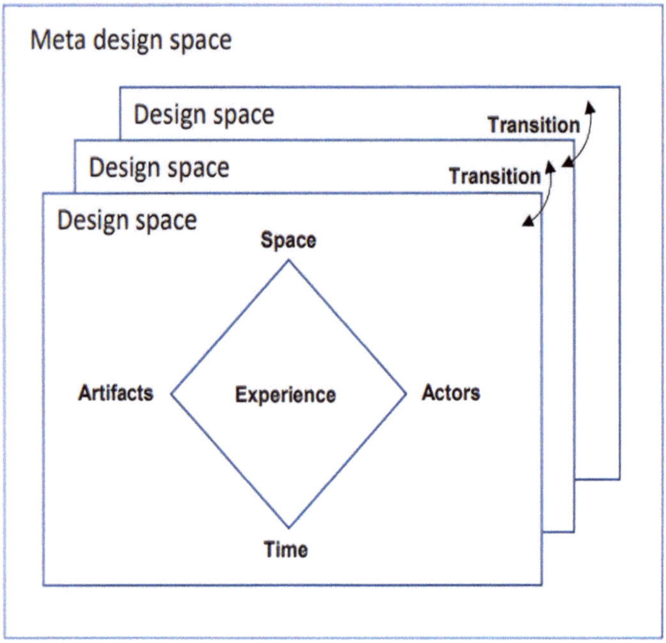

Fig. 4 Metaverse as a meta design space of interconnected design spaces. *Source* [22]

digital realm, and modern computer games are free to change physical constants [23].

In the metaverse, important spatial design choices are related to whether and how the metaverse both controls and enables movement in space. In order to conceptually address these conflicts (spatial constraint versus spatial affordance), we can employ some of the concepts used in the creation of video games, where the term "open world" refers to virtual environments that permit users to freely navigate in an open space. Users can freely roam the metaverse within an open universe (or the region of the metaverse she's in at the moment), while a closed world perspective (deliberately) restricts this navigation—that is, the spatiality of the experience can be chosen by the metaverse designer. Figure 5 illustrates this view.

As the metaverse combines a variety of experiences, it can incorporate all types of closed and open-world strategies. We can observe some evidence for it, such as in Epic Games' Fortnite, a massive game where some game elements offer an open-world experience and the aforementioned concert offers a constrained area for users to attend that concert.

Fig. 5 Designing metaverse spatiality

Fig. 6 Designing metaverse temporality figure

4.2.2 Temporal Tensions (When)

Human experience is temporal. A majority of notions about the metaverse presuppose that experience is real time (contrary to asynchronous). However, the experience's temporality can be determined by the Metaverse's creator. Figure 6 illustrates this view.

Designers of each experience will need to consider the temporal dimension while making decisions. Given that most video games are reactive, one may predict that this dimension will take precedence, however, there are an actually large number of contemplative actions, including virtual reality applications for meditation, sculpting and sculpture viewing, and solving puzzles. Furthermore, the contemplative practices might have health benefits.

4.2.3 Artefactual Tensions (What)

Humans have an artefact experience when they view and engage with artefacts, conjointly in the metaverse and the real world. Despite the fact that they offer a high-fidelity model of a specific real-world occurrence, artefacts in conventional information systems are representational [24, 25], the metaverse does not always follow this rule. Artefacts may be allegorical (such as when a rock that is based on a real rock is presented); however, they are also imaginable. (There are no dragons in reality).

Modern advances in photography enables detailed, high-resolution photographs—but people seek fidelity. Fidelity is a measurement used to assess how well an artefact represents a particular primary occurrence. Designers of the metaverse may purposefully deviate from any real-world experience and offer images that cannot be considered an exact reproduction of an existing region.

It could be too simplistic to view "imagined" as an opposite of "faithful." Instead, we may use the idea of fidelity as a connection to the idea of imaginative thinking. It makes sense to infer that something imagined is a high-fidelity depiction of an imagination if it is presented with a high level of detail.

When we take into account how much of what we are seeing in, consider how high-fidelity representations are imagined in modern video games. For instance, rocks that resemble real rocks yet may not actually exist in the real world (no particular rock that serves as the "original" for that particular depiction can be found in the real world).

As a result, we may describe artefact capacities in the metaverse in terms of fidelity and creativity. Figure 7 depicts the same.

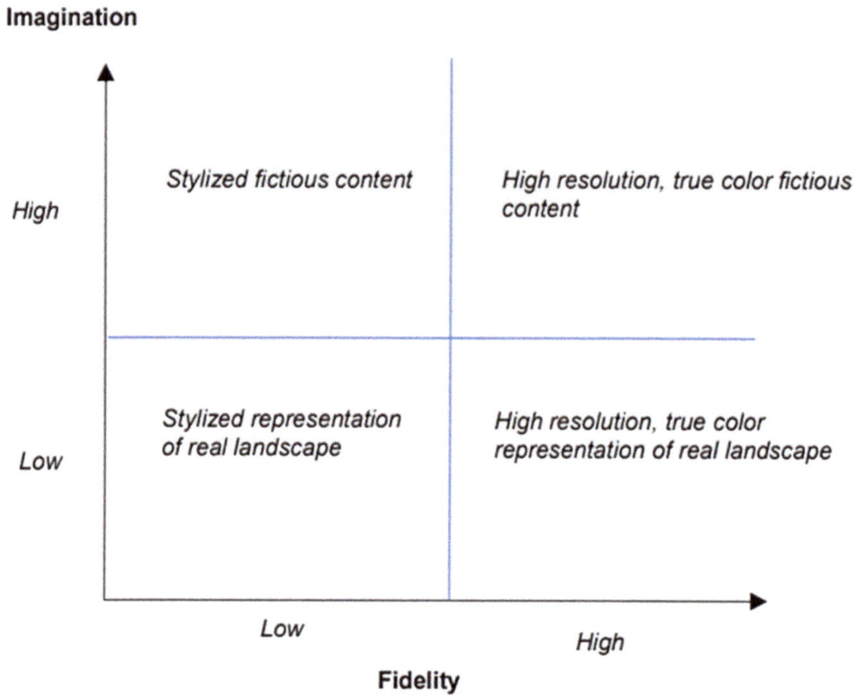

Fig. 7 Designing metaverse artefacts

4.2.4 Actor Tensions (Who)

To engage with the metaverse, users require a computer interface of a particular type, for instance, First-person or third-person avatars are used in Epic Games current release. Owing to their unique properties and capabilities, users can simultaneously interact with other actors and artefacts in the Metaverse's space and time.

The actor's similarity to the human user can be determined by the metaverse designer. (Humans portray themselves on social media platforms) or whether the user is capable of assuming a very different role with possibly different capabilities (an approach used in role-playing games). There is only one physical instance of the unique self in the physical universe, despite the fact that it can take on many other roles and identities. However, in the metaverse, a single physical self can create a variety of instantiations, entities, avatars, or personas. These entities can sometimes have a link or may be completely unrelated. The individual (non-virtual) self may only interact with the metaverse while actively doing something, although online selves may be involved in activities like changing settings or working towards some goals.

In this way, the single self can assign tasks to several selves, and the multiple selves can operate somewhat independently [26]. Figure 8 visualizes this view.

Fig. 8 Designing metaverse actors

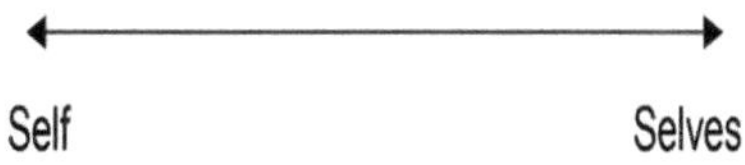

4.2.5 Transition Tensions

Experience dimensions describe the domain of problems where designers work to generate new metaverse experiences. The transition dimension and associated tensions describes the merging of several metaverse experiences and hence jump design spaces, that is, the manner in which one transitions between experiences.

Considering the metaverse as a meta design space, yet another issue is, how does one transition between experiences, or from one design space to another. This fifth design component addresses interconnection, whereas the first four design dimensions focused on experience.

We need three components to characterize a transition: the source, the destination, and the transformations that must be made to go from one experience to another. During any transition within spatiality, temporality, artefacts, and the actor, the transition dimension is applied and therefore operates orthogonal to the other four design dimensions. For example, during actor transitions, the user's roles and skills may vary. Spatial transitions may entail going from a two-dimensional to a three-dimensional space. Temporal transitions may alter the rapidity with which people interact. Artefactual transitions may entail shifting from one environment to another, sometimes with completely different physics and regulations.

Thus, we can talk about two aspects of transitions that are related to.

1. The experience (artefacts, space, temporality, and actors) of moving from one place to another and
2. The transformation's overall design.

Transitions can range from being incoherent to being coherent in terms of the experience. For example, if one is switching from one space to another where each space produces a very different experience, such as going from a two-dimensional game to a three-dimensional game or from a shooter experience to a racing experience, the transition is incoherent. Coherence has been evaluated in terms of both texts and pictures [27, 28].

The transformation might be discontinuous or continuous: a continuous transition is one in which the transformation proceeds step by step, whereas a discontinuous transition is one in which the shift is sudden. Figure 9 visualizes this view.

Taking example of Epic's Fortnite, where players may switch between several experiences, but they must go through a user interface to do it. On the one hand, the same engine is utilized for all experiences and physics, and the appearances are equivalent, resulting in a sense of being coherent. On the other hand, using a user interface forces the user to switch from one experience to another, which gives the user a sensation of discontinuity.

A sense of coherence is crucial to how experiences are put together. For example, a night out might be planned with events that appear to be unrelated, or it can be planned

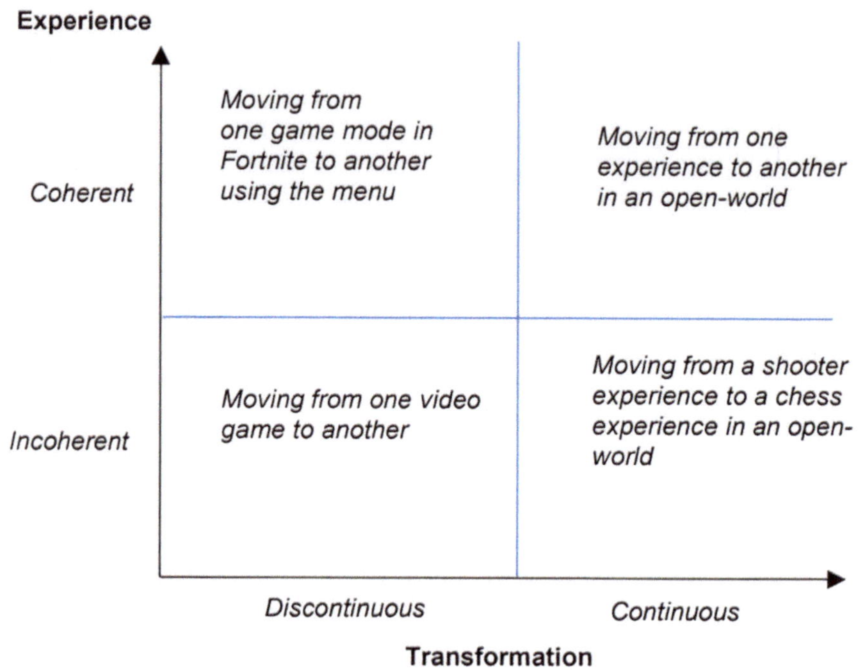

Fig. 9 Designing metaverse transitions

with themes that emerge from each of the experiences. Indeed, an essential part of narrative construction is the creation of linking tissue between events that appear to be distinct. For example, back-to-back episodes in Paris, Berlin, and London might examine virtual art scenes. Several episodes might be united by a shared activity. Similar in literary works, connecting tissue can use metaphors, for example, a shovel in a building game might change into a wand in a fantasy game.

Cities provide for enjoyable evenings since experiences may be combined in logical and coherent ways. Many experiences can be combined in a short amount of time because city infrastructure creates a dense landscape of services [29].

The metaverse may also enable coherent experience design. It might be maintained by virtual zoning and situation-transition infrastructure, just like a city is supported by physical zoning and transportation infrastructure. Although it is possible to envision a network of virtual services that is generated at random, there are benefits to some sort of virtual zoning where experiences that share similar concepts are placed close to one another. Even though transitions may be intended to happen instantly, slower transitions may still be crucial for a number of reasons. Cognitively, we are accustomed to making adjustments to new contexts even before we enter them, thus making adjustments occasionally may be crucial. Technically, switching between virtual worlds may require reconstructing intricate, high-fidelity maps and converting currencies and artefacts for usage in a different setting. To prevent delays and difficulties, it may be important to take some time to complete this task.

More generally, the metaverse will need to offer a wide variety of experiences that can be integrated in order to be productive. The meta experience, such as the virtual night out, may then be built, with coherence being one of the design objectives.

5 AR, VR, and MR as Precursor in Metaverse Design

The Reality-Virtuality Continuum was proposed by Milgram and Kishino [3]. The latest revision of the continuum now includes additional branches of alternate realities that incline further in the direction of physical realities such as depth sensing, machine vision, and artificial intelligence to create a seamless blend between the digital and physical world, notably MR and the innovative holograms, similar to the digital objects from the Star Trek franchise [22] which are projected into real-world environments and can be interacted with by users. Holograms are created using advanced techniques such as interference, diffraction, and polarization, which work together to project a 3D image that appears to be floating in mid-air. However, we focused our discussion on just four major reality types which caught the attention of not only academicians but the business communities. This section starts with the well-known realm of VR and gradually moves on to the future domains of AR and its better versions like MR and holographic technologies. This section also serves as an prelude to the way that XR connects virtual and real worlds.

5.1 Virtual Reality

VR has the defining property of completely synthetic sights. Commercial VR headsets offer standard methods of user interaction, such as head tracking or tactile controllers [30]. User interaction techniques are used to allow people to engage with virtual items while they are positioned in entirely virtual settings. Virtual reality is considered to be at the far end of the Reality-Virtuality Continuum, as it creates fully immersive and simulated environments that are completely separated from the real world [3].

That is, users of VR headsets should fully focus on virtual surroundings and remain disconnected from the actual world. The consumers of the metaverse will provide content for the digital replicas, as already discussed in the prelude. At present, consumers can produce content using commercial virtual environments, e.g. VR painting with Google tilt brush on Meta Quest 2 and Steam. User affordance exploration may be accomplished by consumer interaction with the items in the virtual world, such as tweaking the shape of a virtual item and creating new imaginative creations. These virtual environments allow for real-time collaboration between multiple users. This corresponds to the well-defined needs of the virtual world: a shared sense of space, a similar sensation of presence, a shared sense of time (real-time interaction), communication (through gesture, text, voice, etc.), and a method of sharing information and manipulating digital objects. It is critical to remember that many users in a virtual space, i.e. a subset of the metaverse, should view the same information as other users. Additionally, users can communicate with one another in a consistent and timely manner. Taking into account the metaverse's final stage, users in a virtual shared area should collaborate with any additions or interactions from the physical equivalent, like AR and MR. The essence of constructing the metaverse, through the composition of various virtual shared spaces, must bring together the simultaneous behaviours of objects, avatars representing their users, and their interactions, such as object-avatars, object-object, and avatar-avatar. All processes in virtual environments should be synchronized and should reflect the dynamic events of the virtual world [31]. However, handling and setting up dynamic states and events at scale is a significant challenge, particularly when we take into account large numbers of concurrent users interacting with one another and acting on virtual objects at unacceptable latency, which could negatively affect user experiences.

5.2 Augmented Reality (AR)

Surpassing the virtual worlds, augmented reality (AR) provides alternate experiences to human users in their physical world, with the purpose of improving it. Theoretically, several perceptual information channels, including audio, sights, smell, and haptics can be used to convey computer-generated virtual contents. The first generation of AR system frameworks focused solely on visual upgrades, with the goal of organizing and displaying digital overlays superimposed on top of our actual physical environments.

As evident through very early studies in the 1990s [32], the lack of user movement in a large see-through display necessitates consumers to interact with textual and 2D interfaces using tangible controllers sitting comfortably on their couches.

Significant research efforts have been undertaken since the first study to better user interaction with digital entities in AR. It is crucial to keep in mind that the digital objects, potentially from the metaverse, superimposed before the user's actual surroundings ought to enable human users to seamlessly integrate the simultaneous operations (similar to VR). As a result, one of the main difficulties in integrating users from the real world with the metaverse is ensuring effortless consumer interaction with such digital entities in AR [33]. Freehand interaction techniques show how simple and ready-to-use interfaces for AR user interactions may be created, as seen in the majority of science fiction films, such as Steven Spielberg's Minority Report [34]. Voodoo Dolls [35], a popular freehand interaction technology, is a system solution in which users may use both the hands to pick and operate on virtual items with pinch movements.

HOMER [35] is an additional innovative one-handed handheld augmented reality user interaction that shows the trajectory of rays emanating from a user's virtual hand as they pick and handle augmented reality objects.

Furthermore, AR will locate everything in our living environments, such as annotating directions in an unfamiliar location and pinpointing objects based on consumer contexts [36]. Consequently, we could anticipate that AR will link the metaverse with our urban environment and that digital entity will manifest in obvious and palpable ways on top of a variety of real-world urban areas.

In other words, users of AR simultaneously interact with both the worlds. This necessitates major efforts in detecting and tracking technologies to link the virtual contents apparently shown with the appropriate position in the actual world.

The Touring Machine is regarded as the first research prototype that enables outdoor augmented reality application. The prototype is made up of computing hardware and a GPS unit mounted on a backpack, as well as a head-worn display with map navigation information. Through a touchscreen and a hard-pointed instrument called stylus, the user of Touring Machine can engage with the augmented reality map [37]. In contrast, contemporary AR headsets have shown great advancements, particularly in user mobility. Customers of portable augmented reality headgears can get audible and visual feedback cues identifying AR items, but additional senses like smell and haptics are still underutilized [34]. It is important to note that AR headgears are not the sole way to access the metaverse's contents but are just capable of displaying AR overlays as well as digital entities from the metaverse. AR headgears outperform other strategies in terms of diverting user attention and occupying users' hands. First, human consumers must shift their focus between real-world surroundings and digital content on other kinds of AR tools. AR headgears, on the other hand, allow AR overlays to be presented in front of the user's eye sight [38, 39]. Second, the tangible devices won't require the user to use their hands because the processing units and screens are positioned on the users' heads. These benefits allow users of AR headsets to explore "the metaverse via an AR lens" in real time.

5.3 Mixed Reality (MR)

Unfortunately, no widely accepted definition of MR exists, but it is essential to have a common term which characterizes the alternating reality placed between the two extremes of AR and VR as discussed in Milgram and Kishino's Reality-Virtuality Continuum [3].

According to the research community, MR is an intersection between AR and VR that allows consumers to involve with virtual elements in real-world scenarios. It is important to note that MR objects can interact with other tangible things in a variety of surroundings when they are backed by strong environmental understanding or situational awareness. A physical screwdriver, for example, may spin digital entities of screws with slotted heads in MR, illustrating a key characteristic of digital-physical interoperability. Contrarily, as seen in the applications currently in use [34], typically, AR simply overlays information onto the real world, taking such interoperability into account. A substantial number of papers that depict closer connections and collaborative linkages between the real-world spatial, user interaction, and virtual entities see MR as a stronger form of AR when taking this additional characteristic into account [34, 36, 40].

From the above discussion, despite the fact that we do not have a definitive conclusion about MR, it is the starting point for the metaverse. We believe that the metaverse begins with digital replicas, following which, human consumers begin creating content in the digital twins.

In light of this, digitally created material can reflect physical environments, and users of such digital objects anticipate that they will blend with their real surroundings across time and space. Although we are unable to foresee how the metaverse will ultimately affect our physical surroundings, we can observe that the current MR prototypes include some specific objectives, such as pursuing realistic scenes [24], evoking sensations of presence, and empathetically developing physical-spatial awareness. The aforementioned goals are consistent with the metaverse, which advocates for several virtual worlds to function in parallel.

5.4 Holography, Large Display, and Pico-Projector

This section seeks to speculatively suggest methods for connecting the individually created contents inside the virtual world (ultimately metaverse) to their physical counterparts in the common shared public space. We have no indication that mobile headsets will be the exclusive means of offering metaverse thing into public spaces. To project pixels into our physical world, we may instead use more advanced technology like bigger screens and pico-projectors. Users without mobile headgears can perceive digital objects with a high degree of realism through the use of large screens and pico-projectors [25]. In addition, smartphones with tiny projectors inside of them, e.g. MOVI Phone, with embedded projection technology makes use of laser beam steering that enables multimedia sharing at anytime and anywhere. It is also important to note that smart phones are currently the most common electronic devices.

Finally, we address the possibilities of holographic technology stressing rich communication media on 2D displays and pursuing real volumetric displays (showing pictures or movies) that are indistinguishable from ordinary everyday objects. Currently, there are two forms of holographic technology available: reflection-based and laser-driven holograph. A recent study revealed the possibility of coloured volumetric display on heavy and desk bound devices, despite practical restrictions such as poor resolution, which might influence user perceptions of realism. But the main benefit of reflection-based holography is to produce colourful holograms with colour reproduction that is remarkably accurate to real-world objects [41]. Plasma Fairies, on the other hand, is a 3D aerial hologram that can be perceived by the users' skin surfaces, albeit the devices can only emit plasmonic emission in a 5 cm^3 mid-air zone. We hypothesize that the metaverse can integrate with our living urban and give stakeholders in urban areas a strong sensation of presence if technological advancement enables such volumetric 3D things to appear in the real world frequently. However, the above works suffer from three major flaws in holographic technology: limited resolution, display size, and device mobility. Overcoming these flaws is therefore a crucial step in producing enhanced 3D visuals in the actual world.

6 Challenges in Metaverse Design

This section examines a range of design aspects that influence the metaverse's acceptability. These design aspects include privacy issues, user diversity, fairness, user addiction, cyberbullying, device acceptance, cross-generational design, acceptability of users' digital copies (i.e. avatars), and green computing (i.e. design for sustainability).

6.1 Privacy Issues

Regardless of the novel possibilities enabled by the metaverse ecosystem, it will need to address the issue of potential privacy leakage more quickly, when the issue is so grave in the environment that any solution to address privacy concerns would require redesign from the scratch. The third-party cookie-based advertising ecosystem, where the initial design goal was to provide utilities, is an illustration of this problem. The entire revenue model depended on cookies, which follow users to give customized ads, and it was too late to address privacy concerns. Finally, they were put into effect by privacy laws like GDPR, and Google's decision to remove third-party cookies from Chrome by 2022 effectively put an end to the ecosystem of third-party cookie-based advertisements. The public outcry against Google Glass shows that people can be very wary of new technologies that they see as infringing on their privacy or behaving in a socially unacceptable manner. Following that, a varied solutions presented to safeguard the privacy of onlookers and non-users. Moreover, everything relies on the owners as there exists no legal or technical mechanism in place to verify whether non-users privacy was actually respected. One of the key issues that must be overcome in order to gain social acceptance is creating a verifiable privacy mechanism. The privacy paradox, in which users actively divulge their own information, is another facet of privacy hazard in the context of societal acceptability. Users generally don't pay attention to how their public data are used by third parties, but they do react negatively when the contrast between how their data are actually used and how they are perceived to be used becomes too stark and clear.

For example, almost everyone gives permission on Facebook data sharing. Nonetheless, the Facebook and Cambridge Analytica Data Scandal sparked such a public outcry that Facebook was summoned to congressional hearings in both the United States and the United Kingdom, and Cambridge Analytica went bankrupt shortly after. The elimination of all user data collection is one solution.

However, it will significantly reduce the potential innovations that the ecosystem could enable. Consent-based privacy trading allows consumers to sell their data in exchange of rewards, whether monetary or otherwise, and has also been suggested as a solution by world leaders including former German Chancellor Angela Merkel.

Researchers have already shared their perspectives on the economics of privacy [27], as well as the creation of an efficient market for privacy trade [28]. This strategy will facilitate the data flow required for possible improvements while also fairly compensating people for their data, opening the door for greater social acceptance. As more advanced technologies are developed for creating and accessing the metaverse, it will be important for developers to consider and address these privacy and social concerns in order to gain public acceptance and ensure the safety and security of users in the metaverse. This can involve things like transparent data policies, explicit user consent requirements, and clear guidelines for acceptable behaviour in the virtual world.

6.2 User Diversity

According to a visionary design of human-city interaction [36], different stakeholders should be taken into account when designing mobile AR/MR user interaction in urban settings. Similar to this, the metaverse should be open to all members of the community, including those who are young, old, disabled, and of any colour, gender, age, or religion. Different contents may exist in the metaverse, and we must make sure they are appropriate for a wide range of consumers. Additionally, it is imperative to consider the customized content display for the users and endorse the fairness of suggestions to minimize the biasness of content and thus transform the behaviour and decision-making ability of the recommendation systems [42]. By providing factors of enjoyment, emotional involvement, and arousal, the content in virtual worlds can promote greater acceptance [43]. Creating content with the goal of increasing acceptance under the heading "Design for Diverse Users" would be a difficult task.

6.3 Fairness

The metaverse will be home to a large number of virtual worlds, each with its own set of laws to control user behaviour and activities. Therefore, administering and maintaining such virtual worlds would need a lot of effort. We anticipate that autonomous agents will play a role in virtual world governance, assisted by AI, to reduce the needs of manual labour. It is crucial to note that autonomous agents in virtual environment depends on machine learning algorithms to respond to the dynamic but continual changes of virtual items and avatars. It is a commonly recognized fact that no model can accurately capture every detail of a given real-world instance, and in the same way, a model that is improper or biassed could consistently degrade user experiences in the metaverse. Certain user groups may suffer as a result of the biased services.

Some social groupings may not be well-represented on social networks when user-generated text is summarized using algorithmic techniques. Fairness-preserving summarization algorithms, however, can result in services that are generally of good quality for all social groups [44]. For this reason, the algorithmic fairness should be a basic principle of the metaverse designs, with the metaverse being viewed as a virtual society [45], and maintaining procedural fairness needs a high level of user transparency and result control measures, which are necessary when using algorithms and for computer agents to perform managerial and governance responsibilities. The users' modifications to the algorithmic outputs that they believe are fair and are referred to as outcome controls [46]. It might be harmful if outcomes are not useful to specific people or groups. This suggests that user perceptions are significant for the fairness of such machine learning systems, also called perceived fairness. Leaning towards perceived fairness, however, may fall victim to result favourability bias [22]. Additionally, creators of the metaverse should establish ways to gather the opinions of various community groups and work together to create solutions that promote equity in the metaverse surroundings [45].

6.4 User Addiction

When the metaverse takes over as the most popular place for individuals to spend their time in virtual worlds, excessive usage of digital environment (i.e. user addictions) will become a significant problem. Consumers can exploit the metaverse to aid them in "escaping" from the real world. Earlier research has discovered evidence of addictions to different virtual cyberspaces or digital platforms, including social networks, mobile applications, smartphones, VR, AR, and others. Even if screen time limits had been extensively used, user addictions to cyberspaces might result in psychological problems and mental diseases, such as sadness, loneliness, and user anger. Ever since, the COVID-19 pandemic has caused a paradigm shift away from person to person online gatherings to various virtual ways. As a result, the concern of whether the metaverse increase the addiction among consumers, we examine the current AR/VR systems and, without any supporting data, speculate about probable behavioural changes. First, the virtual reality chat application known as VR Chat may be used as a test case for addiction to the metaverse. In the meanwhile, VR researchers looked at the relationship between such addictive behaviour in VR, its underlying causes, and effective treatments [47]. Additionally, AR games like Pokemon Go may cause large-scale player behavioural changes, including purchasing patterns, group activities in certain physical region, and risky or dangerous activity in the real world. These behaviour changes may have noticeable effects on society. The extended self of a user, which includes the person's mind, body, physical possessions, family, friends, and affiliation groups, is explained by a psychological perspective as encouraging user exploration, combining virtual worlds and the pursuit of rewards, perhaps in an unending reward-feedback loop. We have to make clear that our discussion on the addictions of immersive environments (AR/VR) here is meant to spark discussion and attract research interest.

In the metaverse, the users of the metaverse may engage in a variety of activities that closely mirror those in the actual world caused by super-realism. With the bold assumption that such environments can worsen addictions, such as longer usage times, highly realistic virtual environments also allow people to try things that are impossible in their real lives (such as replicating an event that is immoral in our real lives [48] or experiencing racism [49]). We may be able to better grasp the new aspects of user addiction brought on by the super-realistic metaverse with more research and observation of the wild user behaviour.

6.5 Cyberbullying

Cyberbullying is a form of bullying that involves intentionally sending, posting, or sharing negative or harmful content about someone, whether it's true or not, with the intention of hurting or harassing them [31]. We see the metaverse as an expansive cyberspace as well. As a result, cyberbullying in the metaverse may be another unavoidable social threat to cyberspace.

Authorities will appeal the closure of some virtual environments in the metaverse world, as is customary, in order to stop the current cyberbullying cyberspace, since the metaverse would not be able to function over the long term. Additionally, because there are so many virtual worlds, the metaverse would use algorithms-based cyberbullying detection methods [50]. The fairness of such algorithms will be essential in ensuring that users of the metaverse perceive fairness. After locating any instances of cyberbullying, effective mitigation strategies should be implemented in virtual settings, such as self-disclosures, assistance and support, and online social networks. But unlike social networks, the game-like atmosphere makes it much harder to spot cyberbullying. For instance, the users' inappropriate actions may be ambiguous and hard to determine [51].

Similar to this, 3D virtual worlds within the metaverse can further confuse the situations and hence make large-scale cyberbullying identification challenging.

6.6 Other Factors

First, additional research is required to determine whether the public or bystanders accept certain gadgets, such as mobile AR/VR headsets that connect users to the metaverse. Additionally, the user experience in virtual worlds may suffer as a result of mobile headset user safety issues for both the users and any nearby bystanders. To the best of our knowledge, there aren't many research on the social acceptability of virtual worlds, and neither have we discovered any on digital twins or the metaverse. Also indicating that Gen Z adults like Snapchat, Instagram, and Tiktok when compared with Facebook which are the disparities in cross-generational social networks. Instead, Facebook keeps more members who are in Generations X and Y [33]. Social networks have so far been unable to service all users from all demographics on a single platform. We must plan for the user design of cross-generational virtual worlds in light of the failed scenario, particularly when we take into account the metaverse with the dynamic user cohorts in a unified environment.

Additionally, we must take into account how well consumers will accept avatars, or digital representations of individuals, across time points. What, for instance, is the acceptance of the deceased consumer's family, relatives, or friends to the avatars?

The question posed is extremely imperative to the concept of virtual immorality, which is the concept of digitally preserving someone else's identity and actions. As we continue to develop virtual surroundings made up of both virtual objects and avatars as distinct entities from the physical world, the answer to this question could also influence the direction of digital humanity [52] in the metaverse. For instance, should we permit new users to communicate with an avatar that is two centuries long and represents a user who has likely passed away?

The metaverse, which is thought of as a vast digital universe, will also be sustained by a vast number of computational devices. The metaverse can thus result in significant energy usage and pollution. The designers of the metaverse shouldn't ignore design issues from a green computing standpoint since they shouldn't deny future generations. Eco-friendliness and environmental responsibility may affect how much users love the metaverse, how they feel about it, how many people use it, and even how many people oppose it [53]. Therefore, for the metaverse to be widely adopted, data analytics-based sourcing and metaverse construction based on sustainability indices would be required.

We conclude by briefly mentioning additional elements that could affect how accepting users are of the metaverse, including in-game damage, unforeseen horrors, user isolation, accountability, and trust, identity theft/leakage, virtual misdemeanor, manipulative materials that cause users to act (like persuasive advertising).

7 Conclusion

This chapter laid out a bold new vision for metaverse that outlines the requirements, architecture, standards, status, challenges, and associated research. First, we provided the fundamental concepts of the metaverse and discussed a variety of characteristics that allow the definition of several components of the metaverse. Then, we discussed about the current virtual environment standards used in the metaverse. Because of the use of emerging technologies and the continual expansion and improvement of the ecosystem, the look of our virtual worlds (or "digital twins") will undergo substantial change in the coming years. The metaverse can be viewed as the next generation of the Internet that increasingly provides varied and interconnected immersive experiences. To this end, we have conceptualized the,etaverse as a meta design space that supports transitions and recombination of experiences centred in current and future platform ecosystems. The metaverse is now an empirical phenomenon as software vendors aim to move towards implementing such interconnected immersive experience and have started to create business models around the idea. It may be argued that the topic of the metaverse is simply old wine in new bottles and that debates of a similar kind have already taken place in relation to immersive systems and virtual worlds. The metaverse concept's universality, however, raises the possibility that it may represent the next stage in the evolution of the Internet since it permits the recombination of immersive and emergent experiences in addition to web-based data. The metaverse will involve not only transitions between immersive virtual experiences but transitions between physical and virtual experiences. Simulations allow us to create things in a virtual world that are then transformed back into physical artefacts and processes, thus increasing our overall ability to produce goods and services. The thrust for integration and interconnectedness of tomorrow is not going to look the same as yesterday. A range of experiences will be made possible by the convergence of platform ecosystems and artificial intelligence. The information systems design challenges revolve around the question of how to recombine these experiences. Additionally, the fight over who will rule the metaverse has begun, raising problems of ownership, control, and governance that go beyond just technological ones. Lastly, technology giants such as Google and Apple have grandiose intentions to make the metaverse a reality. Due to the availability of advanced computing devices and intelligent gears, our digitized future will now be more multidimensional, interactive, lively, embodied, and living.

References

1. Joshua J (2017) Information bodies: computational anxiety in Neal Stephenson's Snow Crash. Interdisc Literary Stud 19(1):17–47 (Penn State University Press)
2. Bruun A, Lynge Stentoft M (2019) Lifelogging in the wild: participant experiences of using lifelogging as a research tool. In: 17th IFIP conference on human-computer interaction (INTERACT). Paphos, Cyprus, pp 431–451
3. Milgram P, Takemura H, Utsumi A, Kishino F (1995) Augmented reality: a class of displays on the reality-virtuality continuum. In: Telemanipulator and telepresence technologies, vol 2351. SPIE, pp 282–292
4. Mohammadi N, Taylor JE (2017) Smart city digital twins. In: 2017 IEEE symposium series on computational intelligence (SSCI), pp 1–5 (2017)
5. Grieves MW, Vickers J (2017) Digital twin: mitigating unpredictable, undesirable emergent behavior in complex systems
6. Viljoen S (2020) The promise and limits of lawfulness: inequality, law, and the techlash. Int Polit Econ: Glob eJournal
7. Jiang Y, Zhang C, Fu H, Cannavò A, Lamberti F, Lau YK, Wang W (2021) HandPainter—3D sketching in VR with hand-based physical proxy. Association for Computing Machinery, New York
8. Nebeling M, Lewis K, Chang YC, Zhu L, Chung M, Wang P, Nebeling J (2020) XRDirector: a role-based collaborative immersive authoring System. Association for Computing Machinery, New York, pp 1–12

9. Kumaravel BT, Nguyen C, DiVerdi S, Hartmann B (2019) TutoriVR: a video-based tutorial system for design applications in virtual reality. Association for Computing Machinery, New York, pp 1–12
10. Muller J, Radle R, Reiterer H (2016) Virtual objects as spatial cues in collaborative mixed reality environments: how they shape communication behavior and user task load. Association for Computing Machinery, New York, page 1245–1249
11. Microsoft mesh. https://www.microsoft.com/en-us/mesh. Date: 25 Jan 2024
12. Nvidiaomniverse. https://www.nvidia.com/en-us/omniverse/. Date: 25 Jan 2024
13. Osborne C (2022) ISO/IEC 23005 (MPEG-V) standards. Available https://mpeg.chiariglione.org/standards/mpeg-v. Last accessed: 25 Dec 2023
14. I. SA. IEEE 2888 standards. https://sagroups.ieee.org/2888/. Accessed 25 Jan 2024
15. IEEE 1589 standards. https://standards.ieee.org/ieee/1589/6073/. Accessed 25 Jan 2024
16. IEEE P2048 standards. https://standards.ieee.org/ieee/2048/11072/. Accessed 25 Jan 2024
17. IEEE P7016 standards. https://standards.ieee.org/ieee/7016/11078/. Accessed 25 Jan 2024
18. Hanseth O, Lyytinen K (2010) Design theory for dynamic complexity in information infrastructures: the case of building internet. J Inf Technol 25(1):1–19
19. Clark DD, Wroclawski J, Sollins KR, Braden R (2005) Tussle in cyberspace: defining tomorrow's internet. IEEE/ACM Trans Networking 13(3):462–475
20. Rawat DB, Alami HE. Metaverse: requirements, architecture, standards, status, challenges, and perspectives. arXiv:2302.01125v. Available 24 Jan 2024
21. Muller J, Radle R, Reiterer H (2016) Virtual objects as spatial cues in collaborative mixed reality environments: how they shape communication behavior and user task load. Association for Computing Machinery, New York, pp 1245–1249
22. Wang R, Harper FM, Zhu H (2020) Factors influencing perceived fairness in algorithmic decision-making: algorithm outcomes, development procedures, and individual differences. In: Proceedings of the 2020 CHI conference on human factors in computing systems, CHI'20, New York, NY, USA, pp 1–14
23. Zoshak J, Dew K (2021) Beyond Kant and Bentham: How ethical theories are being used in artificial moral agents. Association for Computing Machinery, New York
24. Lee C, Rincon GA, Meyer G, Höllerer T, Bowman DA (2013) The effects of visual realism on search tasks in mixed reality simulation. IEEE Trans Visual Comput Graphics 19(4):547–556
25. Hartmann J, Yeh YT, Vogel D (2020) AAR: augmenting a wearable augmented reality display with an actuated head-mounted projector. In Proceedings of the 33rd annual ACM symposium on user interface software and technology, UIST '20, New York, NY, USA, pp 445–458
26. Terrace J, Postava EC, Levis P, Freedman MJ (2012) Unsupervised conversion of 3D models for interactive metaverses. In: 2012 IEEE international conference on multimedia and expo, pp 902–907. ISSN 1945-788X
27. Acquisti A, Taylor C, Wagman L (2016) The economics of privacy. J Econ Literature 54(2):442–492
28. Pal R, Crowcroft J, Wang Y, Li Y, De S, Tarkoma S, Liu M, Nag B, Kumar A, Hui P (2020) Preference-based privacy markets. IEEE Access 8:146006–146026
29. Webster C, Garnier F, Sedes A (2017) Empty room, an electroacoustic immersive composition spatialized in virtual 3D space, in ambisonic and binaural. In: Proceedings of the virtual reality international conference—Laval Virtual 2017, VRIC'17, New York, NY, USA. Association for Computing Machinery, pp 1–7
30. Kelly JW, Cherep LA, Lim A, Doty TA, Gilbert SB (2021) Who are virtual reality headset owners? a survey and comparison of headset owners and non-owners. In: 2021 IEEE virtual reality and 3D user interfaces (VR), pp 687–694
31. Chatzakou D, Leontiadis I, Blackburn J, Cristofaro ED, Stringhini G, Vakali A, Kourtellis N (2019) Detecting cyberbullying and cyberaggression in social media. ACM Trans. Web 13(3)
32. Feiner SK, MacIntyre B, Haupt M, Solomon E (1993) Windows on the world: 2d windows for 3d augmented reality. In: UIST '93

33. Fietkiewicz KJ, Lins E, Baran E, Stock WG (2016) Inter generational comparison of social media use: investigating the online behavior of different generational cohorts. In: 2016 49th Hawaii international conference on system sciences (HICSS), Los Alamitos, CA, USA, pp 3829–3838
34. Lee LH, Hui P (2018) Interaction methods for smart glasses: a survey. IEEE Access 6:28712–28732
35. Pierce JS, Pausch R (2002) Comparing voodoo dolls and homer: exploring the importance of feedback in virtual environments. In: Proceedings of the SIGCHI conference on human factors in computing systems
36. Lee LH, Braud T, Hosio S, Hui P (2021) Towards augmented reality driven human-city interaction: Current research on mobile headsets and future challenges. ACM Comput Surveys (CSUR) 54(8):1–38
37. Feiner SK, MacIntyre B, Höllerer T, Webster A (1997) A touring machine: prototyping 3d mobile augmented reality systems for exploring the urban environment. In: SEMWEB (1997)
38. Chaturvedi I, Bijarbooneh FS, Braud T, Hui P (2019) Peripheral vision: a new killer app for smart glasses. In: Proceedings of the 24th international conference on intelligent user interfaces, IUI '19, New York, NY, USA, pp 625–636
39. Zhang T, Li YT, Wachs JP (2016) The effect of embodied interaction in visual-spatial navigation. ACM Trans Interact Intell Syst 7(1)
40. Gardony AL, Lindeman RW, Brunyé TD (2020) Eye-tracking for human-centered mixed reality: promises and challenges. In: Optical architectures for displays and sensing in augmented, virtual, and mixed reality (AR, VR, MR), vol 11310. International Society for Optics and Photonics, p 113100T
41. Martín IV, Sáez JM, Climente MG, Chemisana G, Collados M, Atencia J (2021) Full-color multiplexed reflection hologram of diffusing objects recorded by simultaneous exposure with different times in photopolymer bayfol® hx. Opt Laser Technol 143:107303
42. Schelenz L (2021) Diversity-aware recommendations for social justice? exploring user diversity and fairness in recommender systems. In: Adjunct proceedings of the 29th ACM conference on user modeling, adaptation and personalization, UMAP '21, New York, NY, USA, pp 404–410
43. Holsapple CW, Wu J (2007) User acceptance of virtual worlds: the hedonic framework. SIGMIS Database 38(4):86–89
44. Dash A, Shandilya A, Biswas A, Ghosh K, Ghosh S, Chakraborty A (2019) Summarizing user-generated textual content: Motivation and methods for fairness in algorithmic summaries. In: Proceedings of ACM Human-Computer Interaction, vol 3(CSCW)
45. Woodruff A, Fox SE, Schindler SR, Warshaw J (2018) A qualitative exploration of perceptions of algorithmic fairness. In: Proceedings of the 2018 CHI conference on human factors in computing systems, CHI '18, New York, NY, USA, pp 1–14
46. Lee MK, Jain A, Cha HJ, Ojha H, Kusbit D (2019) Procedural justice in algorithmic fairness: leveraging transparency and outcome control for fair algorithmic mediation. In: Proceedings of ACM Human-Computer Interaction, vol 3(CSCW)
47. Segawa T, Baudry T, Bourla A, Blanc JV, Peretti CS, Mouchabac S, Ferreri F (2019) Virtual reality (VR) in assessment and treatment of addictive disorders: a systematic review. Front Neurosci 13
48. Lewis R, Poleskey MT (2021) Hidden town in 3d: teaching and reinterpreting slavery virtually at a living history museum. J Comput Cult Herit 14(2)
49. Dufresne KV, Stout B (2021) Anchorhold afference: virtual reality, radical compassion, and embodied positionality. Association for Computing Machinery, New York
50. Yan R, Li Y, Li D, Wang Y, Zhu Y, Wu W (2021) A stochastic algorithm based on reverse sampling technique to fight against cyberbullying. ACM Trans Knowl Discov Data 15(4)
51. Kwak H, Blackburn J, Han S (2015) Exploring cyberbullying and other toxic behavior in team competition online games. In: Proceedings of the 33rd annual ACM conference on human factors in computing systems, CHI '15, New York, NY, USA, pp 3739–3748

52. Hou W, Han H, Hong L, Chci WE (2020) A crowd sourcing human-computer interaction framework for cultural heritage knowledge. In: Proceedings of the ACM/IEEE joint conference on digital libraries in 2020, JCDL '20, New York, NY, USA, pp 551–552
53. Zhe S, Wong TN, Lee LH (2013) Using data envelopment analysis for supplier evaluation with environmental considerations. In: 2013 IEEE international systems conference (SysCon), pp 20–24

Ethics and Regulation in Metaverse: Risks, Current Approaches, and Way Forward

Aliya Tabassum, Ezieddin Elmahjub, and Junaid Qadir

Abstract The metaverse is rapidly gaining momentum as the next phase of virtual reality, offering an innovative interactive platform for learning, economics, and social activities. This three-dimensional environment is designed to replicate the real world, and its impact on society is expected to be profound in the coming years. The metaverse has collectively formed a virtual universe, which may become an integral part of our daily lives as technology advances, changing the way we interact with each other and the world. With the evolution and integration of the metaverse into our daily lives, we may face various ethical and legal challenges. In this chapter, we provide a detailed survey on ethics and regulations in the metaverse to understand its current risks and future challenges. We present current approaches and prospective solutions for the challenges faced in this virtual environment. We conclude the chapter with a discussion of appropriate management framework to safeguard the ethical and legal entitlements of metaverse users and co-creators.

Keywords Metaverse · Ethics · Legal · Law · Regulation · Extended reality · Virtual reality · Digital future

---

These authors contributed equally to this work.

A. Tabassum · J. Qadir (✉)
Department of Computer Science and Engineering, College of Engineering, Qatar University, Doha, Qatar
e-mail: jqadir@qu.edu.qa

A. Tabassum
e-mail: aliyatabassum.jntu@gmail.com

E. Elmahjub
College of Law, Qatar University, Doha, Qatar
e-mail: eelmahjub@qu.edu.qa

G. Chhabra and K. Kaushik (eds.), *Understanding the Metaverse*, Blockchain Technologies, https://doi.org/10.1007/978-981-97-2278-5_12

1 Introduction

The term "metaverse" originates from the combination of two key terms "meta" and "universe," indicating a universe beyond our own. The metaverse is a space where individuals can engage with both one another and digital entities within a shared environment, employing advanced technologies such as augmented reality (AR), virtual reality (VR), extended reality (XR), and blockchain [1] as shown in Fig. 1. These technologies are related to each other: augmented reality (AR) overlays digital information, such as text, images, or 3D models, onto the physical world. AR is typically experienced through a device, such as a smartphone or tablet that has a camera and a screen. The device uses its camera to capture the real-world environment, and then applies digital content to that environment in real time. VR submerges individuals in a wholly digital setting, while XR includes both AR and VR, blending the boundaries between the physical and digital realms. Conversely, blockchain has the potential to facilitate secure, transparent, and tamper-proof transactions. As a result, it can play a pivotal role in establishing trustworthy digital identities, virtual economies, and governance systems within the evolving metaverse.

It is essential to recognize that the concept of the metaverse is not a singular entity, but rather a diverse landscape consisting of multiple metaverses managed by different entities and organizations. These metaverses are comprised of a fusion between the digital and physical realms, where virtual worlds feature digital buildings, avatars,

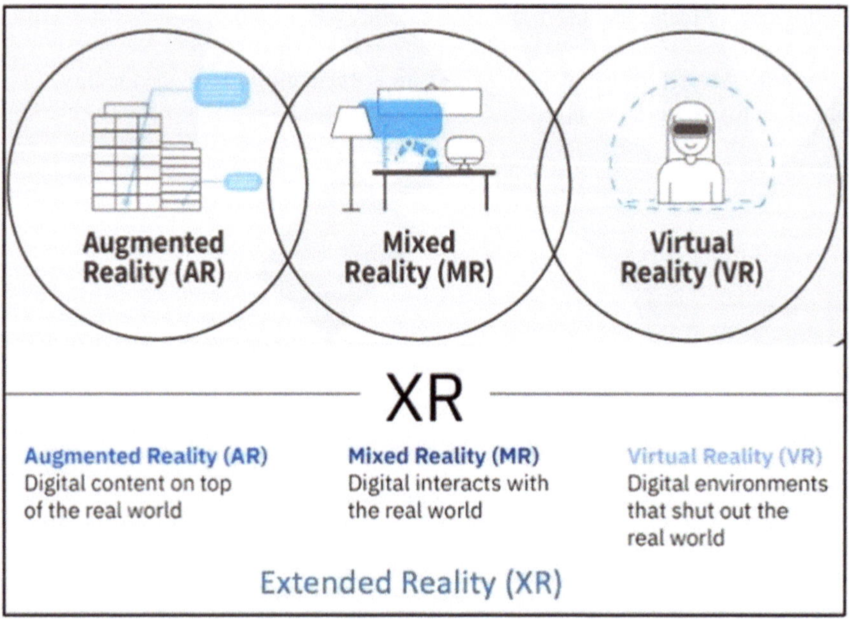

Fig. 1 Various constituent technologies that will play a role in future metaverses

and environments. In these virtual domains, individuals possess the ability to meet, interact, engage in commerce, attend events, and explore various online locations [2]. Furthermore, these virtual worlds encompass their own economic systems that mirror real-world markets, with distinct factors of supply and demand, inflationary dynamics, and even illicit markets [3]. Essentially, the metaverse can be envisioned as an extension of the current online world, but with a heightened level of immersion and interactivity, facilitated by advancements in technology.

The metaverse holds the potential to transform the dynamics of social interaction, transaction, work, and travel. Being a deeply immersive and personalized virtual environment, it allows users to participate in a diverse array of social activities, including chatting, playing games, attending events, and participating in virtual communities. This fosters strong social connections and relationships that are not limited by physical boundaries [2]. Moreover, the metaverse can function as a platform for diverse transactional activities, from buying and selling virtual goods to trading cryptocurrencies and conducting e-commerce transactions. This can create new economic opportunities for individuals and businesses, allowing them to reach a global audience in a decentralized manner [4].

In addition to socializing and transacting, the metaverse can serve as a platform for remote work and collaborative endeavours as well. With the rise of remote work, the metaverse provides an alternative to traditional office spaces, enabling workers to cooperate and engage in joint efforts within a virtual setting. This can reduce the need for physical office spaces, saving costs and providing greater flexibility for employees [5]. Finally, the metaverse can be employed for virtual tourism, offering individuals the chance to discover, experience new destinations and experience different cultures without leaving their homes. This can increase accessibility to travel and empower people to connect with others around the world [2]. The reader is referred to Fig. 2 for a visual representation of some prominent applications of metaverses.

Despite the potential uses of the metaverse, it introduces substantial safety and privacy risks, encompassing concerns such as cyberattacks, hacking, behaviour manipulation, and virtual crimes [6]. This virtual world is susceptible to similar types of crimes that we experience in the physical world, such as theft, fraud, and harassment. Ensuring the safety of users in the metaverse requires measures to prevent and respond to these types of crimes [7]. As individuals participate in social interactions and perform transactions within the metaverse, they disclose personal information. This raises concerns about privacy and data security, especially regarding the potential for data breaches and the improper utilization of personal information [8]. Ensuring the protection of user data and affording control over their information is of utmost importance. The metaverse also creates new ownership and property rights issues. While users buy and sell virtual assets and engage in virtual commerce, questions arise about who owns the assets and what are their rights. The metaverse introduces difficulties concerning intellectual property rights (IPR), including issues like copyright infringement and trademark violations [9]. When users generate and distribute content within the metaverse, there is a challenge to decide who owns the content and what rights they have. There are also possibilities for financial crimes, such as money laundering and tax evasion [10].

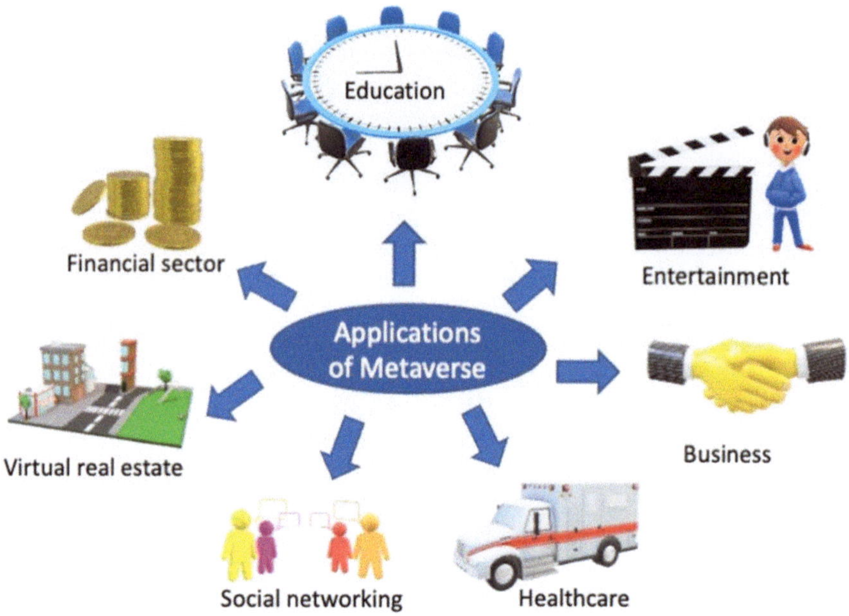

Fig. 2 Prominent application domains of the metaverse

Furthermore, it is important to acknowledge that the creation and advancement of the metaverse are primarily propelled by non-governmental entities, comprising technology firms, game developers, and diverse enterprises. These entities are primarily motivated by profit maximization, as they seek to generate revenue and optimize financial gains through the establishment and progression of the metaverse and its associated applications [11]. Financial incentives drive these private actors to invest resources in metaverse development, aiming to secure returns on their investments. Such returns can manifest in diverse forms, such as user subscriptions, virtual goods sales, in-game advertisements, and other revenue streams. Significant investments have been made by major companies such as Facebook, Google, and Microsoft, who have dedicated billions of dollars to the development of vital VR and AR technologies [11]. As private actors spearhead the metaverse's development, they hold significant influence over its evolution and the range of experiences available to users. This reality raises concerns regarding the potential commercialization of the metaverse and its accessibility and affordability for all individuals.

It is reasonable to assume that private actors will have different priorities and interests than public actors, such as governments and civil society organizations. This will lead to conflicting views around the optimal ethical and legal responses to major issues such as fairness, privacy, safety, and ownership among other things. To ensure that the development of the metaverse benefits to society at large, not limited to private entities alone, it is essential to establish effective ethical frameworks and formal regulations that promote the public interest [12]. This requires collaboration between public and private actors to establish standards and guidelines that promote fairness, transparency, and accountability in the development of the metaverse. From the outset, it is crucial to emphasize that human welfare extends beyond the profit-maximizing and material interests of metaverse designers.

The metaverse may also create new forms of addiction and other mental health issues. Designed for intense engagement and immersion, virtual environments result in individuals dedicating a substantial amount of time within these realms at the expense of their real-world relationships and responsibilities [13]. These have negative consequences for individuals' mental health and well-being.

To tackle such challenges, it is crucial to institute explicit rules and regulations overseeing the utilization of the metaverse. Collaboration among governments, industry organizations, and other stakeholders is essential to formulate policies and best practices that guarantee the safety, privacy, and security of metaverse users. Achieving the full potential of the metaverse as a dynamic and interactive platform for socializing, transacting, working, and travelling will necessitate collaborative efforts.

This chapter tackles these challenges and offers the following contributions.

1. We present a comprehensive examination of the ethical and legal considerations linked to the metaverse. We shed light on the ethical concerns arising from the metaverse's development driven primarily by commercial motives, which may lead to a misalignment with individual and societal well-being. It is crucial to recognize and tackle these issues proactively to guarantee the development of the metaverse in a way that prioritizes the collective well-being of society.
2. We discuss current approaches to mitigate the risks and challenges of the metaverse and proposed solutions to tackle social and economic inequalities, such as digital equality, to ensure that all individuals have equal access to digital technologies and skills.
3. We also explor the potential erosion of privacy and autonomy in the metaverse, where virtual environments collect large amounts of user data for commercial or other purposes, infringing on individuals' privacy and autonomy. We elaborate on the legal and ethical aspects of managing user data to ensure individual privacy and autonomy.

While we acknowledge that each of these topics—ethics, privacy, property, and criminal conduct in the metaverse—possesses sufficient depth to warrant a dedicated research chapter in its own right, we have chosen to present them in this work as an interconnected spectrum of challenges. Our intention is not to exhaustively dissect each area, but to encourage additional interdisciplinary exploration by scholars in law, ethics, and science, recognizing their interconnectedness and mutual influence within the intricate ecosystem of the metaverse.

The chapter is structured as follows: a brief background of the metaverse is presented in Sect. 2, the risks involved with this technology are discussed in Sect. 3, where we elaborate among other risks the problem of surveillance capitalism and behaviour modification. Current and future solutions and methodologies designed to address these issues are explored in Sect. 4. Section 5 delineates unexplored research paths and suggests directions for additional investigation. Ultimately, our findings and observations are summarized in Sect. 6, presenting conclusions derived from the preceding discussions.

2 Background

The inception of VR traces back to the 1960s, attributed to computer scientist Ivan Sutherland, who pioneered the development of the first head-mounted display system. The term "metaverse" was first used by Noel Stephenson in 1992 [3] to define a paradise in order to avert a dystopia in the real world. This idea has been broadened to incorporate tangible activities conducted within a virtual environment utilizing augmented reality, virtual reality, as well as 3D technology, and the Internet of Things (IoT). It is also known as Web 3.0, the subsequent generation of the Internet from its current Web 2.0 state [14]. Etymologically, meta refers to what comes after, beyond, changes, or transcends [2]. The most commonly adopted definition of the term "metaverse" pertains to a virtual realm created through the integration of various technologies where people engage and communicate via avatars, using extended reality as a conduit to connect them. The VR space is substantially becoming an area of interest around the globe. In the year 2001, Edward Castronova [15] performed elementary studies about the virtual worlds and economies by focusing on the online role-playing game EverQuest and forecasted that the VR economy has the potential to impact real-world economies by delivering the perception of community and social interaction that depict the actual world. The way people interact in the metaverse is expected to go beyond using AR or VR headsets and involve the capture of brainwaves to enhance their experiences [16].

1. *Education*: The metaverse has the potential to serve as a platform for distance (remote) education, enabling students from all over the world to attend classes and collaborate with each other in virtual classrooms [17, 18].
2. *Entertainment*: It could be used to generate novel entertainment formats, including virtual concerts, theatrical performances, gaming, and various immersive experiences [19].
3. *Business*: This environment offers new opportunities for businesses to reach customers and engage with them in novel methods, including virtual storefronts, engaging product demonstrations, and interactive customer support [10].
4. *Healthcare*: It is capable to provide virtual healthcare services, such as telemedicine consultations and remote monitoring of patients [17].
5. *Social networking*: The metaverse facilitates new ways of connection and communication through virtual chat rooms, forums, and social networks.
6. *Virtual real estate*: The metaverse enables the creation of virtual real estate, allowing individuals to engage in buying, selling, and leasing virtual properties for various purposes [9].

7. *Financial sector*: It can be an optimal environment to engage in commercial transactions, money transfers, and other forms of banking services [20].

3 Risks and Challenges

In a virtual world where individuals invest a substantial amount of time, there exists the potential for exposure to targeted advertising, propaganda, addiction, and misinformation [21, 22]. The metaverse presents several risks such as potential for targeted influence and manipulation. This technology also raises various ethical concerns regarding the usage of the captured data and the affordability of the technology. The issue of accessibility in the metaverse may be influenced by prevailing technology inequalities, including restricted Internet access, potentially exacerbating divisions and exclusions. The creation of digital avatars using artificial intelligence (AI) algorithms in the metaverse may give rise to disruptions and ethical conflicts. For instance, interactions with AI-based characters possessing human-like personalities and emotions in video games, ethical issues emerge regarding the appropriate treatment of such entities. Moreover, digital twins are created virtually and digitally interconnected through a network of sensors [23]. On one hand, digital avatars can merely be representations or extensions of human creators and therefore do not possess inherent human worth and dignity. But on the other hand, digital avatars can also be imbued with their own unique identities, personalities, and experiences, and they should be afforded the same rights and protections as their human counterparts. Additionally, disputes have emerged regarding toxicity and undesirable behaviours, including bullying or harassment of other players [24]. It is very essential for the creators of metaverse environments to ensure that they are safe for everyone as the metaverse is likely to have a wide range of users. With the change in societal norms, problems surrounding sexuality, virtual affordances, and moderation techniques must also evolve.

In this section, we delve into and delineate various risks linked to the metaverse, encompassing ethical and legal concerns, addiction, cyberbullying, the potential for illicit behaviours like money laundering and terrorism, the challenge of surveillance, behaviour modification, and the automation of AI.

3.1 Ethical Dilemmas in the Metaverse

The evolution of the metaverse raises numerous ethical considerations, involving discussions on various issues, which are deeply embedded in the technical architecture and design of virtual environments. We exemplify a few ethical issues with specific examples below.

1. *Social Disparities in Metaverse Design*: The tech-led development of the metaverse may exploit individuals and communities, particularly those who are

vulnerable or disadvantaged. If virtual spaces and objects are owned by private companies, they might prioritize financial gains at the expense of the well-being of individuals and communities, resulting in some individuals being marginalized or excluded from engaging in the metaverse. Limiting access to the metaverse to those who can afford costly virtual reality equipment intensifies the gap between the affluent and the less privileged [25]. The structural framework of the metaverse holds the capacity to instigate fresh disparities or heighten prevailing digital, social, and economic inequalities. This may result in differential access to digital technologies and skills, amplifying the divide between individuals with advanced technological capabilities and those lacking such access.

2. *Autonomy and Decision-Making in Virtual Realms*: Autonomy refers to an individual's ability to make independent decisions and act freely based on their own values and guided by their motivations, intentions, and decision-making capabilities [26].

   Informed choice, on the other hand, relates to an individual's ability to access and understand relevant information in order to make informed decisions [11]. In the context of the metaverse, autonomy and informed choice are particularly important as individuals are exposed to a wide range of virtual experiences and interactions that may have real-world consequences. Some individuals may encounter virtual environments or interactions that are discriminatory, violent, or otherwise harmful. In such situations, it is essential that individuals are able to exercise their autonomy and make informed choices about their participation in these activities. There are fears that misinformation and propaganda can be worse in the metaverse compared to misinformation on social media and it can undermine autonomy and informed choice when people are trapped in an information bubble [27].
3. *Preserving Worth and Dignity in Virtual Environments*: Virtual environments, while offering innovative possibilities, also introduce ethical challenges concerning the worth and dignity of users. Instances of cyberbullying and virtual sexual assault within the metaverse may infringe upon individuals' dignity and worth, necessitating a careful examination of the ethical implications of such behaviour [28]. These violations in virtual spaces transcends individual experiences, implicating fundamental human rights principles. Careful examination is imperative to uphold human rights within virtual environments, ensuring that users are treated with the respect and dignity they deserve.

3.2 The Risks of Addiction in the Metaverse

The immersive nature of the metaverse raises concerns about potential addiction or dependency issues among users. As individuals immerse themselves in virtual experiences, there is a pressing need to explore strategies to mitigate this risk and promote responsible and healthy engagement with metaverse technologies [29].

Regular interactions with entities within the metaverse may give rise to addiction and emotional dependency. This reliance on virtual entities for information, social interaction, and emotional support raises ethical questions, particularly considering the immersive nature of the metaverse. An important aspect of concern is the impact on the quality of human relationships, questioning the ability to form and sustain genuine human connections [30]. The prospect of addiction in the metaverse introduces not only physical but also psychological risks, raising concerns about potential health problems and mental health issues. This section explores the implications of addiction on both physical and mental well-being within the metaverse environment [31].

3.2.1 Problem of Inclusivity and Bias in the Metaverse

While the metaverse holds promise for creating new forms of social interaction and community, it also presents challenges related to inclusivity and potential perpetuation of biases. The representation of different groups in the metaverse and the treatment of these groups by other users raise ethical complications that need careful consideration [29].

Given the historical and societal biases often inherent in training datasets [32], there's a risk that these biases may persist and even intensify in the recommendations and interactions within the metaverse. Discrimination by algorithms can take various forms, including the exclusion or preferential treatment of user groups based on race, gender, socioeconomic status, or other demographic characteristics. In the metaverse, where AI decisions can yield significant real-world consequences, identifying accountability for biased outcomes becomes an urgent and complex concern. Ensuring equitable access and representation becomes crucial. If certain groups are excluded or marginalized, it contradicts the principle of justice and inclusivity. Regular audits and diverse, representative training data are essential to identify and address biases in AI systems and metaverse, preventing the perpetuation of prejudiced behaviour.

3.3 Ethical Dimensions of Virtual Economies in the Metaverse

Virtual economies within the metaverse introduce complex ethical issues, including copyright infringement and the exploitation of virtual workers. Disputes over ownership and copyright violations of virtual assets, along with the potential for virtual workers to face poor working conditions, emphasize the necessity for robust regulatory frameworks and ethical guidelines to ensure fairness and equity in virtual economic systems [34]. It is necessary to strike a balance between the rights of content creators and the innovative potential of metaverse users.

Apart from virtual assets and economies, huge content is generated which creates a legal conundrum regarding the ownership of these creations. This dilemma necessitates a profound re-evaluation of IPR principles, aiming to reconcile the rights of original creators with the transformative capabilities of AI in the metaverse. Since the metaverse hosts a myriad of virtual assets like digital real estate and in-game items, it also demands legal frameworks capable of handling disputes and facilitating transactions related to ownership rights [35]. The impact of the metaverse on our lives, society, and economies is expected to be substantial. To make the most of these opportunities while minimizing potential negative consequences, proactive consideration of social and ethical challenges is essential.

3.4 Navigating Legal Dimensions in the Metaverse

The metaverse's dynamic landscape brings forth a spectrum of legal intricacies that necessitate careful consideration. Under the broad umbrella of legal concerns, a myriad of issues emerge, spanning intellectual property disputes, contractual frameworks, data protection, and jurisdictional challenges.

This subsection discusses the multifaceted legal landscape of the metaverse, unravelling the complexities associated with each facet. It explores the intricacies of intellectual property within virtual environments, examining disputes arising from virtual asset ownership and copyright concerns. Contractual frameworks governing virtual transactions and engagements also come under scrutiny, shedding light on the challenges of establishing legally binding agreements in the metaverse. Furthermore, the discussion extends to the realm of data protection, addressing the nuances of safeguarding user information within these immersive digital spaces.

1. *Safeguarding Privacy in the Metaverse*: Privacy emerges as a paramount concern within the metaverse, where individuals willingly share personal information and partake in activities they may prefer to keep confidential [36]. For instance, people may take part in virtual activities related to sensitive topics like health or sexuality, and may not want this information accessed or shared without their consent. As users engage in various activities, the potential for unauthorized data access, breaches, and privacy infringements becomes a critical consideration. The metaverse collects and stores a vast amount of user data, including behaviour, preferences, and interactions. Regrettably, concerns and debates regarding data privacy are escalating. Inquiries persist regarding the utilization of data, the entities granted access to it, and the means by which users can assert control over their own data, all of which remain unresolved [8]. Biometric data, encompassing fingerprints, voice, facial features, eye movements, and heart rate can be employed to craft lifelike avatars [26] that resemble the user's appearance or expression, access restricted areas that require identity proof, verify transactions or contracts that involve digital assets or currencies, or monetize user data for advertising or marketing purposes. These forms of privacy risks may

violate established norms of privacy protection such as the General Data Protection Regulation (GDPR-2018). For example, these may involve data processing activities that lack transparency and do not secure explicit user consent, contravening Articles 5 and 6 of the GDPR. Additionally, the collection and utilization of biometric data could infringe Article 9, unless explicit consent has been procured. Users' rights to access and erase their personal data, as stipulated in Articles 15 and 17, may be compromised [37].

2. *Ensuring Safety and Cybersecurity in the Metaverse*: Cybersecurity presents a significant challenge in the metaverse, with the exchange of digital assets and information making it susceptible to cyber threats like hacking and data breaches. These risks pose potential harm to the integrity of virtual assets and personal information, thereby threatening the safety and security of individuals within the metaverse [38]. Addressing legal issues becomes imperative to protect users from these risks and hold accountable those responsible for cybercrimes [38].

   Furthermore, participation in the metaverse may involve engaging in virtual activities that carry risks to others, such as cyberbullying, criminal misconduct, or harassment. Additionally, virtual environments simulating physical experiences, like virtual reality rollercoasters, may introduce potential risks [39]. Consequently, prioritizing user safety is a critical concern in the development of the metaverse.
3. *Safeguarding Virtual Property*: The protection of virtual property stands out as a significant issue within the metaverse, where individuals generate and exchange virtual assets with real-world value, including virtual land, goods, and currency. This activity contributes to the evolution of a virtual marketplace, providing users with opportunities for commerce, creative expression, and the formation of virtual identities. Challenges related to ownership, licencing, and intellectual property emerge, and legal complexities may arise concerning the transfer and sale of virtual assets [20].

3.5 Issues Related to Surveillance and Behaviour Modification

In her insightful work, "The Age of Surveillance Capitalism" [11], Shoshana Zuboff explores the significant impact of major technology companies and their data-centric business models on society. Zuboff posits the emergence of a novel economic paradigm, coined as "surveillance capitalism." This model hinges on the extraction and exploitation of personal data for financial gains, presenting far-reaching implications for democracy, individual autonomy, and the future course of humanity.

The book underscores the practices of tech giants like Google and Facebook, which systematically analyze vast sets of personal data from users. This analytical prowess enables these companies to monitor and predict human behaviour, utilizing this information to manipulate and steer individuals' actions and decisions. The potential consequences extend beyond mere market influence, raising concerns about the undermining of democratic institutions and the erosion of individual privacy and autonomy. Zuboff's exploration sheds light on the intricate interplay between technology, capitalism, and the societal fabric.

A significant challenge in the virtual realm lies in the intricacies of identity, allowing users to craft and manage numerous personas or avatars. This complexity complicates efforts to ascertain a person's authentic identity and pinpoint their actual location in the physical world. Moreover, it creates opportunities for illicit activities such as identity theft, behaviour manipulation, and fraudulent schemes [40]. In such a case it may be challenging to prove the criminal's guilt in a court of law. Additionally, many countries do not have jurisdiction over crimes committed in virtual worlds that are hosted on servers located in other countries [41]. "Behaviour modification" makes predictions to influence behaviour in a way that makes the predictions more accurate. Nowadays, human behaviour prediction models have become sophisticated such that they are capable to predict human behaviour so accurately. Conventional prediction models might not consistently provide accurate forecasts of human behaviour as they often fail to consider the capacity for human behaviour to evolve in response to the predictions being generated.

This method can be applied in various areas such as saving more money by providing predictions of future financial situations or incentivizing healthy behaviours by predicting the long-term consequences of their choices. Similarly, it can be used to improve patient outcomes by predicting the likelihood of a disease or condition and modifying behaviour accordingly. However, the approach has limitations like the potential for unintended consequences, and therefore using predictions to influence behaviour has ethical implications. It may also lead to a reduction in creativity or innovation if individuals become too focused on meeting predicted outcomes. It may also reinforce existing biases or inequalities in society if the feedback is not tailored to the needs of all individuals. If this approach is guided by principles of autonomy, beneficence, and non-maleficence to ensure that individuals are not coerced or harmed in the process of modifying their behaviour, then it can revolutionize society and can be beneficial.

Lanier has underscored the adverse consequences of technology, particularly in the realm of social media. His book delves into how social media companies employ algorithms to manipulate user behaviour, fostering addiction and ensuring continued engagement on their platforms [42]. These platforms have a financial incentive to promote sensational and misleading content, which can erode trust in institutions and undermine democracy. They make it impossible to escape the past by often retaining the media (text or images or videos) and making it public. Sometimes, any embarrassing content may have long-lasting effects on individuals, families, and society. The metaverse, a platform surpassing the immersive capabilities of social media, carries significant implications. Its potential to induce unhappiness through continuous comparison and the anxiety stemming from self-promotion raises concerns. Comparable with social media, the metaverse encourages individuals to present a distorted version of reality, prompting them to exhibit a curated and exaggerated representation of their lives. This dynamic can lead to feelings of inadequacy and foster social comparisons. Lanier encourages individuals to consider deleting their social media accounts and recommends regulatory measures for policymakers to ensure that social media companies operate in the public interest.

3.6 Ramifications of AI Automation in the Metaverse

The prevalence of automation in big tech presents a significant challenge. As machines attain greater intelligence, there exists a concern that they might substitute human workers, resulting in extensive job losses and triggering social unrest [43]. For immersive and customized experiences, the metaverse and conversational AI are used. Recent advancements in AI technology, particularly the emergence of Large Language Models (LLMs), mark significant progress. LLMs, a form of Generative AI, demonstrate the capability to produce text that closely resembles human-authored content without explicit instructions or predefined rules. These advanced algorithms analyze and understand natural language, generating text that is grammatically correct, coherent, and semantically meaningful. While Generative AI is a broader model designed to generate new data, such as images, text, or music, based on patterns or data it has learned from existing examples [18]. The potency of these models raises concerns about their potential misuse. They can be employed to disseminate misinformation, manipulation, and disinformation. Significantly, there have been reported cases where text generated by LLMs was employed to create deceptive news articles or influence public opinion.

Rosenberg [44] has discussed the possible shortcomings that are associated with these technologies by considering a case study as evidence to demonstrate the potential dangers of targeted influence, which include the creation of fake videos using deepfake technology to spread misinformation and the use of virtual personas to manipulate people's emotions of targeted influence, which includes the erosion of trust, misinformation, and manipulation of social and political outcomes [45]. Correspondingly, Lazer et al. [46] highlights the role of bots and automated systems in spreading fake news and propaganda. The potential implications of these models arises ethical issues in virtual worlds.

There are also fears that technologies dependent on AI systems are becoming uncontrollable and may end up causing more harm than good to humans [47]. Controlling these systems is particularly challenging because AI systems can learn and evolve in ways that are difficult to predict or control. The issue at hand is commonly known as the "value alignment problem," as highlighted by various authors including Russell [47] and Christian [48]. This challenge revolves around ensuring that AI systems align with human values and objectives. Analogous to an NP-hard problem, it proves intricate to solve due to the complexity, diversity, and occasional contradictions inherent in human values. In computational complexity theory, NP-hard is a designation for a class of problems that are colloquially deemed "at least as hard as the hardest problems in NP" [49]. Moreover, it is practically difficult to programme human values into AI systems which makes these systems insensitive and degrades empathy. It is because human values are not always easy to articulate, and even when they are, they can be subject to change over time. Additionally, different cultures and individuals may have different values, making it challenging to create a universal set of values that can be programmed into AI systems. In the same way, the future of the metaverse cannot be predicted.

4 Current Solutions for a Pro-Social Metaverse

Several researchers have proposed technical solutions to address the risks and challenges of the virtual world. To illustrate, the use of open-source software to advance more equitable access to the metaverse and reduce the risk of creating closed, exclusive virtual environments [2]. Likewise, to solve the problems of big tech, Weinstein, Reich, and Sahami proposed a system of decentralization [50]. This would involve breaking up the big tech companies and giving more power to individuals and smaller organizations. The authors argue that government has a crucial role to play in rebooting the system. The authors called for greater regulation of big tech and the development of policies that prioritize the interests of people and society over the interests of corporations.

In addition, Russell proposed a novel approach to address the problem of controlling AI systems based on the idea of "corrigibility," which refers to the ability of an AI system to accept and act on corrective mechanisms from humans. For this process, the AI system must be made capable to understand the intentions and values of its human operators. The system must accept feedback and allow human intervention to correct mistakes or steer the system in different directions [47]. The decision-making processes of the AI system should be transparent. By designing AI systems that are corrigible, we can ensure that they remain aligned with human values and goals even as they learn and evolve. This approach minimizes the risks associated with AI and maximizes its potential benefits for humans in the future. Moreover, this approach has implications for how we think about the relationship between humans and machines, and it may require a fundamental shift in our approach to technology and its role in society.

In the subsequent subsections, we propose solutions to tackle the issues elucidated in this study. For a visual representation of these solutions, refer to Fig. 3.

4.1 Ethical Benchmarking

The ethical challenges in the metaverse have the potential to increase already-existing injustices and discriminatory behaviours if they are not adequately governed. According to Rosenberg [51], the regulatory framework for the metaverse should have a strong emphasis on inclusion, diversity, and accessibility to ensure that all users have an equal chance to participate in and gain from the metaverse. Using technical solutions that prioritize transparency, equity, and user-centred design has

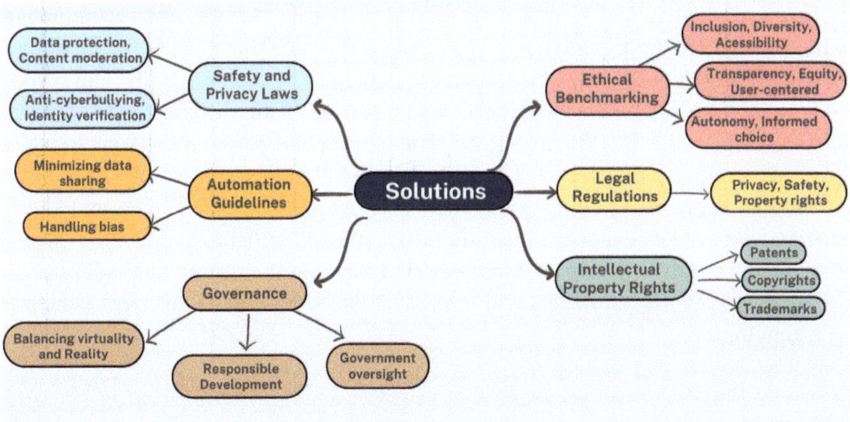

Fig. 3 Overview of current solutions for a pro-social metaverse

the potential to create a metaverse that is both technologically advanced and ethically responsible [52].

Conversely, ensuring informed choice and autonomy is not a straightforward task and is influenced by a number of technical and social factors. For instance, individuals may be subject to algorithmic bias or manipulation, where their experiences within the metaverse are shaped by algorithms that are designed to maximize engagement or profit [6]. Besides, the design of virtual environments and interactions may not prioritize individual autonomy and choice, but rather may be driven by commercial interests or other factors. For example, the use of transparency and user control mechanisms may help to ensure that individuals have access to relevant information and can make knowledgeable decisions about their participation in virtual environments and interactions.

To ensure ethical utilization and prevent exploitation of data, it will be necessary to investigate the types of data collected through these systems, where it is stored, and how it is used. The collection of biometric and physiological data will generate information about user neural activity, necessitating the establishment of rights frameworks to protect mental and biological privacy [53, 54]. Without such privacy protections, new technologies could potentially read our minds, model our identities, and influence our behaviours. As the metaverse introduces an immersive layer to technology, privacy issues could become even more challenging. To address these concerns, groups like the Khronos Group and Open XR are developing new metaverse-oriented standards with a focus on interoperable interfaces. Overall, when creating experiences in the metaverse, it is crucial to consider privacy and ethical frameworks [36].

In an important paper analyzing AI ethics, Jobin et al. [55] provide a detailed analysis of the different AI ethics guidelines that have been developed worldwide from different countries and regions. They use a qualitative content analysis approach to identify and compare the ethical principles, values, and recommendations contained in these guidelines. The analysis revealed that most of the guidelines share common themes and values, such as transparency, accountability, and respect for human rights [26]. However, there were also significant differences in approach and emphasis, reflecting the diverse cultural, social, and political contexts in which the guidelines were developed. The authors stress that there are several challenges in creating effective AI ethics guidelines, such as the need for a clear and shared understanding of key concepts, including what constitutes "ethical AI," the difficulty of balancing competing ethical principles, and the need to ensure that guidelines are not merely symbolic but have a real impact on the development and deployment of AI technologies.

4.2 Necessary Legal Framework and Regulations

The legal complexities linked to the evolution of the metaverse are intricate and diverse. Addressing these challenges will require a combination of technical solutions and legal frameworks that prioritize privacy, cybersecurity, safety, and property rights. There are a number of technical solutions in the literature that are proposed to confront these obstacles. For example, the use of blockchain technology to create more transparent and decentralized virtual environments [1]. This promotes greater security, transparency, and accountability, reducing the risk of harmful or exploitative behaviour in managing virtual assets and transactions, and decreasing the chance of fraud or theft. Additionally, the use of encryption and other security measures may help to protect personal information and prevent unauthorized access. Cryptocurrencies also have negative implications in the metaverse environment. Since it is easier to manipulate the virtual environment using cryptocurrencies. There are growing concerns about the potential enabling force for crypto to augment financial crimes of all sorts [56].

Legal inquiries arise regarding the definition of jurisdiction and the assurance of compliance with local laws and regulations [57]. The central and crucial legal dilemma in the metaverse centres on determining the relevance of property rights and/or intellectual property rights. Property rights are legal systems that govern the right of ownership and the rights of use as well which are necessary to protect virtual assets online. Common property rights are the rights to use a thing and deny others the use of it. The objective of property rights is to establish an equitable, foreseeable, and transparent mechanism for allocating the right to utilize virtual objects among various interests [33]. In the metaverse, all objects are virtual and exist in a virtual usage and storage context, distinguishing them from real property, which necessitates a tangible existence in a specific physical location. While property rights systems can address concerns related to the utilization of intangible objects, they are not suitable for assigning rights of use.

Traditional real property systems may not be directly applicable to virtual assets in the metaverse. However, intellectual property rights offer more relevance and can be applied in the context of metaverse [14]. Unlike general property rights, IP deals with intangible objects, inherently incorporeal, and cannot be physically located, safeguarding ideas and expressions through patents and copyright. While determining applicable copyright laws for audio-visual, artistic, and textual content in the metaverse could be relatively straightforward. Challenges arise in establishing the governing laws for virtual properties or land. The current intellectual property systems aim to safeguard inventions, creative content, symbols, and various other elements. One significant regulatory hurdle involves tailoring current regulations derived from actual laws of property to govern virtual assets in the metaverse.

4.2.1 Security and Privacy Laws

The notion of privacy is intricately tied to the right to protect personal identity. Prominent privacy laws, like the GDPR, delineate the extent and boundaries of digital privacy rights [58]. It is a comprehensive set of tiles that focus on providing strict legal protection for the human right of privacy and greater control for individuals on their personal and sensitive data. However, privacy is often connected with personal integrity (the accuracy and authenticity of data) and personal autonomy (the ability to manage and control one's own data) [59].

- *Data Protection Regulations*: Establishing data protection regulations is crucial to safeguard user privacy and prevent unauthorized or non-consensual sharing of user data [38]. Notably, the GDPR upholds the rights of data subjects, encompassing the right to privacy, access, correction of personal data, objection to processing, and the "right to be forgotten" or erasure. These rights play a pivotal role in upholding human rights in the digital era and impose responsibilities on data controllers and processors [60].
- *Anti-cyberbullying Regulations*: Regulations should be enforced to prevent cyberbullying and harassment within the metaverse. These regulations should include measures to identify and punish cyberbullies by appropriate tracing techniques to identify the guilty.
- *Regulations for Identity Verification*: In order to deter illicit activities such as money laundering, theft, and terrorism, it is imperative to incorporate identity verification regulations to ascertain the true identity of users [60]. Identity verification poses a substantial challenge in the metaverse, especially considering the potential existence of numerous virtual identities.
- *Regulations for Content Moderation*: Additionally, content moderation is challenging due to the ability to create and share user-generated material on a massive scale, which raises concerns about hate speech, harassment, and other harmful content [51]. Content moderation regulations have to be established to prevent the dissemination of detrimental and illicit content within the metaverse. These regulations should include measures to identify and remove such content.

4.3 Preserving Intellectual Property Rights

The metaverse introduces legal intricacies concerning virtual assets, encompassing virtual real estate, digital art, and virtual currency regarding the safeguarding and enforcement of intellectual property (IP) rights [61]. Intellectual property law governs ownership, usage, and transfer of intangible assets and their derivatives, including patents, copyrights, trademarks, which protect inventions, expressions, distinctive signs, and trade secrets. Traditionally, property in the virtual realm can be categorized into two types. Once users in the metaverse possess property rights, the subsequent inquiry arises regarding the adequacy of existing IP rights to safeguard users who exercise their property rights against potential infringements by others. Unlike physical property law, IP law seems perfect for the job. However, there are limits to the operation of the law. Legal challenges obscure the enforcement of IP rights along with the choice of law and enforcement powers because claims are unrestricted in all jurisdictions and involve individuals in many countries. New policies have to be designed to protect the IP of creators within the metaverse and copyright infringement and theft.

Within the metaverse, conflicts may arise among virtual world owners, programmers, and virtual avatars, driven by diverse and sometimes conflicting interests. Barfield underscores the importance of the legal system recognizing the unique characteristics of virtual environments and the diverse array of stakeholders involved [9]. Suggested remedies for resolving conflicts involve the introduction of virtual property rights and the creation of dispute resolution mechanisms that take into account the concerns of all involved parties. Challenges arise in enforcing copyrights and trademarks in virtual spaces, along with difficulties in ascertaining ownership and control over virtual objects and spaces. The issue of virtual identity and the rights of virtual avatars is also addressed, supporting the idea that avatars should have certain property rights regarding their virtual identities.

4.4 Automation Guidelines

Large technology companies are big and powerful, with increasing control over our lives. They have created a system that prioritizes profits over social good and have led to significant social and economic inequalities. Rosenberg [61] has put forth strategies to mitigate targeted influence, encompassing initiatives such as educating individuals about associated risks, implementing technical solutions like AI content moderation, verifying digital identity, and reducing data sharing. Shu et al. [62] developed a framework for detecting misinformation or fake news on social media platforms through the application of data mining techniques. The authors have identified that the large volume of data and the speed at which it is shared are the major obstacles in identifying fake news. Their model involves analyzing the content of social media posts, as well as user behaviour and social context. The article discusses the results of experiments conducted using real-world datasets, showcasing the efficacy of the suggested method. Likewise, an article published [63] outlines strategies for detecting and combating fake content, including using natural language processing techniques and leveraging crowdsourcing to verify the information. The article calls for greater awareness of the risks of LLM-generated fake news and the need for ethical guidelines to regulate the use of these technologies.

When considering LLMs, there are two opposing viewpoints on their development. One is the hands-off camp, represented by figures such as Andrew Ng and Yann LeCun, who believe that LLM development should continue without excessive regulation. They argue that LLMs have enormous potential to revolutionize fields like natural language processing, machine translation, and speech recognition and that their development should not be stifled. One perspective argues that the advantages of LLMs, such as enhanced accuracy and efficiency in language processing, outweigh the potential risks and ethical considerations. Conversely, an opposing stance advocates calling for a six-month suspension on LLM development and release to prioritize safety. They assert that the risks associated with these models are not yet fully understood, and that a pause in their development would allow time to study their potential dangers and implement necessary safety measures. Concerns include the possibility of LLMs being used for malicious purposes, such as the creation of fake news or deepfakes, as well as potential biases in language and datasets [64]. In essence, the discourse surrounding the development of LLMs highlights the necessity for a careful and handling of emerging technologies, including the metaverse, is essential While acknowledging the undeniable advantages of LLM development, it is imperative to conscientiously weigh potential risks and implement measures to ensure their responsible and ethical development and utilization.

Comparably, another article emphasizes the importance of fact-checking and critical thinking in combating fake news, as well as the need for media literacy education to help people distinguish between credible and unreliable sources of information. The authors also discuss the responsibility of social media platforms and news organizations in addressing the problem of fake news [42]. It underscores the importance of addressing the issue of fake news for the health of our democracy and the well-being of society. Most of the problems of the Internet could potentially come to the metaverse by default, as the metaverse is essentially an extension of the Internet, multiple virtual, and augmented reality environments. Similarly, the spread of misinformation could be amplified in a metaverse where users can create and share their own content with ease. Therefore, strict guidelines and regulations have to be enforced in this environment to make it safe and secure for everyone.

5 Unresolved Challenges and Future Avenues

As the metaverse undergoes rapid expansion and integration into our daily lives, the necessity for regulations becomes increasingly apparent. Existing laws and regulations prove insufficient in addressing the distinctive challenges posed by the metaverse, including user-generated content and the tangible consequences of virtual experiences. Ensuring its safe, secure, ethical, and legal operation requires the establishment of tailored regulations. Updating and improving existing regulations are crucial, and international cooperation is essential to regulate the global nature of the metaverse [65].

Having explored the risks associated with the metaverse and the existing regulatory approaches in the previous sections, this section brings attention to open issues and outlines promising future directions that warrant consideration from the community.

5.1 *Establishing a New Social Contract*

This research underscores the necessity of establishing a new social contract among individuals, governments, and corporations to grant individuals greater control over their personal data and ensure accountability for its use by companies. The concentration of power in the hands of a few tech giants, driven by surveillance capitalism, has significant drawbacks and can be susceptible to exploitation for harmful purposes. Qin et al. [40] address the challenges of maintaining law and order in virtual worlds, highlighting the unique difficulties associated with identifying and prosecuting criminals due to the absence of a virtual legal system and physical presence. Collaboration among developers, law enforcement, and users is crucial to create a safer online environment and effectively address these challenges. Moreover, the regulatory framework should include provisions for data protection, content regulation, and dispute resolution [51]. A comprehensive approach combining industry cooperation, self-regulation, and governmental monitoring is essential to implement these policies.

5.2 *Designing the Metaverse: Ethical Considerations and Regulatory Frameworks*

Zallio and Clarkson pose critical questions regarding the design choices made by various tech companies [33], prompting a reflection on how the metaverse can be designed by and for people [65]. To successfully advance this technology, understanding how to optimize opportunities and cultivate a safe, inclusive metaverse that upholds principles of equity and diversity becomes essential.

The existing code of ethics appears inadequate, necessitating the prompt formulation of a new one to align with the expanding metaverse. Therefore, bolstering the supervision of the metaverse and establishing and regularly updating relevant regulations and guidelines are crucial to maintaining a sustainable and equitable environment [66]. Balancing both the ethical and legal considerations is significant to ensure the responsible and sustainable development and usage of the metaverse [57]. Organizations should prioritize creating a metaverse grounded in ethical principles, emphasizing regular updates to regulations and guidelines to align with the metaverse's expansion and ensure its sustainability and ethics.

Moreover, designing the metaverse and its applications should prioritize human welfare over profit-making. This ensures that the metaverse enhances real-life experiences without replacing them, emphasizing the creation of a safe, inclusive, and equitable metaverse.

5.3 Balancing Virtuality and Reality

The impact of immersive virtuality on users' perception and understanding of reality must be carefully considered. Turkle highlights the "Artificial Crocodile Effect," where virtual experiences may distort one's perception of the natural world [67]. While the virtual world cannot replace real-life interaction, it can offer unique benefits in certain contexts. The metaverse's developers must prioritize human-centred and beneficial approaches, designing applications that maintain a meaningful connection with reality while leveraging the advantages of virtual environments. To enhance the situation, it is essential for metaverse developers to adopt human-centred and beneficial approaches, crafting applications that foster a meaningful connection with reality while harnessing the advantages offered by virtual environments.

5.4 Need for Governance

Establishing governance and ethical standards within the metaverse is crucial to prevent it from evolving into a lawless space, prioritizing user well-being and societal interests over commercial considerations. In the absence of proper oversight, the metaverse becomes vulnerable to illicit behaviours, underscoring the necessity for well-defined guidelines and ethical standards to uphold a reliable and secure virtual space. Stakeholder engagement is essential to establish guidelines and criteria that promote the public interest, nurturing a metaverse that is both beneficial and accountable [68]. Critical areas of focus for overseeing and managing include content moderation, data privacy, and user rights. Emphasis should lie in empowering users, giving them control over their data and online identities. Policies need to guarantee fair treatment and real-world consequences for virtual experiences, aligning with established legal frameworks. Robust oversight is imperative to curb crimes related to virtual currencies and assets. Works by researchers like Rosenberg [51] underscore the significance of strategies encompassing collaboration within the industry, autonomous regulation, and supervision by governmental bodies. Implementing and upholding laws concerning data privacy are crucial to protect users' personal information, cultivating reliance and responsibility within the metaverse.

Without proper oversight, the metaverse could become a lawless medium or a "Wild West," where private actors wield unchecked power and users' rights and safety are not adequately protected. Such a scenario could turn the metaverse into a breeding ground for criminal activities, including cybercrime, money laundering, and other illicit activities [68]. Governance and ethical perspectives in the metaverse should prioritize user well-being and broader societal interests over commercial interests. Clear regulations and ethical principles must be established to prevent the metaverse from becoming a lawless environment. Collaboration among various stakeholders is essential to develop guidelines and standards that promote the public interest and ensure a beneficial and trustworthy metaverse for all.

5.5 *Principles for Responsible Development*

As the metaverse undergoes continuous evolution, ensuring its development is safe and responsible becomes paramount, necessitating a framework that prioritizes user wellbeing, safety and societal benefits. Technological advancements democratize access to XR, empowering more individuals to generate and distribute experiences in virtual and augmented reality [69]. The research and development in XR are critical for immersive applications in education, entertainment, and tourism.

To establish effective policies, a blend of collaboration within the industry, self-regulation, and governmental supervision is essential. The responsible development process should adhere to principles such as transparency, accountability, user empowerment, and stakeholder participation, enabling users to understand and manage the effects of technology [51]. Industry management necessitates careful consideration of legal, ethical, and policy aspects, encompassing risk mitigation and regulatory solutions for consumer platforms. Addressing the potential of psychological and health consequences of prolonged exposure to virtual environments is essential. The framework needs to champion sustainable practices and conscientious resource stewardship within the metaverse, incorporating strategies to minimize energy usage and alleviate environmental effects.

Shaping the metaverse responsibly requires a collective effort, considering technical aspects, user experience, governance, and ethical principles. By prioritizing user safety, well-being, and societal welfare, we can unlock the metaverse's full potential while ensuring a sustainable and inclusive future for all.

6 Conclusions

The advent of the metaverse promises a paradigm shift in our interactions, learning experiences, and business engagements within a virtual realm. While the potential for creativity and social connectivity is immense, it becomes imperative to establish a secure and safe environment that allows users to fully harness its advantages. This chapter underscores the pivotal role of ethical and legal frameworks in the metaverse, emphasizing the need for collective endeavours from all stakeholders to effectively address these challenges. The survey presented herein sheds light on the existing ethical and regulatory landscape, offering valuable insights into the imminent risks and hurdles.

The outlined approaches and potential solutions within this chapter play a crucial role in shaping a metaverse that is not only innovative but, more importantly, prioritizes the well-being of individuals and the broader societal welfare over commercial considerations. In essence, the metaverse's success hinges on a thoughtful and collaborative approach that ensures its design and operation align with human-centric values and ethical standards.

References

1. Mozumder MAI, Sheeraz MM, Athar A, Aich S, Kim H-C (2022) Overview: technology roadmap of the future trend of metaverse based on iot, blockchain, AI technique, and medical domain metaverse activity. In: 2022 24th International conference on advanced communication technology (ICACT). IEEE, pp 256–261
2. Mystakidis S (2022) Metaverse. In: MDPI, vol 2, pp 486–497
3. Turdialiev MA (2022) The legal issues of metaverse and perpectives of establishment of international financial center in metaverse. Oriental Renaissance: Innov Educ Nat Soc Sci 2(8):239–249
4. Lee L-H, Braud T, Zhou P, Wang L, Xu D, Lin Z, Kumar A, Bermejo C, Hui P (2021) All one needs to know about metaverse: a complete survey on technological singularity, virtual ecosystem, and research agenda. arXiv:2110.05352
5. Boutenko V, Florida R, Jacobson J, Staff H, Coelho F (2022) The metaverse will enhance—not replace—companies' physical locations. Harvard Bus Rev
6. Rosenberg L (2022) The metaverse: from marketing to mind control. Future of marketing magazine. Future of Marketing Institute, York University
7. Smith CH, Molka-Danielsen J, Rasool J, Webb-Benjamin J-B, UK, KL (2023) The world as an interface: exploring the ethical challenges of the emerging metaverse. In: Proceeding of the Hawaii international conference system sciences, Maui, HI, pp 6045–6054
8. Fernandez CB, Hui P (2022) Life, the metaverse and everything: An overview of privacy, ethics, and governance in metaverse. In: 2022 IEEE 42nd international conference on distributed computing systems workshops (ICDCSW). IEEE, pp 272–277
9. Barfield W (2006) Intellectual property rights in virtual environments: considering the rights of owners, programmers and virtual avatars. In: HeinOnline, vol 39, p. 649
10. Anshari M, Syafrudin M, Fitriyani NL, Razzaq A (2022) Ethical responsibility and sustainability (ERS) development in a metaverse business model. Sustainability 14(23):15805
11. Zuboff S (2019) The age of surveillance capitalism: the fight for a human future at the new frontier of power: Barack Obama's books of 2019. In: Profile Books

12. Vardi MY (2022) ACM, ethics, and corporate behavior. Commun ACM 65(3):5–5
13. Usmani SS, Sharath M, Mehendale M (2022) Future of mental health in the metaverse. General Psychiatry 35(4):100825
14. Kshetri N (2022) Policy, ethical, social, and environmental considerations of Web3 and the metaverse. IT Professional 24(3):4–8
15. Castronova E (2001) Virtual worlds: a first-hand account of market and society on the cyberian frontier. Available at SSRN 294828
16. Frischmann B, Selinger E (2018) Re-engineering humanity. Cambridge University Press
17. Kye B, Han N, Kim E, Park Y, Jo S (2021) Educational applications of metaverse: possibilities and limitations. J Educ Eval Health Professions 18
18. Bozkurt A (2023) Generative artificial intelligence (AI) powered conversational educational agents: the inevitable paradigm shift. Asian J Distance Educ
19. Bibri SE, Allam Z (2022) The metaverse as a virtual form of data-driven smart cities: the ethics of the hyper-connectivity, datafication, algorithmization, and platformization of urban society. Comput Urban Sci 2(1):22
20. Belk R, Humayun M, Brouard M (2022) Money, possessions, and ownership in the metaverse: NFTs, cryptocurrencies, Web3 and Wild Markets. J Bus Res 153:198–205
21. Anderson J, Rainie L (2022) The metaverse in 2040. Pew Research Centre
22. Wang X, Lee L-H, Bermejo Fernandez C, Hui P (2023) The dark side of augmented reality: exploring manipulative designs in ar. Int J Hum–Comput Interact 1–16
23. Far SB, Rad AI (2022) Applying digital twins in metaverse: user interface, security and privacy challenges. J Metaverse 2(1):8–15
24. Wiederhold BK (2022) Sexual harassment in the metaverse. Mary Ann Liebert, Inc., Publishers 140 Huguenot Street, 3rd Floor New ...
25. Dayarathna R (2022) Ethics in the metaverse. 78th 11:79
26. Tabassum A, Elmahjub E, Qadir J (2023) Pathway to prosocial AI-XR metaverses: a synergy of technical and regulatory approaches. In: 2023 International symposium on networks, computers and communications (ISNCC), pp 1–8
27. Washingtonpost. https://www.washingtonpost.com/opinions/2022/08/22/metaversepolitical-misinformation-virtual-reality/. Last accessed 10 May 2023
28. Turkle S (1996) Virtuality and its discontents: Searching for community in cyberspace. American Prospect 50–58
29. Zallio M, Clarkson PJ (2023) Metavethics: ethical, integrity and social implications of the metaverse. In: Intelligent human systems integration (IHSI 2023): integrating people and intelligent systems. AHFE (2023) international conference. AHFE open access, vol 69
30. Barreda-Angeles M, Hartmann T (2022) Hooked on the metaverse? exploring the prevalence of addiction to virtual reality applications. Front Virtual Reality 3:1031697
31. Paquin V, Ferrari M, Sekhon H, Rej S (2023) Time to think "meta": a critical viewpoint on the risks and benefits of virtual worlds for mental health. JMIR Serious Games 11:43388
32. Han S, Shi Z (2024) Unmasking inequality in the metaverse: a study of skin-tone bias in the cryptopunks market
33. Zallio M, Clarkson PJ (2022) Designing the metaverse: a study on inclusion, diversity, equity, accessibility and safety for digital immersive environments. Telematics Inform 75:101909
34. Nazir M, Lui CSM (2016) A brief history of virtual economy. J Virtual Worlds Res 9(1)
35. Papagiannidis S, Bourlakis M, Li F (2008) Making real money in virtual worlds: MMORPGs and emerging business opportunities, challenges and ethical implications in metaverses. Technol Forecast Soc Chang 75(5):610–622
36. Lin J, Latoschik ME (2022) Digital body, identity and privacy in social virtual reality: a systematic review. Front Virtual Reality 3:167
37. Art. 9 GDPR processing of special categories of personal data. https://gdprinfo.eu/art-9-gdpr/
38. Wang Y, Su Z, Zhang N, Xing R, Liu D, Luan TH, Shen X (2022) A survey on metaverse: fundamentals, security, and privacy. IEEE Commun Surveys Tutorials
39. Kim C, Park J (2022) An exploratory study on the production of metaverse ethics education contents for adolescents. In: Technical report, EasyChair

40. Qin HX, Wang Y, Hui P (2022) Identity, crimes, and law enforcement in the metaverse. arXiv: 2210.06134
41. Turdialiev M (2023) Legal discussion of metaverse law. Int J Cyber Law 1(3)
42. Lanier J (2018) Ten arguments for deleting your social media accounts right now. In: Random House
43. Brynjolfsson E (2023) The Turing trap: the promise & peril of human-like artificial intelligence. In: Augmented education in the global age, pp 103–116
44. Rosenberg L (2023) The metaverse and conversational AI as a threat vector for targeted influence. In: 2023 IEEE 13th annual computing and communication workshop and conference (CCWC). IEEE, pp 0504–0510
45. Baeza-Yates R, Fayyad UM (2022) The attention economy and the impact of artificial intelligence. In: Perspectives on digital humanism, pp 123–134
46. Lazer DM, Baum MA, Benkler Y, Berinsky AJ, Greenhill KM, Menczer F, Metzger MJ, Nyhan B, Pennycook G, Rothschild D et al (2018) The science of fake news. Science 359(6380):1094–1096
47. Russell S (2019) Human compatible: artificial intelligence and the problem of control. Penguin
48. Christian B (2020) The alignment problem: machine learning and human values. WW Norton & Company
49. Knuth DE (1974) Postscript about NP-hard problems. ACM SIGACT News 6(2):15–16
50. Weinstein J, Reich R, Sahami M (2021) System error: where big tech went wrong and how we can reboot. Hachette, UK
51. Rosenberg LB (2022) Regulating the metaverse, a blueprint for the future. In: Extended reality: first international conference, XR Salento 2022, Lecce, Italy, 6–8 July 2022, Proceedings, Part I. Springer, pp 263–272
52. Grant TD, Wischik DJ (2020) On the path to AI: Law's prophecies and the conceptual foundations of the machine learning age. Springer
53. Sun J, Gan W, Chao H-C, Yu PS (2022) Metaverse: survey, applications, security, and opportunities. arXiv:2210.07990
54. Sami H, Hammoud A, Arafeh M, Wazzeh M, Arisdakessian S, Chahoud M, Wehbi O, Ajaj M, Mourad A, Otrok H et al (2023) The metaverse: Survey, trends, novel pipeline ecosystem & future directions. arXiv:2304.09240
55. Jobin A, Ienca M, Vayena E (2019) The global landscape of AI ethics guidelines. Nat Mach Intell 1(9):389–399
56. Trozze A, Kamps J, Akartuna EA, Hetzel FJ, Kleinberg B, Davies T, Johnson SD (2022) Cryptocurrencies and future financial crime. Crime Sci 11:1–35
57. Johan S (2022) Metaverse and its implication in law and business. Jurnal Hukum Progresif 10(2):153
58. Egliston B, Carter M (2021) Critical questions for facebook's virtual reality: data, power and the metaverse. Internet Policy Rev 10(4):1–23
59. Koops B-J (2014) The trouble with European data protection law. Int Data Privacy Law 4(4):250–261
60. Rosenberg L (2022) Regulation of the metaverse: a roadmap: the risks and regulatory solutions for largescale consumer platforms. In: Proceedings of the 6th international conference on virtual and augmented reality simulations, pp 21–26
61. Kasiyanto S, Kilinc MR (2022) The legal conundrums of the metaverse. J Central Banking Law Inst 1(2):299–322
62. Shu K, Sliva A, Wang S, Tang J, Liu H (2017) Fake news detection on social media: A data mining perspective. ACM SIGKDD Explor Newsl 19(1):22–36
63. Chen C, Shu K (2023) Combating misinformation in the age of LLMS: opportunities and challenges. arXiv:2311.05656
64. Mark Rossow P (2022) Self-driving cars: what can we realistically expect
65. Rosenberg LB (2022) The growing need for metaverse regulation. In: Intelligent systems and applications: proceedings of the 2022 intelligent systems conference (IntelliSys), vol 3. Springer, , pp 540–547

66. Ning H, Wang H, Lin Y, Wang W, Dhelim S, Farha F, Ding J, Daneshmand M (2021) A survey on metaverse: the state-of-the-art, technologies, applications, and challenges. arXiv: 2111.09673
67. Turkle S (2009) Simulation and its discontents. MIT Press
68. Wu J, Lin K, Lin D, Zheng Z, Huang H, Zheng Z (2023) Financial crimes in web3-empowered metaverse: taxonomy, countermeasures, and opportunities. IEEE Open J Comput Soc 4:37–49
69. De Paolis LT, Arpaia P, Sacco M (2022) Extended reality: first international conference, XR Salento 2022, Lecce, Italy, 6–8 July 2022, Proceedings, Part I, vol 13445. Springer

Intelligent Interactions: Exploring Human–Computer Interaction in the Metaverse Through Artificial Intelligence

Bhavana Kaushik and Tanu Singh

Abstract In the sphere of metaverse, human–computer interaction (HCI) technology and its allied intellectual machine algorithms have fascinating scope of exploration. Intelligent robots are highly used in medical, smart homes, wearable technology, intelligent tutoring, mobile-based learning, privacy, and security domain. HCI refers to the process by which a computer and a person exchange information in a particular way to complete particular tasks. Additionally, a lot of technological companies, including Facebook and Microsoft, have initiated to build metaverse, an alive environment with a new social arrangement. As a result, its future is uncertain. This research work presents complete and comprehensive processing techniques for metaverse and HCI. It also analyzes and confers the features and constraints associated with these methods for compacting visual data, interacting methods, assessment and evaluation method, and assisted living for the better quality of life while dealing with HCI technology. The work examines the use of the concepts of metaverse and HCI and briefly covers their fundamental ideas. This study also tried to include the challenges that arise fundamentally. The chapter also includes the arena that is beyond HCI and metaverse. The metaverse is a promising sector with tremendous expansion possibilities, but there are still several pressing challenges, this paper will argue, drawing on a number of articles and reports from recent years.

Keywords HCI · Artificial intelligence · Metaverse · Wearable technology · Interaction · Challenges · Automation · Augmented reality · Virtual reality

---

B. Kaushik (✉) · T. Singh
University of Petroleum and Energy Studies, Dehradun, Uttarakhand, India
e-mail: bkaushik@ddn.upes.ac.in

T. Singh
e-mail: tanu.singh@ddn.upes.ac.in

G. Chhabra and K. Kaushik (eds.), *Understanding the Metaverse*, Blockchain Technologies, https://doi.org/10.1007/978-981-97-2278-5_13

1 Introduction

The science fiction book Snow Crash [1] presented the idea of the metaverse, in the year 1992 which has piqued the curiosity of the computer world ever since. Together, Meta and Verse represent the concept of "beyond the universe," which is a constructed environment that exists alongside the real world and is reinforced by extended reality and other networked tools. A simulated reality network sphere backed by AR, VR, 3D, and other technologies is the Internet of the future. People can feed back on every action they take in the actual world in this manner. These fundamental ideas allow metaverse to be used in a variety of disciplines, including education, games, business, etc., shown in Fig. 1. Artificial intelligence (AI) technology helps to develop systems, techniques, intelligent methodology, and relevant software for the expansion, expansion, and simulation of human intellect. With the swift and increasing development in the fields of AI are human–computer interaction (HCI), deep learning algorithms, and smart machines has achieved a notable growth in recent years. And if we talk about intelligent robots, they retain near similar human-like aptitude and skill, for example as the ability to listen, speak, read, write, perceive, sense, and realize. Automation and self-directed machines will be specifically rooted in scientifically accompanied environments. Digital world will accompany with intelligence and augment physical reality, which results into the new hybrid smart worlds and therefore, expression will be moved from one interface to another naturally [2]. HCI is reliant upon tailored electronic devices and machinery such as high-resolution interface and communicating microelectronics and often entails tele functioning and video devices. For video devices, the challenges and issues related to image processing are the domain of research. HCI has already enlarged its domain of enquiry, progressed over the time, and has substantially achieved remarkable advances. The need for interaction and communication is increasing with increased complexity in frequent ways, and therefore, the human complement of technology is also changing and has high impact and affect over our everyday life making ourselves more conscious.

This chapter has highly focused on the holistic views, concerns, issues, and challenges that are to be catered while implementing and developing the interaction techniques and interfaces for humans with computers. The basic tenets of the metaverse will be explained, the relationship between HCI and the metaverse's operation will be examined, and a quick forecast about the metaverse's future will be made. The goal of this research is to better the reader's understanding of the metaverse, its uses, and the problems it raises. In Sect. 2, linkage between AI, HCI, and metaverse is shown with detailed literature. User centric framework is discussed in Sect. 3. What is there open for scholars and researchers beyond HCI and metaverse is included in Sect. 4. Section 5 consists of various research challenges present in the area of metaverse. Section 6 discusses the work, and Sect. 7 concludes and summarizes the whole chapter.

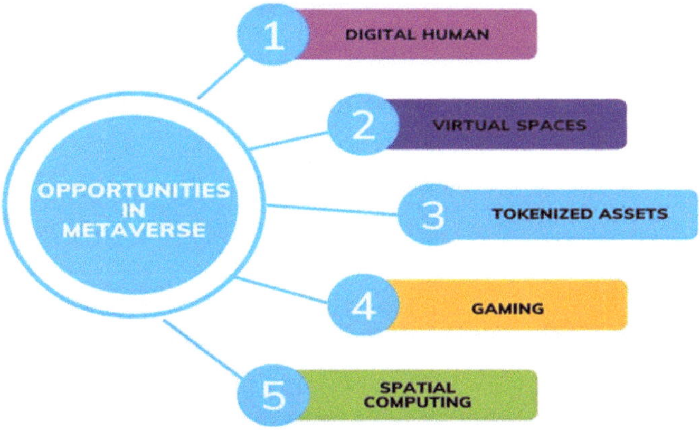

Fig. 1 Opportunities and application area in metaverse

2 Connection of Metaverse with AI and HCI

The metaverse's fundamental idea is to primarily blur the lines between the real world and the virtual one, enabling people to carry out all real-world tasks using technology in virtual space. It started out as a network of virtual worlds on the internet and is built on human–computer interaction. Due to its features, the metaverse can be used for massively multiplayer online activities like online games, video chat, distance learning, remote job, etc. A combination of many of the interactive technologies used today, the metaverse is a concept and technology, the related technologies are described in Fig. 2. Because it involves sending human movements to the machine and then supplying them back into the simulated environment, it necessitates HCI-related technological assistance. Extended Reality (XR), which mixes Virtual Reality (VR), Augmented Reality (AR), and Mixed Reality (MR) technologies, is one of the more important ones for interacting with wearable technology.

Human–computer interaction with smart intelligence techniques has vast development aspects in various areas and major allied tools required for metaverse are shown in Fig. 3. However, there are countless accomplishments in these arenas, but for the enlargement of intelligence fidelity still a huge effort is required. Human beings convey the intentions and emotions of themselves by means of compound indications, such as semantic, pronunciation, facial demonstrations, and gestures. Moreover, many of the physiological signals are also expressed by humans like blood pressure and heartbeat. Most of the pre-defined observation approaches are concentrated on the solitary style, whereas the association between the numerous styles is unnoticed and overlooked. Therefore, multimodal datasets, multimodal datasets categorized hybrid and mixture discernment, and human-like intelligent awareness algorithm grounded on this database will become a significant track for scholars and researchers [3, 4]. When we contemplate humanoid robots or self-driving automobiles, for example, a real-world implementation of an HCI might require significant

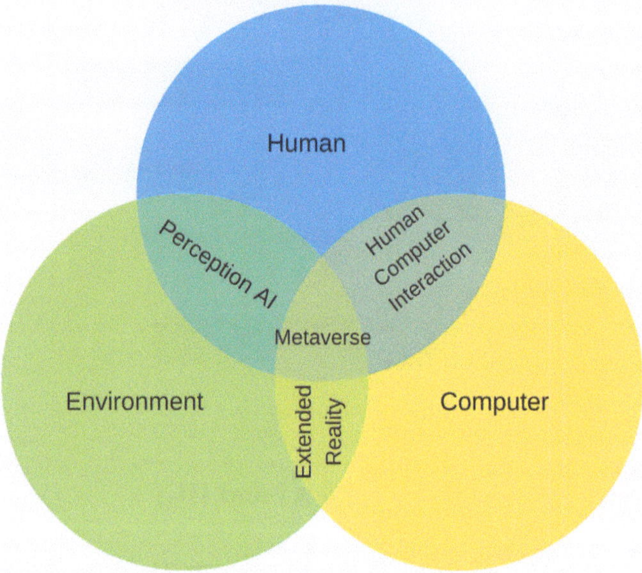

Fig. 2 Metaverse and its connection AI and HCI

resources. Thorough investigations are necessary for the development and assessment of explainable AIs, useful training components, and the fundamental analysis of human–AI interfaces and communications. XR gives a novel perception on the XR-AI combo space and a new testbed for human–AI interactions and interfaces by addressing the question: How can we determine reliable and methodical study processes for human–AI interfaces and interactions?

When looking at HCI from the perspective, it is crucial to comprehend and research whether and how the human user sees the AI she is engaging with. The term "interface" in this context refers to the area where human and machine interaction occurs, comprising all hardware and software elements as well as the underlying interaction concepts and styles. Hands are a typical tool for communication and engagement with the physical environment, and gestures are a popular way for people to express themselves. There are many distinct ways that gestures and speech coexist in communication. They range from those that don't add to the meaning of the speech and just reflect its tone to those that do and represent certain ideas [5]. Three-dimensional modeling, assistive software, data input/authentication, management/direction-finding, and touchless mechanism are the five types of gesture-based software that have been discovered. Instead, then focusing on a wide range of application scenarios, the strategies are grouped according to their immediate use. It is possible, for instance, to design a medium that allows users to interact with 3D picture data while expressing a gesture without using their hands. A doctor might review 3D

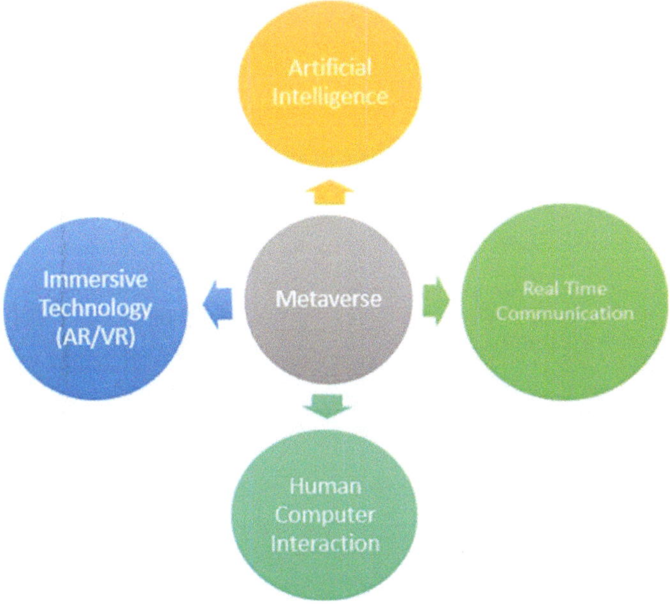

Fig. 3 Metaverse and its allied technologies

scans to guarantee their hands are clean while performing surgery or providing treatment. The interface and gesture requirements were developed to primarily focus on interacting with and manipulating a 3D visualization [4, 5]. This sort of system was classed as platform for manipulation/navigation as they are its specific function. Touchless interaction is an application providing remote machine control but at same time it is a beneficial side effect in terms of interaction. Reference [6] provides an explanation of 3D modeling actions, including shape generation and alteration as well as modification and manipulation. Adjustment is the interaction with that geometry meant to alter its diagrammatic properties, whereas manipulation is the action of moving the geometry around in space, rotating it, scaling it, or translating it. In an empty workspace, creation is the use of hand motions and expression to create a new geometry. Only shapes can be altered with CAD software; the geometry itself cannot be changed or created. In order to improve the development of and interaction with these formations and contours, 3D shaping encompasses both modification and creation and all applications that have investigated 3D design and manipulation of objects in 3D environments. Many organizations do not only require enhancing their levels of artificial intelligence while continuing to develop their better human interaction. Rather, they moreover look for a higher-order artificial intelligence for transforming the intellect in line with corporate policies and strategies. This dynamic metamorphism over time may not be achieved by only relying on human and artificial intelligence. Human–computer interaction requires a meta-intelligence for updating, innovating and transforming the various types of intelligence. Consequently, the firms

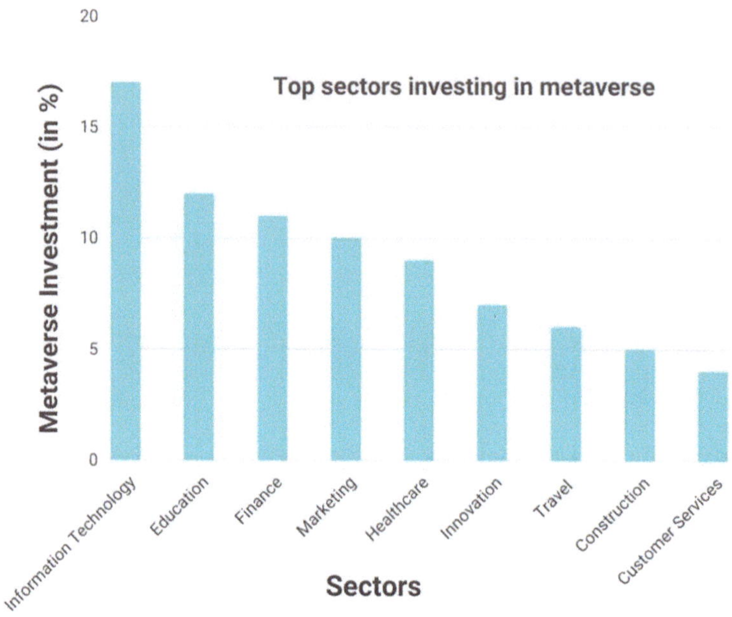

Fig. 4 Top sectors investing in the metaverse, Information Technology is highly investing

need intelligence—comprising meta-intelligence, human–computer interaction, and artificial intelligence. The top industries are investing in metaverse because of its tremendous growth and facts are shown in Fig. 4 [3, 4]. Recombination and renewal of various types of intelligence are two major components of meta-intelligence, which is analogous with the temporal progression of an organization's innovation processes and its technological capabilities [7].

The current system having artificial intelligence is designed to handle only a few fixed set of tasks and is considered as incapacitated. The creation of powerful, generic, or broad applications and systems is anticipated to result from the acceleration of technological development in the field of HCI over the coming two decades [8]. Artificial general intelligence (AGI), a system accomplished of performing the whole spectrum of human rational functions, is the name given to this advanced AI. Futuristic researchers have envisioned super-intelligence—an AI that go beyond all human expertise and reasoning which is far more beyond AGI [9]. It is not intelligence, but wisdom, that is linked with a better status of happiness, well-being, and health, and perhaps it also includes the longevity of society and the individual life [88]. Thus, AW—artificial wisdom is now the future need in the current scenario and this will also serve to reduce the stakes of advanced AIs such as Super-intelligence. "Alexa, play my favorite music track" or "Google call to brother," this the way we nowadays speak to the machines. Why do we do this? The reason is that we are more focused on communicating in these natural and easy ways. All human–computer interaction with equipment and machinery is compulsorily put into machine language, and this is not how the manner human express at all.

3 User-Centric Design Approach for Metaverse and HCI

(i) Understanding User Needs: Understanding the needs and demands of users who will engage with the metaverse is the first stage. This entails gathering data about user preferences, expectations, and goals through the use of user research techniques like surveys, interviews, and focus groups.
(ii) Designing Virtual Environments: Virtual environments that offer a compelling and immersive experience can be created based on user demands and requirements. In order to do this, 2D and 3D models, textures, and animations must be created and then integrated into a user-interactive virtual world.
(iii) Creating User Interfaces: After the virtual environment has been created, user interfaces that enable user interaction must be created. Designing menus, buttons, and other graphical components that let users move about the virtual environment and take activities there is part of this.
(iv) Interactivity: The metaverse heavily emphasizes interaction, and adding interactive elements like physics engines, collision avoidance, and user-controlled avatars can improve the user experience.
(v) Testing and Evaluation: It is critical to examine and assess the metaverse after it has been built and put into action to make sure it satisfies user demands and requirements. This entails carrying out usability tests, receiving user feedback, and revising the design in light of the findings.
(vi) Deployment and Maintenance: Finally, the metaverse can be made available to consumers through either a stand-alone application or as a component of a larger platform. The metaverse must continue to suit the user's needs and be relevant over time, therefore regular upkeep and upgrades are also crucial.

In general, the user-centered design technique for HCI and the metaverse focuses on identifying user needs and creating an immersive and interactive virtual environment that satisfies those goals.

4 Future Work Directions in Metaverse

Public, work, and home, every domain are foreseen to be adaptive, predictive and smart according to the necessities of their dwellers and people at those places, supported by means of artificial intelligence (AI) and engaging massive facts, information and statistics toward training and improving their results and mental ability. In the contemporary upsurge of scientific and technical progress, a nearby upcoming world is envisioned like a place where expertise smart devices are all-pervading, machineries envisage and antedate needs of social being, mechanical organisms would be a fundamental part of ordinary natural life, and capabilities of social being are scientifically embraced. The main center of interest of human–computer interaction had conventionally been the social being and in what way to be certain that algorithm/technique assists the needs of the consumer and in the superlative probable manner. It is considered as a perception that is call for and to also establish the eventual objective of the fresh intelligent technological contemporary era. Not only conscious and determined, but also subconscious and even undetermined interactions will be present in such an environment. Due to its expanded domain of research, HCI has evolved a lot over the years and has completed invincible improvements and progress considerably. Although additional algorithm and techniques fetch new overheads, amplified complication and intensify the need for collaboration and interface in several ways, the humanoid part of knowledge is also evolving and becoming more sensible and aware of the impact that communicating schemes and devices have on ordinary life pf social being: Humans have now become more focused and smart, but also look like to be less hopeful and positive, as well as more worried and specific. As a result of the procedures and plans that center on people, there are some new, significant problems and concerns that need to be addressed, such as formulating and addressing the fundamental problems that support the interaction and communication between people and trustworthy, healthy technology. Harper et al. [10] redirected upon ongoing fluctuations and framed a novel model for analyzing humankind's rapport with technology in the early years of the contemporary era by looking for ways to address social ethics in the expansion of intellectual and smart interaction. In this model, the relationship between machines and people as well as between processors and the physical world is reassessed and revalued. In the following sections of this chapter, each acknowledged and well-known obstacle is examined in the context of concepts and problem definition. With the ambition of making a contribution to the HCI community of academics and scientists, the leading exploration concerns that currently exist, the state of the art, as well as the related developing conditions, are highlighted further [11].

4.1 Meaningful Human Control

In spite the "intelligence" of the technology, person should always be kept back in the iteration by means of regulatory rereading mechanism and reviewing of intellectual and smart self-directed schemes and it's prospective to create instinctive and spontaneous predictions and judgements. The possibility for person learning and attainment of fresh perceptions, understanding, validation, and improvement of the application software as well as compliance with rules are immediate benefits of explainable AI [12]. Considering in the above context, human can be in the round, controlling the iteration, or some additional practice and guidelines, humanoid intervention should be meaningful and for AI it is the main focused short-tenure research investigation primacies [13]. And for successfully and efficiently achieving human supervisor, essential parameter of smart arrangements that to be well-thought-out are understandability, accountability, and transparency, it also helps to ensure in the direction of constructing a connection of confidence and hope between the machine and the user and enhance the enactment of the human-computerization group [14].

4.2 *Emotion Detection and Simulation*

As social beings are empathetic and emotional, articulating, and conveying sentiments are not solitarily common between humans but also with interaction between humans and machines. And as a result, in the manner information and communication technology with HCI fetches and coordinate feelings and expressional behavior, and in what way machines can convey sentiments and demonstrate empathic conduct is a critical concern with regard to humanoid-technology linkage optimization [15].

4.3 *Interactions in AR and VR*

Presenting few more directions in the area of human–computer interactions, virtual reality is the one of the "scientific, philosophical, and technological edge of our era." Demonstration of near lifelike illusionary worlds and then tracking and interacting in same illusionary world with simulated articles and entities, which possibly will truly exemplify physical persons sited anywhere on the earth and universe as well [16]. VR is now available at consumer prices due to recent advances in technology, moreover most market prediction implies that VR will soon have a most important influence [17]. Difficulties and hindrances in conveying near real understandings, that is utterly required to be resolved, take account of inadequate self-representation, consumer inadequacy of lifelike recreation of movement cyber-sickness, and absence of convincing and real communication [18]. Recent trends toward interconnected VR expose new ways for societal practices in virtual reality interface strategy needs, valuation approaches, and the secrecy and integrity apprehensions and stimulate research to address novel user interface [19].

4.4 *Assessment Method and Tools*

It is undoubtedly obvious that novel evaluation ways and means with efficient tools are required, for spontaneously and automatically evaluating the acquired data for self-reporting and assessed metrics for observation [20]. Considering the resulted interpretation, the massive quantity of feature traits and properties that should be assessed in such environments [97], systematic and holistic possibilities for the calculation and estimation of user interface in smart surroundings, taking into account an extensive range of features and traits it is obvious that intelligent new frameworks and model is required [21]. In the forthcoming, brainpower and intellect can remold into a provision and will convert into a novel plan and strategy factual [99].

4.5 Online Social Networks (OSNs)

In virtual world of communal and societal systems confidentiality is a foremost and key point of discussion, and it always in-demand to deliberate on security of personal and confidential data in all technological domains. When OSNs are used by delicate and subtle groups of consumers generally for dedicated commitments like communities of disabled group, students, friends, researchers, and scholars, the privacy becomes more required and demanding [22].

4.6 HealthCare Technologies

Keeping in mind the patients' personal information, fears arise referring to the demand of validity of purpose, and the possibility and probability of misuse of that collected data. Considering the above worry, moral and principal problems contain the consumption of fetched information for juvenile or adult persons, as well as consuming the same for investigation and examination purposes, such as to conduct online surveys, to recruit participants and to use effectively patient-reported data [23]. Accountability, an often neglected yet vital issue to be discussed and controlled. As patients are the most spirited and energetic contributors in the provisioning of healthcare, so as a consequence it raises the problems of probable misinterpretation, blunders, or misquotes [24].

4.7 IoT and Big Data

The Internet of Things is distinguished by means of diversified techniques and process, comprising of smart tangible things which are associated intensely over the network arrangement, permitting massive facilities, and creating smart civil structures, metropolises, and transportation systems [25]. Communications and relation could take place between social beings, machineries, and self-governing agents. And considering this highly interconnected and miscellaneous atmosphere, different types of confidentiality and safety menaces are manifested. HCI research will also uphold regulation actions in the IoT and big data domain.

4.8 Personal Medical Devices (PMDs)

Healthcare community are now looking forward for consumer wearable technology for acquiring and adopting it. But in what way this expertise and mechanism can finely serve treatment continue to persist unclear and it will be resolute by following major concerns, one is to make medical practitioner ready to board the massive amount of patients who will take along their health status data to their medical consultants and doctors, and the another is the high possibility for technical mistakes are high as when sick person try to diagnose the problem without medical training based on information fetched from wearable chip which could be drastically untrustworthy [26]. Therefore, to communicate and interact with wearables would be the research area for HCI researchers to accurately calibrate and calculate the vital data of patients. Research scope is also open in the field of tracking deformable and similar looking objects [27] in the domain of medical area.

4.9 Ambient Assisted Living (AAL)

AAL is in-demand domain pioneering which aimed at supporting elderly and disable users and helping a being's living status. It refers to enhancing their quality of living and help them to stay autonomous and lively to the greatest extent with the use of information communication and technology in an individual's living atmosphere [28]. Future research should emphasis on individual-centered sustenance and direction, by enhancing personalization of support, and keeping in mind various requirements, expectations, and choices of individuals which will need the collaboration of HCI domain [29].

4.10 Mobile Learning

Consumers as students are amicable with portable gadgets, henceforth concern frequently denote to by what means portable and moveable technologies can be fruitfully and proficiently engaged in training a consumer. Including mobile technologies in education may arise with issues as to how learning paradigm should be developed by trainers and how learners' and trainers' philosophy is reconfigured when portable gadgets are recycled effectively around the "conventional edges between official and casual contexts, virtual and physical worlds and planned and explored areas" [30].

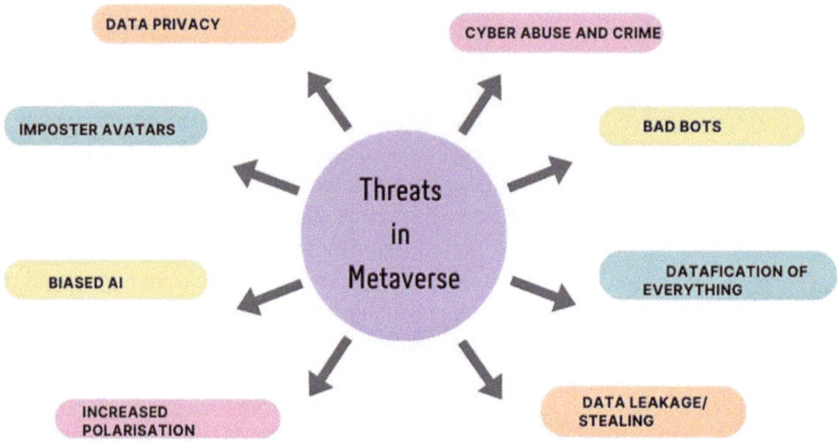

Fig. 5 Threats and challenges in metaverse

5 Challenges and Issues in Metaverse

There will be many challenges in the metaverse, such as unrestricted data collection that compromises our privacy, persistent abuse and harassment, imposter avatars attempting to steal private information (such as your NFTs), widespread security flaws, biased AI, rampant bots and trolls, and even more polarized society, increased inequality, and issues with physical and mental health. A surveillance society that is run by the state or private businesses has also been created by new technologies like big data analytics and artificial intelligence, as well as ongoing data collection brought on by the Internet of Things and social media. Some threats and concerns are identified and shown in Figs. 5 and 6 with statistic details [31–34]. Unless we are able to construct an open metaverse, we may soon be living in a dystopian future where our data dictate our freedom, for better or worse. Intense accumulation of data via AR and VR devices, where the amount of data to be collected, misused, and monetized is projected to be 100× more than it is today [32–35], will aggravate this dystopian future.

6 Discussion

This chapter has discussed main future direction, concerns, and issues that arise the direction contemporary socio-technological real-world human–computer interaction environment, with a view to explore and unleash the soaring accessible collaboration intellect in direction to answer to captivating social and communal demands. The argument has critically promoted an upcoming technical paradigm where brainpower and knowledge will be activated for healthier support and help the wishes of persons

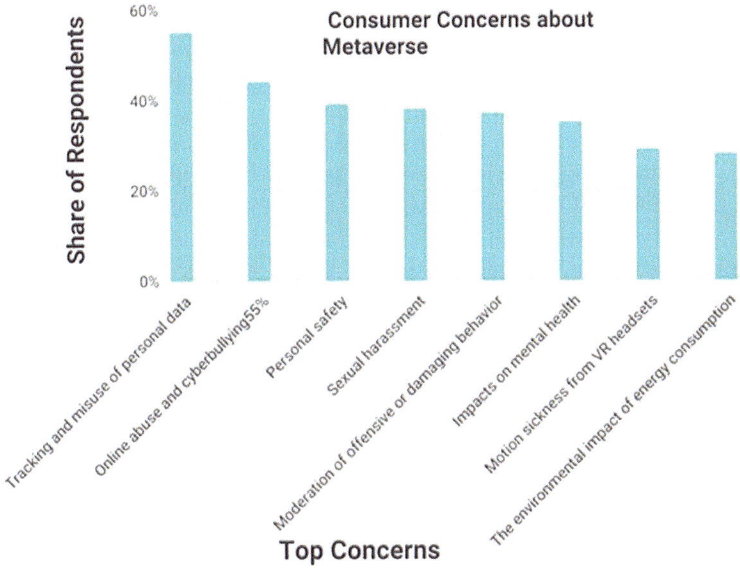

Fig. 6 Top concerns that consumers have about the metaverse

to rightly enlighten and empower them with the enforcement of recent technological development and intelligence. To commence a significant attempt and defend the project of an innovative techniques where the intellect addition will not destabilize individual self-efficacy and power; but as an alternative it converts into an influential instrument that would be developed by HCI researcher community. Various processing techniques for HCI interaction are exploited to present the exceptional features of present techniques with systematic review and enlisting the advantages and disadvantages of all in this chapter. Moreover, the interdependence of social being with smart bionetworks, which goes well beyond practical limitations and involves inter-disciplinary algorithm with the aim to also address all types of societal and ethical enforcing issues is the central issue of research in this domain. This consists of a number of considerations, such as consideration of socio-technological angle, integrating human ethics in design approach and selections, and refining humanistic concerns.

7 Conclusion

The variety of critical concerns also ascend, comprising of like developing for significant humanoid control over machines, safeguarding smart systems' clearness and efficiency, and accounting for intelligent machines intrinsic imperviousness and irregularity. The standard of choice is that the chosen procedure should own most of the desired features of constrained environment, including a reckless and well-organized data treating and handling capability, lesser requirement of storage, high compression proportion as well as high image fidelity. Centered on the investigation of the evidence concise and wide-ranging collected works studies, it is observed that emphasis on the exploring the contents and features of the image and process these contents to achieve high efficiency ratio as well as better image fidelity for diagnosing purposes in HCI imagery. Automatic image segmentation algorithms have been produced in abundance of research, but the problems, such as partial occlusion, and over-segmentation, among many others cannot be overcome. For recent visual data transmission system limitations are inherent and lots of effort and motivation is required to scale up the efficiency and throughput. Intercepting the discussed concerns and inspecting the developing study problems necessitate collaborative efforts underneath a wide-ranging inter-disciplinary space. The collaborated research is to be done by interconnecting elements globally and internationally. In addition to reshaping the contributions and teaching of HCI experts, practitioners, and academics on a worldwide scale, HCI educational teaching will strengthen alliance between academic and research societies and industry. Humans are the primary focus of HCI, and in the current technologically enhanced settings, this will lead to attempt to improve the lives of the elderly and the disabled by raising their quality of life. Future developments in HCI with AI will include whole smart cities and infrastructure, not just home automation or smart workstations.

References

1. Orland K (2021) So what is 'the metaverse,' exactly? Ars Tech, 7 Nov. 2021. https://arstechnica.com/gaming/2021/11/everyonepitching-th
2. Rawat DB, El Alami H (2023) Metaverse: requirements, architecture, standards, status, challenges, and perspectives. IEEE Internet Things Mag 6(1):14–18. https://doi.org/10.1109/IOTM.001.2200258
3. Li G, Kou G, Peng Y (2018) A group decision making model for integrating heterogeneous information. IEEE Trans Syst, Man, Cybern: Syst 48(6):982–992
4. Zhang H, Kou G, Peng Y (2019) Soft consensus cost models for group decision making and economic interpretations. Eur J Oper Res 227(3):964–980
5. Quek F (2004) The catchment feature model: a device for multimodal fusion and a bridge between signal and sense. Eurasip J Appl Signal Process 11:1619–1636
6. Murugappan VS, Liu H, Ramani K (2013) Shape-it-up: hand gesture based creative expression of 3D shapes using intelligent generalized cylinders. Comput-Aided Des 45(2):277–287
7. Lichtenthaler U (2018) Substitute or synthesis? The interplay between human and artificial intelligence. Res-Technol Manag 61(5)
8. Kumar K, Thakur GS (2012) Advanced applications of neural networks and artificial intelligence: a review. Int J Inf Technol Comput Sci 4
9. Bostrom N (2014) Superintelligence: paths, dangers, strategies. Oxford University Press, Oxford
10. Jeste DV, Lee EE (2019) Emerging empirical science of wisdom: definition, measurement, neurobiology, longevity, and interventions. Harv Rev Psychiatry 27:127–140
11. Harper R, Rodden T, Rogers Y, Sellen A (2008) Being human: HCI in 2020. Microsoft, Cambridge, UK

12. Samek W, Wiegand T, Müller KR (2017) Explainable artificial intelligence: understanding, visualizing and interpreting deep learning models. ITU J ICT Discov 1:39–48
13. Russell S, Dewey D, Tegmark M (2015) Research priorities for robust and beneficial artificial intelligence. AI Mag 36(4):105–114
14. Chen JYC, Lakhmani SG, Stowers K, Selkowitz AR, Wright JL, Barnes M (2018) Situation awareness-based agent transparency and human-autonomy teaming effectiveness. Theor Issues Ergon Sci 19(3):259–282
15. Ferscha A (2016) A research agenda for human computer confluence. In: Human computer confluence transforming human experience through symbiotic technologies, pp 7–17
16. Lanier J (2017) Dawn of the new everything: encounters with reality and virtual reality. Henry Holt and Company, New York, NY
17. Steinicke F (2016) Being really virtual. Springer, Cham, Switzerland
18. Bastug E, Bennis M, Médard M, Debbah M (2017) Toward interconnected virtual reality: opportunities, challenges, and enablers. IEEE Commun Mag 55(6):110–117
19. Carvalho RM, Castro Andrade RM, Oliveira KM, Sousa Santos I, Bezerra CIM (2017) Quality characteristics and measures for human–computer interaction evaluation in ubiquitous system. Softw Qual J 25(3):743–795
20. Ntoa S, Margetis G, Antona M, Stephanidis C (2019) UXAmI observer: an automated user experience evaluation tool for ambient intelligence environments. In: Proceedings of the 2018 intelligent systems conference, pp 1350–1370
21. Holmquist LE (2017) Intelligence on tap: artificial intelligence as a new design material. Interactions 24(4):28–33
22. Parmaxi A, Papadamou K, Sirivianos M, Stamatelatos M (2017) E-safety in web 2.0 learning environments: a research synthesis and implications for researchers and practitioners. In: Proceedings of the 4th international conference on learning and collaboration technologies, pp 249–261
23. Denecke K, Bamidis P, Bond C, Gabarron E, Househ M, Lau AYS, Hansen M (2015) Ethical issues of social media usage in healthcare. Yearb Med Inform 24(01):137–147
24. Kluge EHW (2011) Ethical and legal challenges for health telematics in a global world: telehealth and the technological imperative. Int J Med Inform 80(2)
25. Ziegeldorf JH, Morchon OG, Wehrle K (2014) Privacy in the Internet of Things: threats and challenges. Secur Commun Netw 7(12):2728–2742
26. Piwek L, Ellis DA, Andrews S, Joinson A (2016) The rise of consumer health wearables: promises and barriers. PLoS Med 13(2)
27. Kaushik B, Koundal D, Goel N, Zaguia A, Belay A, Turabieh H (2022) Computational intelligence-based method for automated identification of COVID-19 and pneumonia by utilizing CXR scans. Comput Intell Neurosci 2022:12, Article ID:7124199
28. Moschetti A, Fiorini L, M. Aquilano M, Cavallo F, Dario P (2014) Preliminary findings of the AALIANCE2 ambient assisted living roadmap. In: Proceedings of the 4th Italian Forum on ambient assisted living, pp 335–342
29. Kachouie R, Sedighadeli S, Khosla R, Chu MT (2014) Socially assistive robots in elderly care: a mixed-method systematic literature review. Int J Hum-Comput Interact 30(5):369–393
30. Burden K, Kearney M (2016) Conceptualising authentic mobile learning. In: Churchill D, Lu J, Chiu T, Fox B (eds) Mobile learning esign, pp 27–42
31. Wang Y et al (2023) A survey on metaverse: fundamentals, security, and privacy. IEEE Commun Surv Tutor 25(1):319–352. https://doi.org/10.1109/COMST.2022.3202047
32. Kaushik B, Kumar M, Jalal AS, Bhatnagar C (2018) A context based tracking for similar and deformable objects. Int J Comput Vis Image Process (IJCVIP), pp 1–15
33. Wu J, Lin K, Lin D, Zheng Z, Huang H, Zheng Z (2023) Financial crimes in Web3-empowered metaverse: taxonomy, countermeasures, and opportunities. IEEE Open J Comput Soc 4:37–49. https://doi.org/10.1109/OJCS.2023.3245801

34. Pramanik PKD, Pal S, Choudhury P (2018) Beyond automation: the cognitive IoT. Artificial intelligence brings sense to the internet of things. In: Sangaiah A, Thangavelu A, Meenakshi Sundaram V (eds) Cognitive computing for Big Data systems over IoT. Lecture Notes on Data Engineering and Communications Technologies, vol 14. Springer, Cham
35. Dunnett K, Pal S, Jadidi Z, Jurdak R (2023) The role of cyber threat intelligence sharing in the metaverse. IEEE Internet Things Mag 6(1):154–160. https://doi.org/10.1109/IOTM.002.2200003

Metaverse Explosion and Its Consequences for the Travel Industry, AI, and VR

Shakeel Basheer, Sandeep Walia, and Sheezan Farooq

Abstract A new era of discovery and adventure has begun as tourism, AI, and transportation intersect. This book chapter examines how these four principles will affect tourism and how they interact. In the chapter, the metaverse, a virtual world that uses immersive technologies for social interaction and exploration, is examined. The metaverse allows people to visit distant places without really going there, expanding tourist opportunities. AI is undoubtedly advancing the holiday industry. AI can handle massive volumes of data and offer tourists individualized suggestions, simplifying and enriching vacation planning. AI-enabled chatbots can instantly answer client questions and solve problems. Virtual tour guides, who can provide travelers a unique experience wherever, are also being developed using artificial intelligence (AI). Digital tour guides can customize their services based on each person's interests, preferences, and physical ability. The chapter concludes with these technologies' travel industry effects. These advancements have many benefits, but some worry about job losses and local economies. The chapter finishes by considering how to employ this creative technology to benefit travelers and the communities they visit. This book chapter explains the tourist sector, AI, and vacationing relationship. It explores how the metaverse may affect the travel and tourism industry and presents opportunities for expansion.

Keywords Artificial intelligence · Travel · Tourism · Virtual space · Virtual reality · Metaverse

S. Basheer (✉) · S. Walia
School of Tourism Management and Airlines, Lovely Professional University, Phagwara, India
e-mail: Shakeelbasheer40@gmail.com

S. Walia
e-mail: sandeep.24651@lpu.co.in

S. Farooq
Department of Computer Science, Islamic University of Science and Technology, Pulwama, India

G. Chhabra and K. Kaushik (eds.), *Understanding the Metaverse*, Blockchain Technologies, https://doi.org/10.1007/978-981-97-2278-5_14

1 Introduction

The advent of ICTs has shaken up the travel sector and altered the method in which businesses operate [33]. In the 1980s, global distribution systems (GDSs) like Amadeus and Sabre were developed, making it easier for travel agencies to consolidate data from several vendors and order travel-related services for their clients [23]. The expansion of global networks and mobile devices has led to the creation of an intelligent infostructure, which has altered the tourism industry. Technology platforms have helped the tourism industry by allowing providers to reach more customers directly, cutting out middlemen, nevertheless, the sharing economy has presented new disruptive concerns, for instance, [23]. Competition in the hotel and transportation sectors has intensified and business models have shifted as a result of platforms like Airbnb and Uber. Therefore, it is crucial for those involved in the tourism business to keep up with and effectively implement new technological developments in order to maintain a competitive edge [34].

Smartphones and mobile commerce, with their explosive growth, have the potential to shake up the travel sector [3]. Smartphones allow for the real-time and contextual co-creation of value [7]. The advent of AR and VR has amplified this possibility, opening up new horizons for tourist administration, advertising, and interactive stakeholder participation [43]. Travelers can now enhance their in-person experiences at a place with virtual ones they can preview beforehand [31]. These possibilities have been expanded by the Metaverse, which has been in the spotlight ever since Facebook rebranded as Meta and refocused its commercial strategy [29]). Smartphones, wide-ranging networks, and intelligent infostructures have all contributed to the industry's dramatic shifts in recent years [29]. These technological developments have opened up new potential for the tourism industry, but they have also raised competitiveness and introduced new obstacles [19]. In order to be competitive and produce superior experiences for travelers, those involved in the tourism business must keep up with technology advancements and effectively use them.

According to this study, the Metaverse is "the seamless integration of the real and virtual worlds for a wide range of purposes, including but not limited to work, education and training, health, curiosity exploration, and socialization" [20]. There is substantial evidence of widespread acceptance in the game industry despite its conceptual nature [9]. The future of the Metaverse is now being defined, structured, organized, and depicted by researchers from a variety of fields working together [11]. Newcomers to the Metaverse can start by creating an avatar and then venturing out to discover the digital world [9]. The development of the Metaverse is far from complete, and developments in the future may have unanticipated consequences for the travel sector [34]. The full potential of digital environments will be realized when they support lifelike interactions in the digital realm. To wit the advantages of digitization were made evident during the COVID-19 epidemic, but the shortcomings of flat, muddled, and unreal encounters were also made manifest (Young Lee Assistant Professor, 2021).

By 2030, all physically possible devices will be digitally connected, and all digital actions will have tangible consequences in the real world and vice versa [25]. The metaverse relies on the integration of several technologies that were previously seen as separate, such as cloud and edge computing, AI, blockchain, IoT, VR/AR, and so on. Numerous people may attest to its existence. [17] claims that popular brands like Fortnite, Meta, and Roblox are altering how we interact in our leisure time. Metaverse natives are one group, while metaverse colonists make up the other [16]. The former are the young people who have grown up with online interactive games and are more at ease interacting with others in virtual worlds than they are in the actual world. Based on previous research [41], this second camp maintains that we can divide reality into digital and physical spheres.

2 Innovation and Market Shifting as a Result of COVID-19's Metaverse Impact

The concept of the metaverse became widely discussed after the COVID-19 pandemic (Tik-Tsuen WONG, n.d.). With the rise of telecommuting and online education, it became necessary to devise effective methods or channels for facilitating online communication [15]. Consumers and businesses alike have given it more attention since the outbreak. In 2020, tech companies said that they would begin developing and investing in this technology. Metaverse invested a billion dollars on technology investments in 2021 [21]; therefore, it was a strong year for the company [13]. Metaverse technologies were reportedly sparked by the COVID-19 pandemic, and experts believe that post-pandemic conditions will pique consumer interest. As a result of the epidemic, virtual online communities, such as interactive gaming landscapes and the expanding use of mixed reality [1], have emerged as key lifestyle destinations for users who have been imprisoned due to lockdowns and closures. Many sectors may benefit from the COVID-19 pandemic's positive effect on the metaverse commerce. Nonfungible Tokens (NFTs), blockchain, and other digital assets are helping to propel the metaverse into the mainstream. The author, [44]. It is expected that in the future, the metaverse will naturally foster spatial patterns throughout the education industry, much like how campus digital twins have improved the remote education experience. [36] Pursuing knowledge will be reimagined as an interactive experience enriched by digital records in the metaverse [24] rather than just words, images, and lectures available on demand.

In addition, it would lay the groundwork for the development of digital social experiences in the era after COVID [14]. For some users, the metaverse is just a digital setting in which they can participate via controlling a virtual character called a "avatar" [12]. Market revenue is expected to expand significantly in the coming years as a result of rising demand for the implementation of blockchain-based metaverse systems for digital trade assets [32]. Finally, the chapter discusses how metaverse technology has affected the tourism industry as a whole [38]. Despite the benefits of technological progress, others worry about the potential negative effects on employment and local economies [5]. The chapter concludes with suggestions for maximizing the positive impact of these technology developments for tourists and locals alike [21]. This chapter of the book provides an in-depth analysis of how AI, the Metaverse, and tourism interact with one another. It draws attention to the possible benefits of new technology for the tourism industry and considers their potential drawbacks and consequences.

3 The Metaverse: An Expanding Digital Universe

Millions of people all around the world interact in a shared digital space called the "Metaverse" [37]. The idea of the Metaverse is to create a digital space where people can interact socially, recreationally, scholastically, and professionally [37]. Popularized by science fiction, the concept of the Metaverse is becoming increasingly tangible as technology advances [26]. The Metaverse is built on top of VR and AR systems, which allow users to interact with a simulated world as if it was real. In the coming years, we may expect the technology powering the Metaverse to become even more immersive and user-friendly [26]. The growth of cutting-edge technologies like AI and blockchain is driving this improvement. One of the most significant aspects of the Metaverse is its capacity to bring together people from different parts of the world in a common virtual space. This allows users to communicate with one another in real-time regardless of where they physically are. This paves the way for fresh methods of communication and teamwork, and it could have far-reaching effects on our daily lives and professional spheres in the future [2, 3] (Fig. 1).

The Metaverse may also become an economic superpower. As people spend more time online, they may start buying and selling virtual goods and services. For businesses and startups interested in making money in the Metaverse, this might mean new opportunities. However, issues with privacy and security are exacerbated in the Metaverse. The likelihood of having one's personal information hacked or stolen rises in proportion to the amount of time one spends online [27]. There are also worries about the potential for addiction and social isolation as people spend more time in the virtual world. Despite these concerns, the Metaverse is predicted to play a larger role in our lives in the near future. As technology improves, the possibilities of the Metaverse will grow. It could be used for all sorts of purposes, including communication, learning, and recreation. It will be fascinating to see how the Metaverse develops in the years to come and whether or not it ends up having a significant impact on the way we currently live and work.

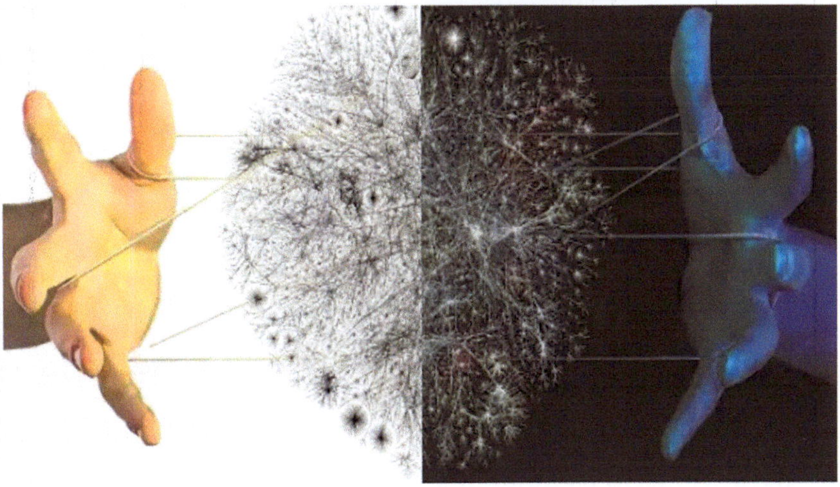

Fig. 1 Potential of the Internet. *Source* https://www.livemint.com/technology/tech-news/the-internet-30-big-hope-to-big-bother-1554156689591.html

4 "Disruptions and Opportunities in Tourist Behavior and Experience Due to the Metaverse"

Augmented and virtual realities, blockchain, and AI are just a few of the developing technologies being used to create the metaverse [30]. In this setting, users' avatars can interact with one another in real-time, bringing the virtual and real worlds closer together [44]. The idea of the metaverse is gaining traction in the travel industry as a means to provide novel experiences for tourists. Metaverse in tourism includes the development of virtual tours as one of its dimensions (Zhu et al. n.d.). Virtual reality and 360-degree videos allow potential tourists to get a feel for a place before committing to a trip there. Those with mobility issues or financial constraints may benefit greatly from this. Historical locations, cultural events, and natural wonders that are out of the way can also be highlighted through the use of virtual tours. The development of virtual hotels is an additional facet of the metaverse in the hospitality industry. Travelers can have the full experience of staying at a hotel without leaving their rooms by staying in a "virtual hotel." Travelers can enjoy the hotel's ambiance and services without actually staying there by using a virtual reality headset or an augmented reality app. Similarly, cutting-edge hotel architecture and technology can be showcased through the usage of virtual hotels [28].

Similarly, the metaverse can be utilized to design and build fully immersive tourist attractions like VR theme parks. Combining AR/VR/gamification technologies allows for the creation of such theme parks. Through the use of their avatars, visitors to virtual attractions, games, and characters can engage in real-time activities that are sure to leave a lasting impression. The development of online markets is just another facet of the metaverse that has relevance to the tourism industry. Local goods, souvenirs, and handicrafts can be shown in online marketplaces so that tourists can shop for genuine items without leaving their houses. By patronizing locally owned enterprises, sustainable tourism may be promoted through these types of markets. Last but not least, the metaverse can be used to provide meeting places for tourists. Through the use of online community forums, vacationers can meet new people and talk about their trips with old acquaintances. Virtual concerts, festivals, and conferences can be hosted through these social networks. In conclusion, the metaverse has enormous potential for the development of novel tourist experiences. Companies in the tourist industry may provide customers with interactive experiences, virtual tours, hotels, marketplaces, and social gathering places by leveraging cutting-edge technology like virtual reality, augmented reality, and blockchain. The metaverse has the potential to grow in importance as a tourist destination as technology advances.

5 Sixth, Tourism's Potential for Growth and Innovation

Virtual tourism experiences: The metaverse could allow users to visit destinations virtually, without having to physically travel there [45]. This could include virtual tours of landmarks and attractions, as well as interactive experiences that allow users to explore the destination in more detail. Interactive cultural experiences: The metaverse could provide interactive experiences that allow users to engage with the local culture of a destination, such as virtual cooking classes, language lessons, and traditional dance performances. Social connections: The metaverse could enable users to connect with other travelers and locals in real-time, allowing them to share tips and recommendations, and create new social connections [40] Immersive marketing: The metaverse could be used to create immersive marketing campaigns that showcase destinations in a new and engaging way. This could include interactive virtual events, virtual reality experiences, and more. Sustainability education: The metaverse could be used to educate travelers about sustainable tourism practices, including responsible travel and environmental conservation efforts (Fig. 2).

6 How Travel is Being Enhanced by Artificial Intelligence Chatbots

Chatbots that use artificial intelligence (AI) are becoming more and more common in a variety of industries, including travel and tourism. The usage of AI chatbots in metaverse tourism is an intriguing new application [10]. A virtual environment where users can communicate with one another and partake in a variety of activities like gaming, interacting with others, and exploring is referred to as the metaverse. As augmented reality and virtual reality technologies have grown in popularity, individuals all around the world are traveling to the metaverse more and more. Computer

Fig. 2 Tourism's technological potential. Reference: https://www.hlp.city/what-is-smart-tourism-and-why-does-it-matter/

programs called chatbots are made to mimic human speech [18]. To comprehend and reply to customer inquiries, they make use of machine learning (ML) and natural language processing (NLP) methods. Tourism organizations may improve user experience and give customers a more personalized and interesting journey by incorporating AI chatbots into the metaverse. Instant help and support for users are one of the key advantages of employing chatbots in metaverse tourism. Chatbots can be designed to guide users through the metaverse, offer recommendations, and respond to often asked inquiries [27].

Making the most of their time in the metaverse and navigating the virtual world more simply can be achieved with this assistance. Using chatbots to customize the user experience is another benefit of metaverse tourism. Chatbots are able to offer personalized recommendations and suggestions that are tailored to the user's interests and preferences by analyzing user data and preferences. Users who might not have otherwise investigated new places of the metaverse may find them thanks to this. Additionally, chatbots can be utilized to raise the metaverse's entertainment value. It is possible to create chatbots to converse with people, tell jokes, and offer fascinating information and facts, for instance [27]. Users may feel more a part of the virtual world and be inspired to explore for longer as a result. Ensuring that chatbots can effectively interpret and reply to user queries is a challenge when deploying them in metaverse tourism. Since NLP and ML algorithms are always changing, it might be challenging to guarantee that chatbots can accurately interpret user inquiries. Additionally, sophisticated or nuanced questions requiring human assistance could be beyond the capabilities of chatbots. Businesses can spend money on continuous training and development for their chatbots to solve these issues. This may entail gathering user input and applying it to gradually enhance the chatbot's functionality. Additionally, businesses can employ human moderators to supervise chatbot interactions and step in when needed.

In summary, chatbots driven by artificial intelligence possess the capability to transform the metaverse tourist sector by offering real-time support, customization, and amusement. This will enhance the metaverse's appeal and accessibility for individuals worldwide. While there are some drawbacks to employing chatbots in metaverse tourism, businesses may get over these and provide a more seamless experience with continued training and development.

6.1 AI-Powered Guides Enhance Online Travel Experience

Virtual tour guides have grown in popularity in recent years, particularly since real travel has become more difficult due to travel restrictions and safety concerns. Travelers can now enjoy a more customized and immersive experience thanks to the emergence of virtual tour guides made possible in large part by AI technology. The capacity to personalize the experience for each passenger is one of the main advantages of using AI to create virtual tour guides. Virtual tour guides can generate a tailored itinerary by analyzing a traveler's interests, preferences, and previous travel experiences through the use of machine learning algorithms. For instance, the virtual tour guide can suggest particular museums, historical locations, and cultural events to visit if the traveler has an interest in history and culture. Artificial intelligence (AI) is also utilized in virtual tour guides by means of natural language processing (NLP) technology. This makes the experience more dynamic and engaging by enabling the virtual tour guide to comprehend and reply to inquiries and requests from visitors in a natural way. For visitors who might not speak the language, NLP technology can also be utilized to deliver translations in real-time, which facilitates communication and helps visitors grasp the customs and language of the area.

Travelers can explore realistic and immersive virtual settings made possible by AI. High-quality 3D models of actual places can be produced by virtual tour guides employing computer vision and image recognition technologies. This makes it possible for visitors to experience these places virtually, with lifelike sights, sounds, and interactions. Making sure that the experience is accessible and user-friendly for all visitors, regardless of their level of technological skill, is one of the issues faced by virtual tour guides. AI can be utilized to design user-friendly interfaces that are simple to browse in order to overcome this difficulty. Travelers can easily access the experience on their preferred device as virtual tour guides can be created to work on a variety of devices, including smartphones, tablets, and virtual reality headsets. Lastly, tourists can also receive real-time warnings and recommendations from AI. For instance, the virtual tour guide can modify the timetable and provide substitutes if a visitor is going to be late for an appointment. Similarly, the virtual tour guide can offer real-time navigation and assistance to a traveler in need of directions if they are in an unknown place.

In conclusion, the development of virtual tour guides has been greatly influenced by artificial intelligence (AI) technology. Through the use of natural language processing, computer vision, image identification, and machine learning algorithms, virtual tour guides may offer tourists customized and engaging experiences. AI can also be utilized to develop user-friendly user interfaces, offer real-time alerts and recommendations, and improve accessibility and usability for travelers.

6.2 Using AI to Build Virtual Marketplaces

Online marketplaces are growing in popularity because of how convenient they are. Virtual marketplaces are growing more intelligent and user-friendly with the introduction of AI technology, giving customers a better online shopping experience. This essay will go over how artificial intelligence plays a part in the development of online markets. To begin with, AI is utilized to customize the buying experience. Through the examination of a client's past purchases, search terms, and online activity, artificial intelligence systems are able to suggest goods that are probably going to catch their attention. This makes it more probable that a sale will be made and that the client would visit the store again in the future. Additionally, personalization fosters consumer loyalty and fosters trust between the consumer and the marketplace. AI is applied to pricing optimization. AI algorithms are able to ascertain the best price for a product by examining trends in the market, the prices of competitors, and consumer demand. This contributes to maintaining the market's profitability and competitiveness. Additionally, it aids in avoiding under pricing, which can lead to lost income, and overpricing, which can lead to lost sales. AI is applied to enhance inventory control. Artificial intelligence (AI) algorithms can assist marketplace owners in optimizing their inventory levels by analyzing sales data and forecasting future demand. In this method, the possibility of lost sales is decreased because products are guaranteed to be in stock whenever clients wish to buy them. Additionally, it aids in avoiding overstocking, which can lead to missed sales and squandered inventory.

Artificial intelligence (AI)-powered chatbots may respond to consumer questions and offer assistance around-the-clock, guaranteeing that clients receive timely and effective support—a measure that can boost client happiness and loyalty. Additionally, chatbots can manage repetitive duties like order tracking, freeing up customer support agents to answer more difficult inquiries. AI is finally applied to enhance security.

In summary, artificial intelligence (AI) technology is essential to the growth of virtual marketplaces. AI algorithms can help marketplace owners provide customers a better buying experience by optimizing pricing, boosting customer service, streamlining inventory management, customizing the shopping experience, and bolstering security. Virtual marketplaces are anticipated to grow even more intelligent and user-friendly as AI technology develops, offering even better purchasing experiences (Fig. 3).

Fig. 3 Tourism's future. Reference: https://www.way2smile.ae/blog/ai-in-travel-industry/

7 Both Visitors and Locals' Profit from the Application of New Technologies

The way we travel has been completely transformed by new technology, which have also made previously unthinkable opportunities and experiences possible [39]. Technology is becoming a necessary component of travel, from the time we begin itinerary planning until we arrive home [22]. The significance of these technologies, meanwhile, extends beyond just tourists [6]. They can also help local communities when applied appropriately, which promotes more ethical and sustainable travel. The sharing economy is one of the most important ways that technology helps local communities and vacationers alike. [8] can now travel with greater authenticity and experience local culture and communities thanks to platforms like Airbnb and Uber. Travelers can now rent private homes, flats, or even spare rooms from local hosts as an alternative to staying in impersonal hotels. In addition to offering visitors a more engaging experience, this helps the local economy by giving hosts in the area a source of revenue. Similar to this, ride-sharing services like Uber and Lyft have increased travelers' ease and convenience when traveling between cities and given local drivers a way to make money. Additionally, by lowering the number of cars on the road, these services have contributed to lower emissions and a cleaner environment. The usage of travel applications is another way that technology is helping local communities. To help tourists make the most of their time at a destination, these apps offer real-time information about nearby restaurants, sights, and events. They also give nearby companies a chance to present their products to a larger market, which raises their profile and boosts sales.

Tourism applications and the sharing economy are not the only technological tools being utilized to promote sustainable tourism. As an illustration, a few hotels and resorts are conserving resources and lowering their carbon footprint by implementing smart technology. This covers water-saving devices, intelligent heating and cooling systems, and energy-efficient lighting. Hotels may minimize their environmental impact and cut operating costs by utilizing this technology. In a similar vein, certain locations are encouraging ethical tourism behaviors with the help of technology. For instance, the Icelandic Tourist Board has created an app and website that teaches visitors about responsible travel strategies, such appreciating the environment and local customs. This helps to protect the destination's natural and cultural assets in addition to helping the local populations. But it's important to recognize that technological advancements don't necessarily translate into environmentally friendly travel. For instance, some detractors contend that using ride-sharing services like Uber and Lyft may worsen traffic in urban areas. Similar worries have been raised over the influence on regional property markets due to the growing popularity of short-term rentals via websites like Airbnb.

In conclusion, when applied properly, technology can help both tourists and local populations. Applications for travelers, the sharing economy, and environmental programs are just a few instances of how technology may be used for the benefit of society. It's critical that we continue to be aware of how technology affects local communities as we navigate the always shifting travel landscape and work to use it in a way that encourages ethical and sustainable travel. Using chatbots is one possible way artificial intelligence is being used in the metaverse. Computer programs known as chatbots are designed to mimic human-user conversation. Chatbots could be deployed in the metaverse to help tourists by recommending restaurants and other activities and by responding to inquiries about nearby sights. In addition to assisting visitors in making the most of their stay abroad, this might support regional companies. Making virtual assistants is another method AI may help local groups in the metaverse. AI-powered apps known as virtual assistants may aid users with a range of chores, including ordering takeout and making appointments. Virtual assistants could be utilized in the metaverse to aid local firms in running their operations more effectively. A virtual assistant, for instance, may aid a restaurant owner in handling orders and reservations, freeing up their time to concentrate on giving their patrons a wonderful dining experience.

AI may potentially be utilized to develop metaverse experiences that are more immersive. Avatars with AI capabilities, for instance, might be utilized to build realistic characters that communicate with users in real-time. Because the avatars may adapt to each user's unique requirements and tastes, this could help provide a more personalized experience for them. Artificial intelligence (AI) has the potential to produce more lifelike metaverse landscapes with authentic animals and weather patterns. A possible advantage of AI and the metaverse for nearby areas could be a rise in tourism. As more individuals delve into the metaverse, they might find themselves drawn to traveling to the actual places that served as the inspiration for the virtual worlds they encountered. This might result in higher tourism-related income for nearby companies, bolstering the local economy. However, using AI in the metaverse carries certain potential risks as well. Chatbots and virtual assistants driven by AI, for instance, may be used to collect user data without authorization. Furthermore, it's possible to fabricate false personas using AI-powered avatars in order to influence users.

We must create strict privacy laws and guidelines pertaining to AI use in the metaverse in order to mitigate these concerns. Developers should also be open and honest about the data they gather and the purposes for which they use it. Ultimately, people ought to have the authority to manage their own data and be able to refuse data gathering. In conclusion, consumers may have a more customized and immersive experience thanks to the metaverse and AI, which could also help local communities. Travelers can interact with lifelike characters and receive personalized advice via chatbots, virtual assistants, and AI-powered avatars; local businesses can employ AI to streamline operations. It will be crucial to handle any risks—like worries about data privacy—that could arise from the application of AI in the metaverse. All things considered, the metaverse, and AI have the ability to build a more interconnected and advantageous environment for tourists as well as local populations.

8 The Difficulties and Ramifications of These Technologies in Tourism

There are several emerging technologies that are expected to have a significant impact on the tourism industry. These technologies have the potential to transform the way people plan, book, and experience their vacations. However, they also bring new challenges and implications that the tourism industry must address. Here are some examples: virtual reality (VR) and augmented reality (AR): VR and AR technologies offer tourists the opportunity to experience destinations in a more immersive and engaging way. However, the use of VR and AR may also reduce the demand for physical travel, as people can experience destinations virtually from the comfort of their own homes. Artificial intelligence (AI): AI has the potential to revolutionize the tourism industry by personalizing travel recommendations, improving customer service, and streamlining operations. However, there are concerns about the potential loss of jobs as AI takes over tasks traditionally performed by humans. Internet of Things (IoT): IoT enables the integration of various technologies and devices, making it possible to create smarter and more connected tourism experiences. However, IoT also raises concerns about data privacy and security, as personal information is shared across devices and networks. Blockchain: Blockchain technology offers a secure and decentralized way to manage transactions and data. In the tourism industry, it can be used to improve transparency in the booking process, reduce fraud, and streamline payments. However, there are also concerns about the complexity of implementing blockchain solutions and the potential for unintended consequences. 5G Networks: 5G networks provide faster and more reliable internet connectivity, making it possible to access and share information more quickly and efficiently. However, there are concerns about the potential for 5G networks to overload existing infrastructure and the potential health effects of increased exposure to electromagnetic radiation.

Overall, these technologies have the potential to transform the tourism industry in numerous ways. However, it is essential to address the challenges and implications associated with their adoption to ensure that the benefits outweigh the risks. The tourism industry must be prepared to adapt to these emerging technologies while also maintaining a focus on providing high-quality and authentic travel experiences for tourists. The concept of the metaverse has become increasingly popular in recent years, with the COVID-19 pandemic accelerating the shift toward digital experiences. The metaverse can be defined as a collective virtual shared space that is created by the convergence of virtual and physical reality, and it has the potential to transform the way we interact, work, and play. One area that is poised to benefit from the emergence of the metaverse is tourism. As people become more comfortable with digital experiences and the potential of the metaverse becomes clearer, the tourism industry can use this new platform to create immersive and engaging experiences for travelers. In this article, we will outline a research agenda for building blocks for metaverse tourism.

9 Conclusion

The development of the Metaverse offers the travel and tourism sector a rare chance to completely transform and rethink how it interacts with travelers. The Metaverse presents opportunities and difficulties for tourism locations and organizations alike by seamlessly fusing the real and virtual worlds. The Metaverse is a disruptive trend that, while still in its early phases, has the potential to significantly affect how competitive travel locations and businesses are. Tech-savvy users who can partake in virtual tourism experiences prior to making purchase selections are expected to enjoy the Metaverse as consumers grow more used to digital services, technologies, and the digital economy. Future Metaverse platforms will need to be able to handle complicated requirements and make it easier for users to interact with destinations and organizations in order to flourish in this new era of blended tourism experiences and improve the user experience. In certain situations, the Metaverse can serve as a substitute for travel while also arousing curiosity and awareness that could result in greater actual travel. In conclusion, by providing a seamless travel experience for tech-savvy people, the Metaverse offers the tourist industry a viable chance to get over COVID-19 constraints and improve travel experiences.

10 Recommendations and Suggestions

The growth of the Metaverse provides a once-in-a-lifetime chance for the tourism industry to be completely revolutionized. The way the industry interacts with customers, in general, could be drastically altered by this new development. The Metaverse presents both exciting new opportunities and daunting new problems for travel destinations and businesses by fusing the real and virtual worlds into one seamless experience. Organizations in the tourism industry are urged to take use of the

Metaverse by creating fully interactive virtual tourist attractions. Travel options can be better understood and more excitingly anticipated by tech-savvy consumers who can virtually experience destinations before making decisions to travel. This is in step with the trend toward digitalizing services and the digital economy more generally. Metaverse platforms will need to improve in the future so that they can meet increasingly sophisticated user needs and provide convenient, frictionless connections to real-world places and groups. This is essential in today's era of hybrid vacations since it improves the consumer experience overall.

The Metaverse also has the potential to replace real-world travel in some instances, providing both ease and variety. On the other hand, it has the potential to pique interest, leading people to want to see more of the world outside of their hotel. In conclusion, the Metaverse is a powerful resource that may help the travel sector adapt to the changes brought on by COVID-19 and improve customer satisfaction. Adopting this technology is crucial for maintaining a competitive edge and keeping up with passengers' ever-changing expectations.

References

1. Alamoodi AH, Mohammed RT, Albahri OS, Qahtan S, Zaidan AA, Alsattar HA, Albahri AS, Aickelin U, Zaidan BB, Baqer MJ, Jasim AN (2022) Based on neutrosophic fuzzy environment: a new development of FWZIC and FDOSM for benchmarking smart e-tourism applications. Complex Intell Syst 8(4):3479–3503. https://doi.org/10.1007/s40747-022-00689-7
2. Basheer S, Farooq S, Hassan V, Malik YM, Reshi MA (2023) Augmented reality and virtual reality in cultural heritage tourism enhancing visitor experiences, pp 13–34. https://doi.org/10.4018/978-1-6684-9957-3.ch002
3. Basheer S, Walia S, Farooq S, Shah MA, Mir FA (2023a) Exploring the metaverse, pp 195–204. https://doi.org/10.4018/978-1-6684-8898-0.ch012
4. Basheer S, Walia S, Farooq S, Shah MA, Mir FA (2023b) Exploring the metaverse: the future of tourism through AI and virtual reality. In: Influencer marketing applications within the metaverse. IGI Global, pp 195–204. https://doi.org/10.4018/978-1-6684-8898-0.ch012
5. Basheer S, Walia S, Mehraj D, Reshi M (2023) Creating transformative travel unveiling the link between meaningful experiences and sustainable tourism, pp 14–29. https://doi.org/10.4018/979-8-3693-0650-5.ch002
6. Billore S, Anisimova T (2021) Panic buying research: a systematic literature review and future research agenda. Int J Consum Stud. https://doi.org/10.1111/ijcs.12669
7. Bolger RK (2021) Finding wholes in the metaverse: posthuman mystics as agents of evolutionary contextualization. Religions 12(9). https://doi.org/10.3390/rel12090768
8. Buhalis D (2020) Technology in tourism-from information communication technologies to eTourism and smart tourism towards ambient intelligence tourism: a perspective article. Tour Rev 75(1). https://doi.org/10.1108/TR-06-2019-0258
9. Buhalis D, Leung D, Lin M (2023) Metaverse as a disruptive technology revolutionising tourism management and marketing. In: Tourism management, vol 97. Elsevier Ltd. https://doi.org/10.1016/j.tourman.2023.104724
10. Chuang TC, Liu JS, Lu LYY, Tseng FM, Lee Y, Chang CT (2017) The main paths of eTourism: trends of managing tourism through Internet. Asia Pac J Tour Res 22(2):213–231. https://doi.org/10.1080/10941665.2016.1220963
11. Damar M (2021) Metaverse shape of your life for future: a bibliometric snapshot

12. Elhoseny M, Haseeb K, Shah AA, Ahmad I, Jan Z, Alghamdi MI (2021) Iot solution for AI-enabled privacy-preserving with big data transferring: an application for healthcare using blockchain. Energies 14(17). https://doi.org/10.3390/en14175364
13. Farooq S, Farooq B, Basheer S, Walia S (2023) Balancing environmental sustainability and privacy ethical dilemmas in AI-enabled smart cities, pp 263–286. https://doi.org/10.4018/979-8-3693-0892-9.ch013
14. Filieri R, D'Amico E, Destefanis A, Paolucci E, Raguseo E (2021) Artificial intelligence (AI) for tourism: an European-based study on successful AI tourism start-ups. Int J Contemp Hosp Manag 33(11). https://doi.org/10.1108/IJCHM-02-2021-0220
15. Gretzel U, Fuchs M, Baggio R, Hoepken W, Law R, Neidhardt J, Pesonen J, Zanker M, Xiang Z (2020) E-tourism beyond COVID-19: a call for transformative research. Inf Technol Tour 22(2):187–203. https://doi.org/10.1007/s40558-020-00181-3
16. Gursoy D, Malodia S, Dhir A (2022) The metaverse in the hospitality and tourism industry: an overview of current trends and future research directions. J Hosp Market Manag 31(5):527–534. https://doi.org/10.1080/19368623.2022.2072504
17. Han Y, Oh S (2021) Investigation and research on the negotiation space of mental and mental illness based on metaverse. In: International conference on ICT convergence, 2021-October. https://doi.org/10.1109/ICTC52510.2021.9621118
18. Hennig-Thurau T, Gwinner KP, Walsh G, Gremler DD (2004) Electronic word-of-mouth via consumer-opinion platforms: what motivates consumers to articulate themselves on the Internet? J Interact Mark 18(1):38–52. https://doi.org/10.1002/dir.10073
19. Jovanović A, Milosavljević A (2022) VoRtex metaverse platform for gamified collaborative learning. Electron (Switz) 11(3). https://doi.org/10.3390/electronics11030317
20. Koo C, Chung N (2021) Understanding of human nature, smart tourism and metaverse. J Internet Electron Commer Res 21(6). https://doi.org/10.37272/jiecr.2021.12.21.6.1
21. Ku ECS, Chen CD (2015) Cultivating travellers' revisit intention to e-tourism service: the moderating effect of website interactivity. Behav Inf Technol 34(5):465–478. https://doi.org/10.1080/0144929X.2014.978376
22. Kulkov I, Hellström M, Tsvetkova A, Malmberg J (2023) Sustainable cruise tourism: systematic literature review and future research areas. Sustain (Switz) 15(10). https://doi.org/10.3390/su15108335
23. Kye B, Han N, Kim E, Park Y, Jo S (2021) Educational applications of metaverse: possibilities and limitations. J Educ Eval Health Prof 18. https://doi.org/10.3352/jeehp.2021.18.32
24. Li F, Ruijs N, Lu Y (2022) Ethics & AI: a systematic review on ethical concerns and related strategies for designing with AI in healthcare. AI 4(1):28–53. https://doi.org/10.3390/ai4010003
25. Maranisya U, Sutanto S (2022) Efektivitas metaverse tourism Sebagai Sarana Promosi Wisata Alam Hiu Paus Di Taman Nasional Teluk Cenderawasih. Syntax Literate; Jurnal Ilmiah Indonesia, *7*(3)
26. Masri NW, You JJ, Ruangkanjanases A, Chen SC, Pan CI (2020) Assessing the effects of information system quality and relationship quality on continuance intention in e-tourism. Int J Environ Res Public Health 17(1). https://doi.org/10.3390/ijerph17010174
27. Mehraj D, Qureshi IH, Singh G, Nazir NA, Basheer S, Nissa VU (2023) Green marketing practices and green consumer behavior: demographic differences among young consumers. Bus Strat Dev. https://doi.org/10.1002/bsd2.263
28. Mehraj D, Ul Islam MI, Qureshi IH, Basheer S, Baba MM, Nissa VU, Asif Shah M (2023) Factors affecting entrepreneurial intention for sustainable tourism among the students of higher education institutions. *Cogent Business & Management*, *10*(3). https://doi.org/10.1080/23311975.2023.2256484
29. Monaco S, Sacchi G (2023) Travelling the metaverse: potential benefits and main challenges for tourism sectors and research applications. Sustain (Switz) 15(4). https://doi.org/10.3390/su15043348
30. Moon J-W, An Y (2022) Uses and gratifications motivations and their effects on attitude and e-tourist satisfaction: a multilevel approach. Tour Hosp 3(1):116–136. https://doi.org/10.3390/tourhosp3010009

31. Narin NG (2021) A content analysis of the metaverse articles. J Metaverse 1(1)
32. Nguyen HM, Dang LAT, Ngo TT (2019) The effect of local foods on tourists' recommendations and revisit intentions: the case in Ho Chi Minh City, Vietnam. J Asian Financ, Econ Bus 6(3). https://doi.org/10.13106/jafeb.2019.vol6.no3.215
33. Noh H-K (2022) Metaverse-related issues in tourism using news Big Data. J Tour Leis Res 34(2). https://doi.org/10.31336/jtlr.2022.2.34.2.151
34. Park S, Kim S (2022) Identifying world types to deliver gameful experiences for sustainable learning in the metaverse. Sustain (Switz) 14(3). https://doi.org/10.3390/su14031361
35. Park SM, Kim YG (2022) A metaverse: taxonomy, components, applications, and open challenges. IEEE Access 10. https://doi.org/10.1109/ACCESS.2021.3140175
36. Pillai R, Sivathanu B (2020) Adoption of AI-based chatbots for hospitality and tourism. Int J Contemp Hosp Manag 32(10). https://doi.org/10.1108/IJCHM-04-2020-0259
37. Sahli A (2015) Revisiting perceived risk and trust in E-tourism context: toward an extended technology acceptance model. J Mark Res Case Stud, 1–14. https://doi.org/10.5171/2015.516086
38. Samala N, Katkam BS, Bellamkonda RS, Rodriguez RV (2022) Impact of AI and robotics in the tourism sector: a critical insight. J Tour Futur 8(1). https://doi.org/10.1108/JTF-07-2019-0065
39. Schuckert M, Liu X, Law R (2015) Hospitality and tourism online reviews: recent trends and future directions. J Travel Tour Mark 32(5):608–621. https://doi.org/10.1080/10548408.2014.933154
40. Shafiee MM, Tabaeeian RA, Tavakoli H (2016) The effect of destination image on tourist satisfaction, intention to revisit and WOM: an empirical research in Foursquare social media. In: 10th International conference on E-commerce in developing countries: with focus on e-tourism, ECDC 2016. https://doi.org/10.1109/ECDC.2016.7492964
41. Suh W, Ahn S (2022) Utilizing the metaverse for learner-centered constructivist education in the post-pandemic era: an analysis of elementary school students. J Intell 10(1). https://doi.org/10.3390/jintelligence10010017
42. Tik-Tsuen Wong A (n.d.) E-tourism: how customers intention to use be affected? Acad Mark Stud J
43. Um T, Kim H, Kim H, Lee J, Koo C, Chung N (2022) Travel Incheon as a metaverse: smart tourism cities development case in Korea. In: Information and communication technologies in tourism, 2022. https://doi.org/10.1007/978-3-030-94751-4_20
44. Yoo K-H, Gretzel U (2012) Use and creation of social media by travellers impact of technology view project handbook of e-tourism view project, Chapter 15 Use and creation of social media by travellers. https://www.researchgate.net/publication/304381560
45. Zhang Y, Lee H (2022) Wine tourism experience effects on co-creation, perceived value and consumer behavior. Cienc e Tec Vitivinic 37(2):159–177. https://doi.org/10.1051/ctv/20223702159
46. Zhu Y, Wang Y, Song B, Feng Q, Lin H, Tang J (n.d.) The interactive effects of extrinsic and intrinsic motivations on service recovery performance in the hospitality industry: the mediating role of self-efficacy Os efeitos interativos das motivações extrínsecas e intrínsecas no desempenho da recuperação de serviço na indústria hoteira: o papel mediador da autoeficácia. Tour Manag Stud 19(3):7–22. https://doi.org/10.18089/tms.2023.190301

MIX
Papier aus verantwortungsvollen Quellen
Paper from responsible sources
FSC® C105338

If you have any concerns about our products,
you can contact us on
ProductSafety@springernature.com

In case Publisher is established outside the EU,
the EU authorized representative is:
Springer Nature Customer Service Center GmbH
Europaplatz 3, 69115 Heidelberg, Germany

Printed by Libri Plureos GmbH
in Hamburg, Germany